THE EVOLUTION OF THE BIOENERGETIC PROCESSES

The Evolution of the Bioenergetic Processes

BY

E. BRODA

Institute of Physical Chemistry, University, Vienna

PERGAMON PRESS

Oxford · New York · Toronto

Sydney · Paris · Braunschweig

U. K.	Pergamon Press Ltd., Headington Hill Hall, Oxford OX3 0BW, England
U. S. A.	Pergamon Press Inc., Maxwell House, Fairview Park, Elmsford, New York 10523, U.S.A.
CANADA	Pergamon of Canada Ltd., 207 Queen's Quay West, Toronto 1, Canada
AUSTRALIA	Pergamon Press (Aust.) Pty. Ltd., 19a Boundary Street, Rushcutters Bay, N.S.W. 2011, Australia
FRANCE	Pergamon Press SARL, 24 rue des Ecoles, 75240 Paris, Cedex 05, France
WEST GERMANY	Pergamon Press GmbH, 3300 Braunschweig, Postfach 2923, Burgplatz 1, West Germany

First edition 1975

Library of Congress Cataloging in Publication Data

Broda, Engelbert.
Evolution of the bioenergetic processes.

1. Bioenergetics. 2. Evolution. I. Title.
QH510.B76 1975 574.1'9121 75-6847
ISBN 0-08-018275-5

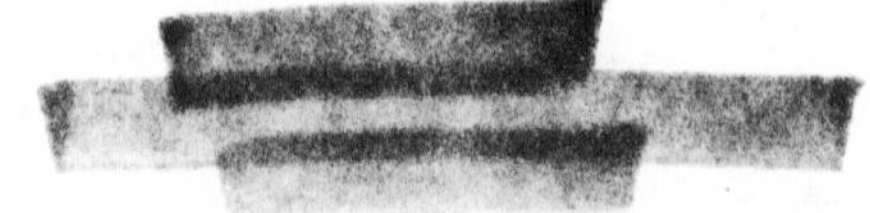

Printed in Great Britain by A. Wheaton & Co., Exeter

CONTENTS

FOREWORD

THIS book has grown from an essay "The Evolution of Bioenergetic Processes", published in 1970, and from even earlier articles and lectures. The treatment has now been expanded and brought up to date. Again I should like to thank the colleagues with whom I could discuss the previous survey, or parts of it, and whose names are given there. Particular thanks are due to C. B. Van Niel, the great pioneer in microbial energetics, and to F. Egami, O. Hoffmann-Ostenhof and M. D. Kamen for working through the whole of the previous manuscript. Parts of the present manuscript were read by P. Broda, H. Ruis, P. Schuster, R. Singer and J. Zemann, and the whole manuscript was ploughed through most carefully by G. A. Peschek. Cordial thanks to them, too.

The central part of the present book is, of course, constituted by the discussion of the path from fermentation to photosynthesis and respiration, and of the relationship between the bioenergetic processes in prokaryotes and in eukaryotes. I have preferred to cite the most recent articles, and have given special attention to good secondary literature. This selection of references should be most helpful to the reader. May the many colleagues forgive me whose contributions, often important, thus have not been emphasized sufficiently.

Chapters 1 to 6, in which a brief account of the conditions and possibilities of biopoiesis is given, should be looked at as merely introductory. Especially, the paragraphs on free energy and entropy are just a reminder. No derivation of thermodynamic relationships will here be expected. The Chapter 22, on the multicellular organisms, as well as the Chapters 23 to 25, where largely "inductive" evidence is presented, are supplementary. In all these accessory chapters, quotations have been kept down to a minimum, and are mostly meant to direct the reader to more detailed information.

The book could have been embellished with fine electron optical pictures of bacterial cells, mitochondria, chloroplasts, etc. I have, however, resisted this temptation, as such pictures would not have contributed to the discussion, and are abundant in other articles and books.

References in the text to other chapters of this book appear in italics in brackets.

My thanks for permission to reproduce illustrations are due to the publishing firms and journals: Georg Thieme Verlag, Stuttgart (Figs. 7.1, 9.1, 11.1, 12.5), Elsevier Publishing Co., Amsterdam (Fig. 8.1), John Wiley & Co., New York (Figs. 8.2, 12.3), American Journal of Botany, Norman, Okla. (Fig. 12.2), Zeitschrift für Naturforschung, Tübingen (Fig. 12.6), Cambridge University Press, London (Fig. 16.1), Gustav Fischer, Verlag, Stuttgart (Fig. 18.1), W. H. Freeman & Co., San Francisco (Figs. 18.2, 18.4), Scientific American, New York (Figs. 18.3, 23.1), Yale University Press, London (Fig. 19.1), Macmillan Publishing Co., New York (Fig. 20.1), Hutchinson Educational Ltd., London (Fig. 20.2) and Wadsworth Publishing Co., Belmont, Calif. (Fig. 23.4).

I want to express my thanks to Pergamon Press and their staff for their friendly patience.

Vienna, May 1974 E. BRODA

1

ENERGY IN THE BIOSPHERE

(a) Life and Work

Life implies work. Organisms and their parts move against the resistance of a medium, and thereby they perform mechanical work. Less obviously, but more generally, all organisms do internal work. For the synthesis of "biomolecules" chemical work is needed, for the setting up and the maintenance of concentration differences osmotic work, and of differences in electrical potential, electrical work. Occasionally light is emitted by organisms, and this also requires the performance of work. Energy is the capacity to do work.

As long as it was not known that heat is a form of energy, and therefore the First Law of thermodynamics had not been discovered, progress in the analysis of the bioenergetical processes was necessarily slow, patchy and contradictory, though already in the eighteenth century Crawford, Lavoisier and Laplace had introduced calorimetric methods into biology (see, for example, Mendelsohn, 1964). Only when the principle of the conservation of energy had been established (after 1840, with the living body as R. J. Mayer's main test object!) could reliable energy balance sheets be drawn up.

There are forms of energy which in given conditions cannot be used for work. According to the Second Law, established after the pioneering work of the isolated genius, Sadi Carnot, by William Thomson (later Lord Kelvin) and by Clausius, no device is possible that in a cycle of operations does nothing else but consume heat and do the equivalent amount of work. Another well-known expression of the Second Law is that in an isolated system entropy can never decrease. Entropy was established by Clausius as a thermodynamic concept, and was later recognized as a measure of molecular disorder by Boltzmann. Boltzmann's well-known equation, as written on his gravestone in Vienna, is

$$S = k \log W \tag{1.1}$$

where S is the entropy, W the probability of the state of the system, and k Boltzmann's constant.

In 1852 Thomson still thought it prudent to restrict his statement of the Second Law to "inanimate material agency". However, a violation of the Second Law by living things has never been observed. Therefore, already on empirical grounds the validity of the Second Law no less than that of the First Law for living as well as for nonliving matter is now generally accepted. In this respect, Nature is certainly unitary.

But not all processes permitted by the Second Law are available to living beings. The methods for the performance of work that are most common in modern industry are not used by organisms: they do not employ differences in temperature, whether set up by combustion or by nuclear reactions. It will be seen that the source of all work done by organisms is chemical energy directly. They are "chemodynamical" machines [*1d*]. Furthermore, only very few classes of chemical reactions are put to use in bioenergetics.

(b) Energy and Entropy in the Biosphere

The whole Earth and its surface layer are now, broadly speaking, in a steady state. The part of the surface layer that is available to organisms is called the biosphere. This term was coined in 1875 by the geologist of Vienna, E. Suess, and extensively used by Vernadsky (see Vernadsky, 1929). The biosphere is also in a steady state. Thus the energy content of the biosphere, alone of interest here, does not vary significantly with time. The intake of energy is just compensated by loss.

For the energy intake, only light—in the infrared, visible and ultraviolet ranges—need be considered. The maximum of the solar energy spectrum lies in the visible, namely, in the green near 500 nanometres (nm), corresponding to 57 kcal per mole of light quanta, i.e. per einstein. The total energy of incoming electromagnetic radiation of very short (γ- and X-rays) and of very long wavelengths (radiowaves) is negligible. The same is true of the energy of particulate cosmic radiation. The heat production of the Earth itself, from radioactive processes, must be equal to the net outward energy flux across the crust. It is only 0.02% of the energy flux that comes to the Earth from the Sun and therefore is lost again into space (see, for example, Hubbert, 1971). Thus compared with the solar heat flux, the heat production of the Earth is minute.

The energy loss of the Earth and also of the biosphere occurs entirely through radiation. The terrestrial radiation consists mainly of infrared rays, with a maximum near 11,000 nm, i.e. at 2.2 kcal/einstein. The difference between the wavelengths of incoming and outgoing radiation reflects the difference between the temperatures of Sun and Earth.

In the steady state, the entropy content of the Earth must also be constant. The idea of an entropy of radiation, associated with its energy, was introduced by Boltzmann†, and later exploited by Planck to derive his famous radiation law, and therefore quantum theory. So entropy is taken in in the form of solar radiation, and carried away by the terrestrial electromagnetic radiation. But much entropy is also produced all the time in the Earth through irreversible processes. The rate of production of entropy must be equal to the net loss. Thus in balance the Earth emits entropy, or, as Schrödinger (1945) put it, absorbs "negative entropy" ("negentropy"; Brillouin, 1949). Compensation between entropy production and net entropy emission must also hold for the different parts of the Earth singly, as long as they are in steady states. Thus it is true for the surface layer, and, more particularly, for the biosphere.

Schrödinger has applied his striking concept of negentropy to living matter itself. Organisms or ensembles of organisms in steady states can indeed be said to feed on negentropy. While they give off just as much energy as they take up, the entropy associated with the energy given off must exceed that associated with the energy taken up. Thus the chance of organisms for existence is that they interpose themselves into the gradient of entropy between incoming and outgoing energy.

The core of Schrödinger's idea may perhaps be traced to a magnificent statement by

† Boltzmann's first paper on the relationship of heat radiation to the Second Law appeared in *Wied. Ann.* **22,** 31 (1884). There Boltzmann referred, however, to an even earlier paper by the Italian, Bartoli (1876), which had come to his knowledge a short time before. Yet Boltzmann's rather than Bartoli's considerations became influential.

Boltzmann (1886), whom he much admired, though Boltzmann was not quoted by Schrödinger in this context:

> Der allgemeine Lebenskampf der Lebewesen ist daher nicht ein Kampf um die Grundstoffe—auch nicht um Energie, welche in Form von Wärme, leider unverwandelbar, in jedem Körper reichlich vorhanden ist—sondern ein Kampf um die Entropie (more exactly; Negentropie E.B.), welche durch den Übergang der Energie von der heißen Sonne zur kalten Erde disponibel wird. Diesen Übergang möglichst auszunützen, breiten die Pflanzen die unermeßlichen Flächen ihrer Blätter aus und zwingen die Sonnenenergie in noch unerforschter Weise, ehe sie auf das Temperaturniveau der Erdoberfläche herabsinkt, chemische Synthesen auszuführen, von denen man in unseren Laboratorien noch keine Ahnung hat.

Entropy is absorbed by organisms in the form of "food" (chemical substrate of bioenergetic processes) and of light. The entropy content of the components of food can, in principle, be determined by standard physico-chemical methods. Attempts have also been made to apply the concept of the entropy of radiation in a quantitative way to the light relations of organisms, more particularly to photosynthesis (Duysens, 1952, 1958, 1959, 1962; Spanner, 1964; Knox, 1969; Rueda, 1973). The essence is that a definite temperature is ascribed to the effective light radiation so that the maximum work that can be done by a given amount of light can be derived by means of the fundamental equation for the Carnot cycle [*1c*]. This work sets a limit to the yield in photosynthesis. Such computations may in certain circumstances be used to exclude proposed mechanisms that would violate the Second Law.

(Of course, it must not be forgotten that the actual processes must obey not only thermodynamics. Energy can be utilized only in parcels prescribed by the relevant chemical equations. The limits in this way set by stoichiometry are naturally narrower than those given by thermodynamics.)

However, the accent on entropy in this connection tends to obscure the fact that usable energy is really what organisms need. It will be seen later that the free energy function G is optimally suited to bioenergetics. Even when entropies must be computed for the evaluation of free energies, it is advisable to apply explicitly to bioenergetic considerations free energy.

Perhaps a crude analogy is allowed. A town in a steady state converts useful materials, not only food, into waste at a more or less constant rate. The mean value of the unit amount of material is thereby decreased, just as the value of energy is decreased when entropy goes up. Consequently, we may define a magnitude "negworthlessness" that decreases owing to the town's activities. So one could correctly say that the town feeds on negworthlessness. Would it not be better to say that it feeds on worth?

(c) Heat as a Source of Work?

Organisms have often been compared to heat engines (see, for example, von Helmholtz, 1869). In fact, both in living cells and in man-made engines stored chemical energy is used to produce movement, in a wide sense. More particularly and obviously, both engines and respiring cells convert fuel (food) to CO_2 and water.

However, the engines burn the fuel first, and work is done only by exploiting the temperature difference between the heat source and the environment. According to Carnot, the maximum amount of work A from a given amount of heat Q is related to the temperature difference by

$$A/Q = (T_1 - T_2)/T_1 \tag{1.2}$$

where T_1 is the upper temperature of the working substance, i.e. also of the heat supply reservoir, and T_2 the temperature of the environment (the coolant). Consequently, only part of the energy obtained in the combustion of the fuel can be recovered as mechanical (useful) energy. For instance, in an engine working between 500°C (773°K) and 30°C (303°K), the theoretical yield is about 60%; in practice, it is much less.

It cannot be too strongly emphasized that in the isolated system Carnot engine plus heat supply reservoir plus coolant the entropy of the whole system remains constant at every step throughout (ideal conditions). It is a common misconception that at some steps entropy increases and at further steps there is a compensating decrease. In this case, the "further steps" would contradict the Second Law. What really happens is merely that heat and therefore entropy is transferred between the working substance and the heat supply reservoir or the coolant. This statement can clearly be extended to chemical processes. In ideal conditions, the sum of the entropies of the reacting chemical system and the environment remains constant during every time interval.

In ideal conditions, the process is carried out "reversibly". This means that conditions are only infinitesimally removed from equilibrium. At any instant the driving force is just sufficient to overcome the resistance, and therefore an infinitesimal change in parameters could reverse the direction of the process. This parameter may be a temperature difference in a heat engine or a concentration difference in a chemical system. In reversible processes avoidable losses are zero.

The possibility that organisms are mere heat engines was soon doubted. For instance, basing himself on Carnot and Joule, Thomson (1852) stated:

> . . . it appears nearly certain that, when an animal works against resisting force, there is not a conversion of heat into external mechanical effect, but the full thermal equivalent of the chemical force is never produced; in other words, that the animal body does not act as a thermodynamic engine; and very probably that the chemical forces produce the external mechanical effects through electrical means.

Curiously, it follows:

> Whatever the nature of these means, consciousness teaches every individual that they are, to some extent, subject to the direction of his will. It appears, therefore, that animated creatures have the power of immediately applying, to certain moving particles within their bodies, force by which the motions of these particles are directed to produce desired mechanical effects.

Whether such will is also shown by plants and bacteria?

Nutritionists have observed since the last century that surprisingly high yields of useful energy can be obtained from food (see, for example, Kleiber, 1961). Such tests are best carried out on man in a treadmill or a similar device. Only mechanical energy is considered. The Carnot equation can be used to compute the temperature difference corresponding to the efficiency, as was done first by Fick (1882). With a yield of 25%, as observed, the temperature difference works out as 105°. Hence if the body were a heat engine, local temperatures of at least 310° (body temperature) + 105° = 415°K (142°C) would be needed. Such temperatures occur nowhere in organisms.

(d) Organisms as Chemodynamical Machines

Clearly work from food is not produced by way of heat. Rather, the organisms are now considered as chemodynamical machines in a more direct sense. Chemical energy is interconverted, or is transformed into other forms of energy, without passing a heat stage. The (small) gradients of temperature that exist within organisms are never applied to work. In practice, the organisms work isothermally (and isobarically), which is most fortunate for computations!

It has been said that in bioenergetics the concept of free energy is, and should be, used explicitly. Useful introductions into the application of the free energy function in bioenergetics have been written, for example, by Klotz (1957, 1964), by Ingraham and Pardee (1967) and by Lehninger (1970, 1971). The molar free energy G is defined as

$$G = H - TS \tag{1.3}$$

where H is the molar energy content, T the temperature and S the molar entropy content. Strictly, in an isobaric system H should be called "enthalpy" and G "free enthalpy". But the volume changes, allowed in isobaric conditions, are small with reactions in living systems so that the difference between enthalpy and energy vanishes.

(Moreover, most reactions occur in solutions, and the relevant functions should then be prefixed by the word "partial". But this may be taken for granted, and therefore omitted.)

So the energetics of a reaction are described in terms of the differences between products and reactants by

$$\Delta G = \Delta H - T\Delta S. \tag{1.4}$$

The "free energy change" ΔG has a negative value when the reaction is written in the direction in which it proceeds spontaneously, i.e. as an "exergonic" reaction. In the opposite direction, the reaction is "endergonic". The maximum work that can be got from a chemical reaction is given by ΔG. It is obtained when the reaction is performed reversibly (ideally). But whenever a reaction has an irreversible component, some of the energy is dissipated as heat, and additional entropy is generated.

While ΔH is mostly measured by calorimetry, ΔG (and, therefore, $T\Delta S$) is normally derived from measurements of chemical equilibria. ΔG is connected with the equilibrium concentrations of products (c_{p1}, c_{p2} . . .) and reactants (c_{r1}, c_{r2} . . .) and their actual concentrations (starred values) by

$$\Delta G = -RT \ln \frac{c_{p1}c_{p2}\cdots}{c_{r1}c_{r2}\cdots} + RT \ln \frac{c^*_{p1}c^*_{p2}\cdots}{c^*_{r1}c^*_{r2}\cdots} \tag{1.5}$$

provided the thermodynamic activities† can be replaced by the concentrations. For physicochemical standard conditions (all $c_p{}^*$ and $c_r{}^* = 1$, also implying a hydrogen ion concentration 1, i.e. a pH value 0), the value of ΔG is called ΔG_0. For physiological standard conditions (otherwise the same, but hydrogen ion concentration 10^{-7}, i.e. pH = 7), ΔG is by definition $\Delta G_0'$. The $\Delta G_0'$ values in this book are generally those of Burton (1957), but more recent values have for many reactions been listed by Bassham and Krause (1969).

The equilibrium constant K is defined as

$$K = \frac{c_{p1}c_{p2}\cdots}{c_{r1}c_{r2}\cdots} \tag{1.6}$$

Therefore, for standard conditions

$$\Delta G_0 \text{ (or } \Delta G_0') = -RT \ln K. \tag{1.7}$$

† Be it recalled that the thermodynamical activities are so-to-speak idealized concentrations. The system behaves thermodynamically (in respect to free energy changes in reactions) as if the concentrations did not have the values given by naïve chemical analysis, but the values indicated by the thermodynamical activities.

As ΔG_0 (or $\Delta G_0'$) is a measure of the tendency for a reaction to occur, it is also known as the affinity of the reaction. The affinity is the greater, the more negative ΔG_0 ($\Delta G_0'$) is, i.e. the more exergonic the reaction is in standard conditions. It cannot be stressed too strongly that the actual exergonicity or endergonicity of a reaction in given conditions is determined not by the standard values (ΔG_0 or $\Delta G_0'$), but by the values ΔG or $\Delta G'$, i.e. the work obtainable from a reaction is influenced by the actual concentrations (more exactly: activities) of the reactants and the products.

Other reactions depend on solid structures within the cells, e.g. mitochondria. Their exact treatment requires concepts of multiphase thermodynamics and (mechanistically) of solid-state chemistry. In the present book, approximations will do.

(e) Dynamic States of Organisms

In the ideal case of reversible reactions, no entropy is generated. Entropy is merely shifted, in blocks of magnitude $T\Delta S$ (eqn. 1.4), between the working system and its environment. Production of entropy occurs in the irreversible components of the real processes [*1b*].

Reactions cannot proceed fully reversibly if they have finite velocities. Now speed is needed to a greater or lesser extent by the organisms in all their life functions. Metabolism must not be too slow. Therefore, full reversibility cannot be attained, and entropy must be generated. Hence entropy is certainly produced when organisms do external work.

But why does even a resting cell, while seemingly not doing work, produce entropy? Why does basal metabolism (metabolism at rest) exist universally? In bacteria, with which we shall be much concerned, the energy of basal metabolism is known as "energy of maintenance" (see Dawes and Ribbons, 1964; Pirt, 1965; Forrest and Walker, 1971).

The explanation lies in the "dynamic state of body constituents", established by Schoenheimer (see Schoenheimer, 1942; von Hevesy, 1948). Investigations with isotopic tracers, first with deuterium, have shown that the concentrations of most biomolecules even in resting organisms are maintained not statically, but dynamically. This means that the metabolic processes continue in the state of seeming rest, with compensation of break-down (catabolism) and build-up (anabolism). Energy is needed for these processes.

Not all biomolecules are in a dynamic state. For instance, the genetic material, DNA, is not, nor are in mammals some of the proteins, e.g. haemoglobin or casein (see Broda, 1959, 1960, 1968). Equally, at least some of the proteins of bacteria are not maintained dynamically (Hogness *et al.*, 1955; for further references, partly contradictory, see Broda, 1959, 1960).

Experience shows that the antagonistic catabolic and anabolic processes always proceed along differing pathways. This is understandable. Catabolic reactions are generally exergonic so that the corresponding anabolic reactions cannot proceed spontaneously, at least not in the same conditions. Therefore, different pathways, which will permit the investment of additional chemical energy, are used for anabolism. Typically, more energy in useful form is needed for anabolism than is obtained in catabolism. Often no useful energy is obtained at all in the catabolic reactions of dynamic states, and the energy is dissipated as heat.

A good example for biomolecules in dynamic states are the serum proteins of vertebrates. Their concentrations in the blood are kept constant by means of specific cycles. Thus serum albumin is synthesized, with input of energy, in the ribosomes of liver cells from aminoacids through the well-known DNA-directed and messenger RNA-mediated process for protein production [*2b*]. The required energy comes from the hydrolysis of ATP, the universal energy transfer agent [*6a*]. In the steady state, circulating albumin falls, at the same rate, victim to hydrolysis by proteinases.

Other reaction cycles serve energy metabolism. For example, in the citric acid (Krebs or tricarboxylic acid) cycle [*13c*] oxaloacetic acid is consumed and regenerated while organic substrate is converted to CO_2 and "metabolic hydrogen" [H]. The latter is usually burned with atmospheric oxygen through the respiratory chain [*13d*].

Oxaloacetic acid is sometimes called a "catalyst" for this conversion, and a catalytic role, namely, for the hydrolysis of ATP, can be attributed to serum albumin in the previous example as well. It is an important feature that in contrast to ordinary catalysts these biomolecules pass a whole series of well-regulated chemical steps before they are, eventually, regenerated, or more precisely, before an equal amount of the same substance is reproduced.

The cyclic processes that are at the core of the dynamic state may be looked at as a phenomenon that is qualitatively new and characteristic for life; probably they are the most characteristic feature of life, as we know it. The orderly sequence of reactions presupposes, of course, reasonable stability of reactants and products. That is why Schrödinger (1938) has called life a "quantum phenomenon near absolute zero".

Commonly, properties that are the basis of proliferation and evolution, especially replication and mutation, are named as criteria of life (see, for example, Horowitz and Miller, 1962). It is true that life could not continue and proliferate without replication and could not evolve without mutation. But a living cell is in any case, even when incapable of proliferation, sharply distinguished from a piece of non-living matter.

(f) Active Transport

Another kind of dynamic state occurs in connection with "active transport" (Ussing, 1954, 1960, 1967; Andersen and Ussing, 1960; Schoffeniels, 1964, 1967; Stein, 1967; Caldwell, 1969). This is transport of solutes across membranes against the "natural" direction, i.e. the direction in which the solutes would move (diffuse) if there were no special mechanisms to reverse the direction—an extraordinary achievement of life, and of life only. Broadly speaking and forgetting about subtle points, the natural direction is given by the free energy change, which must be negative:

$$\Delta G = -RT \ln \frac{a_1}{a_2} \tag{1.8}$$

a is the thermodynamical activity, again often to be approximated by the concentration (c). In diffusion, ΔG is negative. In active transport, ΔG is positive, and "osmotic" work must be done on the system at the expense of free energy. If there is an electrical potential difference between the compartments, an additional term must take care of this in respect to ions.

Through active transport, large differences in the concentrations of organic substances or ions may be set up between cells and their environments, or between different cell "compartments". When the difference is large enough, spontaneous back-flow through the membrane may compensate for active transport, and a steady state is observed. In higher organisms, "transcellular" active transport across layers of cells is also found. In the hands of Ussing, the surviving frog skin has become, from the point of view of bioenergetics, the best-studied object in "membranology" (Teorell, 1967).

Evidently, the solutes are pumped across the membranes by special mechanisms; in simple cases, they may return by ordinary diffusion through pores (pump-and-leak). The pumps are generally found to be highly, though not necessarily absolutely, specific for particular solutes (substrates). The membranes are asymmetric. The direction of work of a given pump is then determined.

In active transport, as in the dynamic maintenance of chemical non-equilibria (case of serum albumin), work must be done all the time. It has, for example, been stated that 20–45% of the total energy needs of resting normal mammalian tissue is for active transport (Whittam, 1964).

In conclusion, we emphasize that quite generally steady (dynamic) states are maintained on the basis of metabolism (Fig. 1.1). The advantage of a dynamic against a static state in answering rapid calls is known to every gangster who keeps the engine of his car running while he robs the bank.

The exceptions may be seen in this light. It is understandable that in the primitive bacteria dynamic states in respect to anabolism and catabolism are not so well developed while active transport is pronounced. The bacteria concentrate on feeding and proliferation rather than on individual survival. For mammals, it would be a waste to maintain haemoglobin or casein, proteins to be expended or exported, in dynamic states. With DNA, a dynamic state would be even dangerous; it is precisely the task of DNA to remain safe and undisturbed, and not to run risks. But on the whole, dynamic states have proved precious. No organism is known to do without them, while nowhere outside a living cell have dynamic states ever been observed. The complicated mechanisms to maintain them, which require delicate control, can have evolved only gradually through long times by trial and error. For considerations about the evolution of transport through membranes see Tosteson (1971).

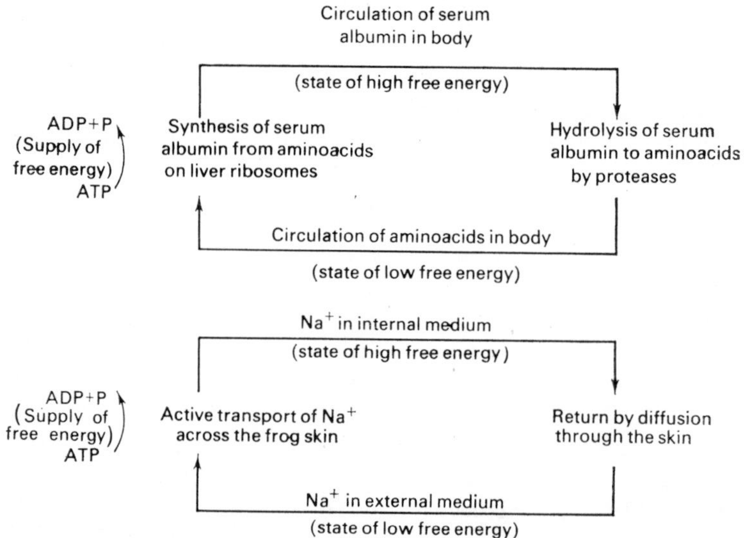

FIG. 1.1. Examples for the dynamic state of body constituents. *Top.* In the vertebrate organism, serum albumin is degraded and resynthesized at equal and constant speeds. *Bottom.* Sodium ion is transported across the frog skin and concentrated. It diffuses back as a consequence of the concentration gradient set up. In the steady state, the two velocities are equal.

From the point of view of the thermodynamics of open systems, non-equilibrium structures in living and other systems in steady states that are maintained through continuous loss of free energy (dissipation of energy) are called "dissipative structures". Dissipation of energy is, as we have seen, a basic feature of systems in dynamic states. It implies generation of entropy. Dissipative structures have been most widely discussed by Prigogine (see Prigogine and Nicolis, 1971; Prigogine and Glansdorff, 1971; Prigogine *et al.*, 1972). The analysis is capable of yielding general properties of such systems, and on its basis some kinds of systems may be excluded from existence. Yet the results of the analysis, as is always true of thermodynamics, clearly are compatible with many different molecular or supermolecular mechanisms, and therefore do not define or predict them.

The idea of the non-equilibrium maintained by constant input of free energy can, of course, be applied not only to single organisms and its parts, but also to collectives of organisms, and even to whole parts of the biosphere. Examples are the non-equilibria of free nitrogen [*16c*] and of free oxygen [*25c*]. Indeed Schrödinger's concept of the thermodynamics of life [*1b*] can be looked at in this light.

2

BIOENERGETICS AND EVOLUTION

(a) Classes of Bioenergetic Processes

Among the metabolic processes, we shall consider the processes for the supply of energy in useful form, the bioenergetic processes. They can be broadly divided into three classes: fermentation [*7*], photosynthesis [*8*] and respiration [*13*]. The division between fermentation and respiration may seem artificial, but is in fact well justified nevertheless [*7a*].

Relations between these processes will be considered from an evolutionist point of view, and plausible transitions between the classes will be looked for. These transitions must indeed have occurred if life has a monophyletic origin. The basic unity of life has been emphasized by Kluyver (Kluyver and Donker, 1926; Kluyver, 1931) as the principal conclusion from "comparative biochemistry" (see Baldwin, 1937; Van Niel, 1949b; Kluyver and Van Niel, 1956; Florkin and Mason, 1960; Florkin, 1966; Dayhoff and Eck, 1969). For accounts of Kluyver's thought see La Rivière (1959), Van Niel (1966) and Benemann and Valentine (1971).

Evolution requires the adaptation and the improvement of the bioenergetic processes. Experience with extant organisms shows that on a cellular level these processes are most diverse among the most primitive groups (see Kluyver and Van Niel, 1956; DeLey, 1971). It appears that the diversity goes back to the early times when Nature still "experimented" in this field, and had not yet "decided" in favour of the now familiar scheme that is dominated by the compensating antagonistic processes of photosynthesis and of respiration. So the important transitions must have taken place in the remote past.

In contrast, in the evolution of higher organisms, when the bee and the whale and the rhododendron appeared, the bioenergetic equipment of the cells underwent relatively little further change. Among the higher organisms, the diversity in morphology, anatomy and macrophysiology far surpasses that in basic biochemistry and microphysiology, including bioenergetics.

It will be assumed here that the properties of the biosphere changed to an important extent only as a consequence of the activities of expanding and evolving life itself. In particular, it were the living organisms themselves that produced the atmospheric oxygen (Spencer, 1844). Consequently, the somewhat fashionable ideas about a great influence on evolution of occasional cataclysmic increases in the intensities of cosmic or other rays from space (see, for example, Russell and Tucker, 1971; Ruderman, 1974) or of processes in the Earth's crust (see Hallam, 1974) will not be considered. Telling arguments against the validity of such ideas have been put forward by Simpson (1968; see also McElhinny, 1971).

We have to reconstruct, then, the pathway of bioenergetic evolution from the pattern of the bioenergetic processes in the extant organisms. The task of reconstruction (not specifically in respect to bioenergetics) has been compared (Orgel, 1973) to the task of tracing the evolution of technology from the early beginnings, when mankind invented the wheel and the lever, merely from knowledge of contemporary machinery.

Evolution is unlikely to have involved large steps between intermediate types. In a related context, the term "principle of continuity" has been used (Orgel, 1968). Moreover, one remains aware that the capacity for any process can have established itself only when the external physico-chemical conditions for its operation were given. For instance, respiration in the ordinary sense (oxygen respiration) was impossible in an atmosphere devoid of free oxygen, an "anoxygenic" atmosphere, as termed by Rutten (1962). Consequently, our guideline will be that every step in physiological evolution must be (a) biologically useful, (b) thermodynamically possible, and (c) mechanistically plausible.

(b) Genotype and Phenotype

Ultimately, all expressions of life are based on metabolism. This is catalysed and directed by enzymes. While not all proteins are enzymes, all enzymes are proteins. The proteins are built up within the cells through "translation" in accordance with the instructions laid down in messenger RNA, and these instructions in turn are equivalent to, and "transcribed" from, those encoded in DNA. This is the concrete, modern, form of the famous "one gene—one enzyme" hypothesis of Beadle and Tatum (see Beadle, 1945).

In this way, information flows from nucleic acid, the "template", to protein. This fact has been called the central dogma of molecular biology (see Crick, 1970). Proteins and nucleic acids are known as "informational" macromolecules. It should be recalled that not all RNA is messenger RNA. "Transfer RNA" and "ribosomal RNA" serve protein production in other ways.

The difference between the roles of the two kinds of informational macromolecules must be explained through their characteristic chemical properties (see Schuster, 1972; Miller and Orgel, 1974). On the one hand, nucleic acids are capable of faithful replication through base pairing, best known for the case of the DNA double helix. On the other hand, they are not catalysts. This may be due to the rigidity of the nucleic acid chains which will not allow them to adapt their spatial shapes to the manifold requirements of specific catalysts.

In contrast, the protein molecules can take on innumerable shapes and very finely adjust them to specific needs. Their backbone is more flexible, and moreover many kinds of side groups are available, namely twenty in the contemporary proteins. Thus an extraordinary multiplicity of catalytic functions can be fulfilled by the proteins. On the other hand, no general direct mechanism for replication of proteins by copying is known. The reason is that amino-acids do not interact specifically (1:1), as nucleotides (bases) do. Interaction between protein chains rather occurs on the level of the tertiary structure, and is of a different nature.

On the basis of the composition of DNA, the "genotype", each kind of cell has definite metabolic capabilities, a metabolic potential. The genotype of a cell remains, as a rule, constant. Apparently it changes only through mutations and through horizontal gene transfer [*2c*]. The genes are transmitted in division to daughter cells, and at certain levels of complexity parasexual and sexual processes for the reassortment (reshuffling) of the genes are observed.

In given conditions and with a given history of the cell, the metabolic capabilities may be expressed or not. Thus, the "phenotype" is defined by the reactions that actually occur in the cell during its life. The phenotype of a cell can quite easily change with time. The case caterpillar–butterfly provides a drastic example. Within the multicellular organisms the division of labour clearly affects the extent to which the biochemical, also bioenergetic, capabilities are developed and exploited, or neglected and lost. The processes actually occurring within the various, differentiated, kinds of cells of, say, a tree or an adult man are, of course, quite diverse.

Higher organisms contain far more DNA in their cells than can be accounted for directly, through protein production. It is thought that much of the DNA is concerned in some way with regulation, the scope of which increases with the complexity of the organism (see for example, Kohne, 1970; Britten and Davidson, 1971; Davidson *et al.*, 1973; also *Science 182*, 1009 [1973]).

In respect to the mechanisms of evolution, no strict terminological precision will be maintained in this book. Simpler and at times also more colourful expressions may be used. For instance, it may be said, without pedantry, that organisms evolve in response to changed circumstances. I am not afraid of the teleological flavour. May the reader forgive me! Whoever prefers the term, may replace "teleology" by "teleonomy" (Pittendrigh, 1958). It goes without saying that evolution is considered to proceed, as generally assumed, through mutation, reassortment and selection. Concerning modern views on the role of mutation in microbial evolution, see Drake (1974), and of selection, Kubitschek (1974).

(c) Gene Transfer and Evolution

Genes are transferred from one generation to the next ("vertically"). In higher organisms, this is (with or without sexuality) practically the only possibility for transfer. Therefore, the concepts of descent and of the phylogenetic tree are unambiguous.

Among primitive organisms, notably the bacteria, "horizontal" gene transfer between cells is also observed and it may have important consequences for phylogeny and taxonomy (Anderson, 1966; Richmond and Wiedeman, 1974; Sneath, 1974). It is quite a threat to the species concept among the bacteria. Bacterial cells have been considered as composite mosaics (Pollock, 1969). One author (Hedges, 1972) has even proposed that

> perhaps it would be best to think of the bacteria as constituting one gene pool from which any "species" may draw genes as these are required . . . A (bacterial) taxonomy depicting well-defined species or groups of related species would be inherently unnatural whilst the real situation would be better represented by a picture of a continuous gene pool in which the "species" represent sets of genetic information which happen to be adaptive.

At least four relevant mechanisms are known (see Jones and Sneath, 1970): genes are transferred as soluble DNA molecules ("transformation"), they are transferred on cellular particles (plasmids), a whole part of the bacterial chromosome is transferred ("conjugation"), or, finally, DNA from a host is "transduced" to an acceptor cell with or instead of the DNA of a bacteriophage. Gene transfer is possible not only intraspecifically, but also interspecifically and indeed intergenerically. "It is not beyond all possibility that genes may be transferred across the whole taxonomic spectrum of bacteria, though doubtless they would need to be transmitted in steps from one genus to nearly-related genera, in a chain-like fashion."

For the time being, it may at least be maintained that interspecific transfer is the more difficult and the more rarely observed the farther the species are removed. For instance, a transfer between Gram positive and negative genera [*17a, b*] has not been established so far (Jones and Sneath, 1970). DNA from widely differing organisms would not find the proper milieu in which its products could act as enzymes. If it nevertheless breached the "nucleic acid intrusion barrier", it could play havoc with the metabolism of the host (Pollock, 1969). To some extent, transfer is, in Nature or in the laboratory, a question of suitable technique. After all, viruses do succeed in breaking through the barriers and in injecting their DNA.

If the most radical view really were true, it would be futile to build up a phylogenetic tree on the basis of information about bioprocesses, e.g. bioenergetic processes. At least, such a tree would not look like a proper tree, but would show all kinds of cross-connections. One would have to accept quite erratic changes, as groups of genes are transferred from one group of bacteria to the other. If there were really full promiscuity, one would have to tolerate concepts that appear quite bizarre. For instance, one could not exclude that a group of (primitive, strictly anaerobic) clostridia suddenly turned into aerobic respirers!

However, no observations, experimental or otherwise, force us to such drastic conclusions as yet. Evolution by "accretion" may be rare (Mandel, 1969; Pollock, 1969). While gene transfer among different kinds of bacteria is no doubt important, it still appears that the most fundamental properties of the cell are not involved. Notably, the bioenergetic processes (fermentation, photosynthesis, respiration) require large numbers of well-integrated genes. The whole organization of the cell must fit the processes to obtain useful energy and to perform work. For the time being, it will therefore be reasonable to admit that the bioenergetic processes can be modified in detail by interspecies and intergenus gene transfer, but to maintain that the main lines of evolution are still determined by ordinary, vertical, heredity.

(d) Processes and Substances

Life is a thing "that happens", as Gowland Hopkins used to say. Its chemical processes, as an integrated whole, are its essential feature, and are described in "dynamic biochemistry". The biomolecules (nucleotides, proteins, carbohydrates, lipids, photosensitizers, minerals, etc.) may be looked upon as pillars to hold the fabric of the processes, and the pillars are held together and maintained by the fabric. The processes depend on the substances, and the substances depend on the processes.

It follows that bioenergetic evolution could be described in terms of substances rather than processes. Given sufficient knowledge, the two descriptions must be equivalent. But while the activities of a given cell, more particularly bioenergetic processes, may often be well described in terms of dynamic biochemistry, a sufficient description in terms of biomolecules and their spatial order will be practically impossible for a long time to come.

Yet evolutionist dynamic biochemistry needs collaboration with "chemotaxonomy" (Rochleder, 1854; Hegnauer, 1962; Swain, 1963, 1966, 1971, 1973, 1974; Alston, 1966, 1967), i.e. comparative studies of the chemical composition of the species. Chemotaxonomy has led even before the rise of molecular biology to what has later been called a "molecular approach to phylogeny" (Florkin 1944, 1966, 1974). Where information about the detailed structure of the biomolecules and about their supermolecular aggregates exists, the potential contribution to evolutionist knowledge is tremendous. Among specific products of cells, i.e. biochemical markers (Aaronson and Hutner, 1966), we may mention pigments (carotenoids! see Goodwin, 1971b; Swain, 1974) and components of the cell envelope (Swain, 1974); the latter are also responsible for the Gram reaction [*17a, b*].

The blueprint for the synthesis of biomolecules rests in the last analysis in the DNA [*2b*]. Indications about evolutionary kinship may therefore be sought in the comparison of "GC ratios". Chargaff (see Chargaff, 1955) had observed that the molar fractions of guanine and cytosine on the one hand, and of adenine and thymine, on the other hand, in any given preparation of natural DNA are identical. This result is understandable, indeed necessary, in the light of the later Crick–Watson double helix model, established in 1953. Adenine pairs with thymine, and guanine with cytosine. Hence the base composition of any DNA species is defined by a single parameter. This has been determined for many groups of organisms, and has among bacteria been shown to vary between 23 and 74% GC (see, for example, DeLey, 1968). This implies, of course, 77–26% for AT (adenine and thymine). Related groups are supposed to have GC ratios that do not differ widely.

(A warning note has, however, been struck (Singer and Ames, 1970). The GC content has been found to be positively correlated to ultraviolet (UV) exposure in the ecosystem where particular kinds of the bacteria normally live. This may be due to the well-known higher sensitivity of thymine against photochemical attack (formation of dimers). Possibly, then, the GC content is not wholly indicative of kinship.)

Stronger information about similarities and dissimilarities between DNA molecules from different species ("DNA homology") is obtained in hybridization (base pairing) experiments. Various techniques for the determination of homology and a large number of results exist (Doty *et al.*, 1960; Ravin, 1960; Marmur and Lane, 1960; Schildkraut *et al.*, 1961; Marmur *et al.*, 1963; Sneath, 1964; Hoyer *et al.*, 1964; DeLey, 1968b; Mandel, 1969; Swain, 1974).

A maximum of information would be provided by the base sequences in DNA. Unfortunately, because of the technical difficulties (see Mandeles, 1972; Salser, 1974) the entire base sequence of a DNA molecule has not yet been obtained. But portions of such molecules are translated, via messenger RNA, into the protein language, and so they define the aminoacid sequences in the protein molecules. The protein contains less information than the correspondent part of the DNA molecule, as several differing triplets of nucleotides may code [*5c*] for one and the same aminoacid. Nevertheless, the importance of sequence studies with proteins for judging phylogenetic relationship is tremendous (Crick, 1958; Anfinsen, 1959; Zuckerkandl *et al.*, 1960; Zuckerkandl and Pauling, 1962, 1965; Dayhoff *et al.*, 1974; Wu *et al.*, 1974). The determination of full aminoacid sequences has been technically feasible for many years (see Fitch and Margoliash, 1970; Margoliash *et al.*, 1971; Dayhoff, 1971; Jollès and Jollès, 1971).

The question whether the secular changes in protein composition were, on the basis of selection, biologically useful, as traditionally believed, or neutral (Kimura, 1968a,b; King and Jukes, 1969) has not been settled (see, for example, Zuckerkandl, 1963; Mandel, 1969; Fitch and Margoliash, 1970; Richmond, 1970; Margoliash *et al.*, 1970; Kimura and Ohta, 1974). Discussions of this important question from the point of view of organismal rather than molecular biology are owed to Simpson (1964) and Ayala (1974).

A favourite object for sequencing has been cytochrome c (Margoliash, 1963, 1964; Fitch and Margoliash, 1967, 1970; Nolan and Margoliash, 1968; Margoliash *et al.*, 1970; Dickerson, 1971, 1972). Other proteins used in similar research include fibrinopeptide, histone IV and haemoglobin (Dayhoff, 1969b; Dickerson, 1971). Selected proteins from higher plants have also been sequenced (Boulter, 1970, 1973; Boulter *et al.*, 1970a,b, 1972; Watts, 1970).

Unfortunately, most work with cytochrome c has been done with higher organisms. Bacteria, most interesting from the point of view of bioenergetics, have been largely neglected. One reason may be that it is not always obvious which cytochrome of a given bacterium ought to be correlated with cytochrome c in higher organisms. The other proteins mentioned do not even occur in bacteria. Only recently the sequence method with cytochrome c has been applied to the photosynthetic bacterium *Rhodospirillum rubrum*, and with a kind of transfer RNA to other bacteria (McLaughlin and Dayhoff, 1970). Probably the ancient protein ferredoxin [*7d*], which may occur in all organisms, will emerge as the most important object of sequence analysis for phylogeny (Lipmann, 1965; Eck and Dayhoff, 1966; Cammack *et al.*, 1971; Hall *et al.*, 1971a, 1973 a, b; Dayhoff, 1972; Wickramasinghe, 1973; Dayhoff *et al.*, 1974; Hall *et al.*, 1974, 1975).

(Incidentally, Yamanaka (1972, 1973) has attempted to use the differential reactivity of cytochrome c with cytochrome oxidase and also with nitrite reductase from various sources for tree construction, with bewildering results!)

Through sophisticated computing methods (Dayhoff, 1969a, b), the results of sequencing are used to draw up partial phylogenetic trees. Moreover, if information from palaeontology about the time elapsed since certain branchings is used for calibration, average rates of evolutionary change can be calculated; they differ for different informational macromolecules. On the basis of knowledge of rates the time of the separation of vertebrates from invertebrates has been estimated as 0.6 gigayears (Gy)† before now. The higher organisms (eukaryotes (*18*): plants except blue-green algae, animals, fungi) diverged from the bacteria (prokaryotes; *18*) 2.7 times further back than the groups of eukaryotes diverged from each other (McLaughlin and Dayhoff, 1970). Reference to information about the evolution of biological energy conversion from protein structure will be made later [*8g*].

It has been pointed out that with bacteria horizontal gene transfer may be a source of errors also in phylogenetic studies through sequence analysis (Hedges, 1972).

(e) Bioenergetics and Classification

For classification, biologists have traditionally relied mainly on the macrostructure of the organisms and their external organs (morphology), and on macrophysiological processes, notably reproduction and locomotion. This approach is certainly successful with higher plants, fungi and animals, but for primitive organisms the additional criteria of microphysiology and biochemistry are needed. Claude Bernard, A. Dohrn, O. Jensen, H. Molisch and Gowland Hopkins were among the first to have stressed the importance of these criteria (see Molisch, 1907; Lwoff, 1944; Florkin and Mason, 1960). Among the champions of the physiological approach has also been Lwoff (1944):

> Les schizomycètes (bacteria, E.B.), en raison de la variété de leur métabolisme . . . devraient constituer un matérial remarquable pour l'étude de l'évolution physiologique. La comparaison des processus biochimiques en rapport avec leur évolution devrait fournir de documents de première importance . . .
>
> Les caractères morphologiques des bactéries ne permettent pas une distribution satisfaisante en ordres et en familles. Les seuls critères dont dispose le morphologiste sont la forme, la présence des cils et les modalités de leur insertion, la présence ou l'absence des spores. Aucun de ces caractères présente de stabilité suffisante.

† 1 Gy = 10^9 years; also known as 1 aeon or eon. As an admirer of Boltzmann, I am reluctant to assign a definite value to an aeon (see Broda, 1955).

The relative merits of "morphology" and "physiology", to introduce short terms, for classification have been discussed many times (see Murray, 1962; Sneath, 1962; Mandel, 1969). Broadly speaking, physiology is the more useful the more primitive the organism, as is expressed by Lwoff. On the other hand, physiological classification is obviously hardest to carry through or even practically impossible for higher organisms where enormous numbers of morphologically distinct species are known. Significant metabolic differences are subtle and generally unknown. But it should not be forgotten that in the last analysis morphology, including micromorphology, and physiology are interdependent.

Many years ago, Kluyver and Van Niel (1936) have concluded that for the classification of bacteria "morphology remains the first and most reliable guide", and this view has still been repeated by Stanier and Van Niel (1941) and Van Niel (1946, 1955). Nevertheless, already at that time these authors have underlined the importance for classification of the physiological, and more particularly, catabolic processes. These processes mostly serve energy supply. Stanier has more recently stated that metabolic unit processes are the functional properties most appropriate for taxonomic purposes.

The physiological approach has been emphasized more strongly, in respect to the eubacteria (true bacteria), in the modern text by Stanier *et al.* (1966):

> Their diversity is expressed only to a limited extent by differences in the structures of the cells since the structural simplicity of these organisms severely restricts the number of different categories that can be established purely on the basis of structural attributes. The diversity of the eubacteria, rather, receives its major expression in the unparalleled range of specialized physiological and biochemical types which exist among them . . . Most bacteria can be adequately defined only by stating what they do, not simply by how they look . . .

Yet as an illustration of the persisting difficulties the authors point to the specific case of the photosynthetic purple bacteria [*9, 11*], united by their distinctive energy-yielding metabolism and, in connection with it, by their common possession of a specific array of pigments. There are, however, three different structural types: rods or spirals that reproduce by binary fission, and one oval organism that reproduces by budding. For each of these three types structural counterparts are found among the non-photosynthetic eubacteria. Possibly in this particular case the conversion hypothesis [*14e*] may show the way out. Generally any apparent contradiction between the results of morphology and of physiology–biochemistry, including bioenergetics, should be looked at as a challenge and an invitation to deepened analysis.

In the end those criteria may prevail that allow the best approach to a natural (phylogenetic, i.e. evolutionary) classification of the organisms, as opposed to a phenetic classification, based on the properties of the organisms today. The goal of a fully phylogenetic classification is, of course, so remote that as yet no unanimity has been reached whether it should be approached at all. This latter problem has recently been debated again by Bisset (1962), Cain (1962), De Ley (1962), Heslop-Harrison (1962), Sneath (1962) and Heywood (1966).

The phenetic classification has been given a modern form in "numerical taxonomy" (see Sneath and Sokal, 1973; Sneath, 1974), where characters are counted, and degrees of relatedness subsequently obtained by computing. The weighting of characters is omitted, although the tremendous intrinsic interest of phylogenetic relationship is, of course, not disputed, and this probably requires weighting. The goal in this approach is, then, to classify by criteria that are meant to be objective, and every well-informed observer is supposed ultimately to come to the same conclusions. However, it has been pointed out that already the choice of the characters to be used involves subjective judgment (see for example Gould, 1974).

In morphology, in physiology or in biochemistry, convergence may lead to deception in respect to phylogenetic relationships. A striking example of (presumably) convergent evolution is that of the pathway for the breakdown and utilization of benzoate and related compounds (Canovas *et al.*, 1967). Both in *Acinetobacter* and in pseudomonads, groups believed to be very distant, the same intermediates (about a dozen) are observed, while the mechanisms of regulation differ. It is thought that an identical pathway of degradation has in this instance been forced upon the organisms by the chemical properties of the substrates; apparently no alternative pathway was chemically feasible. However, the horizontal transfer of a whole gene complex [*2c*] with subsequent changes in the regulation mechanisms would be an alternative possibility.

Kinship has been compared to plagiarism (Orgel, 1973). When two patent applications from different people claim rights to the same invention, and the texts are also in essence identical, i.e. when at least one of the applicants has copied the idea, this corresponds to the situation where not only the metabolic products, but also the pathways leading there are identical. Kinship is likely in such a case. In contrast, fundamental difference in the pathways indicates convergence. But the criterion may fail, as the benzoate example shows.

The power of convergence in the cases, say, of fins (fishes, dolphins, penguins) or of wings (birds, bats, extinct reptiles) is generally known. Perhaps convergence in behaviour is less appreciated. Who recalls that, e.g. foxes and geese, both of which are devoted husbands, wives and parents, are phylogenetically connected merely through the surely less virtuous reptiles?

3

EARLY CONDITIONS ON THE EARTH

(a) Formation of the Planets

It is generally thought now that the planets formed along with the Sun by accretion of components of the "solar nebula", of gas, dust and meteorites (see Kuiper, 1952; Urey, 1952a, 1959; Birch, 1965; Larimer and Anders, 1970; Larimer, 1973). Some authors (Ringwood, 1960, 1966 a, b; Cameron, 1970; Oversby and Ringwood, 1971; Arrhenius, 1974; see Anders, 1971; Anderson *et al.*, 1974) think that accretion was so fast that the Earth was heated to high temperatures by the gravitational energy released. In the molten Earth, heavy components (iron) could rapidly move towards the centre. In this "fall" further gravitational energy was set free as heat.

Alternatively (Urey, 1952a, 1953, 1957, 1959, 1962; Kuiper, 1952; Vinogradov, 1961), the Earth remained fairly cool at first; only subsequently its interior was slowly heated up and melted by radioactivity. Radioactivity is, of course, the process that maintains the interior of the Earth hot at present, but in the early time after accretion far more radioactive matter was present in the Earth than is left now. Various kinds of intermediate hypotheses are also possible.

Preplanetary matter consisted largely of hydrogen; in the solar system as a whole, hydrogen is still, although it has long served as the Sun's fuel, far more abundant than all other elements combined. Probably helium was the next common element in preplanetary matter. The interpretation of the differences between the planets is the task of "comparative planetology" (Young, 1973).

The heavy and distant cold planets, notably Jupiter and Saturn, still have what may be a primary atmosphere, rich in hydrogen (see, for example, Urey, 1959; Rasool, 1968, 1972; Oró, 1972; Lewis, 1973; Newburn and Gulkis, 1973; Hunt, 1974). But during the accretion of the smaller and warmer (Earth-like) planets, hydrogen, helium and other primordial gases escaped into space. Evidence for this loss is the near-absence of the inert gases neon, krypton and xenon from the present terrestrial atmosphere (Aston, 1924; Russell and Menzel 1933; Brown, 1949, 1952; Suess, 1949, 1962). The original (cosmic) abundances of these inert gases are normal. They are about as common as other elements in their neighbourhood in the periodic table. But not being chemically bound in solids, and remaining in the primary atmosphere instead, they were defenceless against loss during accretion.

(b) Age Determinations

Physicists and geologists used to disagree how long a time span could be allotted to the Earth. Thus in 1899 Lord Kelvin complained that the geologists demanded practically unlimited time for the processes envisaged by them. But he, as a physicist, thought he could not provide more than about 20 million years for the time since the consolidation of the surface of the Earth (see Badash, 1968; Brush, 1969).

Only after radioactivity had been discovered as a long-term heat source, and after reliable age determination methods based on nuclear transmutation (see Hamilton, 1965; York and Farquhar, 1972) had been introduced by Rutherford in 1904–5, a long time scale has definitely been accepted by all. The oldest rocks now known have been found in South Africa (3.5 Gy; Hickman, 1974), Greenland (3.7 Gy; Moorbath *et al.*, 1973) and Minnesota (3.8 Gy; Goldich and Hedge, 1974).

The first useful radiometric method was the well-known "overall" uranium–lead method. Basically, the amount of Pb accumulated in a rock is compared, on the assumption that the rock has remained undisturbed all along, with the amount of parent U in the same rock. This crude method can be applied to rocks whose initial ("primeval") Pb content was relatively small; this is true especially for U ores. The method has later been greatly refined by separate consideration of the different isotopes of U and Pb, measured with the mass spectrometer. It is of great value that the Pb isotope 204 is not radiogenic so that its abundance indicates the original Pb content of the mineral.

Other important methods based on radioactive decay are the potassium–argon and the rubidium–strontium method. In the fission track method (Fleischer *et al.*, 1965 a, b), the number of tracks in a suitable mineral, as counted with the microscope after etching, is compared with the uranium content of the mineral. The tracks in terrestrial minerals are due to the spontaneous fission of U238; the fission fragments carry far more energy than the α-rays from the uranium so that with a suitable technique they alone are registered.

Methods for the estimation of the age of the whole Earth are based on the content of radiogenic lead in whole ("rock") lead (Gerling, 1942; Holmes, 1946; Houtermans, 1946, 1953; Patterson, 1956; see York and Farquhar, 1972). In the most modern form, the argument runs roughly as follows. At the beginning, terrestrial Pb had the same isotopic composition as now Pb has in those meteoritic minerals that are practically free of U and Th. (To keep the argument simple, we shall forget about Th.) Now under certain assumptions the share of radiogenic Pb in any given sample of rock Pb (from uranium-free minerals) depends (1) on the time interval between the formation of the Earth and that of the particular rock, and (2) on the U content in the part of the Earth from which the radiogenic Pb came before the mineral formed. In principle, two equations are available for the two unknowns. First, the age of the rock (from now!) is measured by standard radiometric methods. Secondly, the ratio of the amounts, in the rock, of radiogenic ^{206}Pb (derived from ^{238}U) and radiogenic ^{207}Pb (derived from ^{235}U) is, as they come from the same U pool, uniquely determined by the time interval between the formation of the Earth and that of the rock. From a series of such analyses, with different rocks, ages of the Earth around 4.6–4.7 Gy have been computed. A recent value is 4.66 Gy (Sinha and Tilton, 1973). Even with the assumption of a hot start of the Earth, the solid crust cannot be much younger, as cooling must at first have been fairly rapid.

The greatest known ages of meteorites are similar to that of the Earth (Gerling and Pavlova, 1951; Urey, 1957; Wänke and König, 1959; Anders, 1962, 1963, 1964; Kirsten *et al.*, 1963; Wood, 1968; see Sinha and Tilton, 1973). The age of the oldest lunar rocks (Tatsumoto *et al.*, 1972; see Anderson and Hinthorne, 1973) is also the same order. In view of the concordance of the results, this age may be taken as that of the planetary (solar) system as we know it now.

It is remarkable that the oldest rocks found on Earth are so much younger than those on the Moon or in meteorites. A violent early event may have consisted in the capture of the Moon by the Earth, followed by intense heating and volcanism (Singer, 1968a; Cloud,

1968a, 1972; Alfvén and Arrhenius, 1972; Turcotte *et al.*, 1974). This event would have greatly helped the rapid segregation of the core. It would also have destroyed all evidence about the early crust, hydrosphere and atmosphere. It is not clear why in this case the ancient parts of the Moon crust were not obliterated. Also survival cannot have been easy for any of the terrestrial organisms.

(c) The Secondary Terrestrial Atmosphere

After the Earth had more or less reached the present size and structure, and most of the primary atmosphere had been lost, a new atmosphere was formed by outgassing (volcanism, hydrothermal phenomena and related processes), a "secondary" atmosphere (Chamberlin, 1897; Chamberlin and Salisbury, 1905; Urey, 1952a, b, 1959; Rubey, 1951, 1955; Holland, 1962; see Kuiper, 1952). It is mostly thought that outgassing was a gradual process, but on Singer's theory [*3b*] impulsive heating led to a sudden release of gas. Gravity now being strong, the secondary atmosphere has been mostly held back by the Earth.

The importance of outgassing is demonstrated in a semiquantitative way by the presence of argon 40. This is one of the decay products of the natural radionuclide potassium 40 (von Weizsäcker, 1937) and must, to judge by its quantity, largely have come from the mantle rather than from the crust of the Earth (Damon and Kulp, 1958; Schwartzman, 1973). But the secondary atmosphere evolved not only owing to outgassing, but also owing to escape of light gases from the Earth, and later owing to the activities of life.

Some think that the secondary atmosphere initially consisted, in addition to water vapour, partly of other volatile hydrides: methane, ammonia, and hydrogen itself (Urey, 1952 a, b, 1957, 1959; Miller and Urey, 1959). This atmosphere was reducing. It may also have contained carbon dioxide and nitrogen.

According to a number of other authors (Rubey, 1955, 1964; Abelson, 1966; Cloud, 1968 a, b) the secondary atmosphere was devoid of CH_4 and NH_3 (almost) from the beginning. Rather, the atmosphere is thought to have consisted (almost) all along, in addition to H_2 and to H_2O, mostly of CO_2, CO and N_2. The exergonic reaction

$$CO_2 + 4H_2 = CH_4 + 2H_2O;\ \Delta G_0 = -31 \text{ kcal} \tag{3.1}$$

and other, similar, reactions may not have been rapid. Of course, even a CO_2-dominated atmosphere was reducing as long as there was CO and H_2. An atmosphere consisting merely of CO_2, H_2O, N_2, but free of O_2, an atmosphere which probably came into existence only later, might be called neutral.

A kind of compromise solution, argued in much detail quantitatively, has been offered by Holland† (1962, 1964; see Kenyon and Steinman, 1969). Methane, H_2 and, to a lesser extent, H_2O, H_2S and NH_3, prevailed as long as elementary iron was within reach of the crust, in the upper part of the mantle. Stocks of metallic Fe kept the gases reducing that emerged to the surface. Only after the formation of the iron core had been completed, maybe 0.5 Gy after the birth of the Earth, CO_2, H_2O and N_2 gradually took the place of CH_4, H_2 and NH_3. It is quite likely that volcanoes continued to emit reducing gases for a long time so that a stationary, though small, concentration of hydrides persisted.

The CO_2 pressure probably was never very large. Otherwise the acid seas would have eroded rocks far more strongly than they did. Urey (1952 a, b; see Rubey, 1955; Holland,

† I am indebted to H. D. Holland for much oral and written information. E.B.

1965) argues that the reaction of CO_2 with silicates in presence of water, expressed schematically by

$$CaSiO_3 + CO_2 = CaCO_3 + SiO_2 \tag{3.2}$$

prevented accumulation of CO_2 ("Urey equilibrium"). Here $CaSiO_3$ stands for all kinds of silicates. The equilibria always lie far to the right. $CaCO_3$ also gradually precipitated in the oceans, which at that time may have contained a fair amount of silica (Holland, 1965).

Free hydrogen was lost into space from the top layers of the atmosphere, the hot exosphere (Urey, 1959; see Miller and Orgel, 1974). Unfortunately, the rate is not known with any certainty. It has been argued that because of the higher thermal conductivity and also of the higher infrared emissivity the gravitational escape of hydrogen from an early CH_4—H_2 atmosphere was much slower than from the present atmosphere, dominated by N_2 and O_2 (Rasool and McGovern, 1966). Whether escape was rapid or slow, certainly in addition to the primeval H_2 also the H_2 was lost that was formed in the decomposition of hydrides, including CH_4, NH_3 and H_2O, by UV light (Wildt, 1942; Dole, 1948; Harteck and Jensen, 1948; Kuiper, 1952; Urey, 1952 a, b; Ferris and Nicodem, 1972).

Whatever the history of the atmosphere was in very early times, the ratio of oxygen to hydrogen in the atmosphere was gradually driven to increasing values. Yet the atmosphere may still have remained reducing for a long time. This fact was crucial for chemical evolution [*4b*], and it also determined the path of early biological evolution. After the advent of life, fermentation must have contributed H_2 (Yčas, 1972). Indeed, this is still so in limited parts of the biosphere. As far as any free oxygen formed at all by action of light, it combined readily with reduced components of the crust, including hydrides, sulphides and ferrous iron [*25a*]. The atmosphere remained anoxygenic.

(The influence of life on the terrestrial atmosphere has become determining for oxygen [*25a*], for nitrogen [*16c*] and for carbon dioxide [*25c*]. Life has become a strong factor for homoeostasis [see Lovelock and Margulis, 1974].)

A minor component of the secondary atmosphere has all along been helium 4, always produced in α-decay (see Kockarts, 1973). However, most of this radiogenic helium, like free hydrogen, has been lost (Goldschmidt, 1938; Wasserburg, 1964). Argon 40, which is not lost, has been mentioned already.

This may be the place to ask the reader to excuse the absence of any discussion of isotope effects in this book. While important and often intriguing results in palaeobiology have been obtained, the author feels that the time for the inclusion of the data into a book of this kind has not yet come. The power of these data and also the sources of error ought to be discussed *ab initio* by one of the leading experimentalists in this field. Be it recalled that measured isotope effects are determined not only by the nature of the relevant chemical reaction and the temperature, but also by the degree to which the reaction in the given conditions went to completion. Further, there are trivial sources of error that are not easily appreciated. However, it is with regret that this important source of information has been left out here.

(d) Conditions on Earth-like Planets

Why do the atmospheres of the Earth-like planets Venus and Mars differ so much from that of our planet, and why are their climates so inhospitable? In respect to Venus (see Urey, 1959; Ingersoll and Leovy, 1971; Lewis, 1971), one view (Rasool and DeBergh, 1970) is that owing to the lesser distance from the Sun and the consequent higher initial surface temperature outgassed water could not, in contrast to the Earth, condense at any stage. So the abundant CO_2 (now about 75 atmospheres) did not establish the Urey equilibrium with silicates. Remaining in the atmosphere and efficiently holding back heat, the CO_2 made the planet a "runaway greenhouse".

The absence of water vapour, except for small amounts, has been explained by photochemical decomposition by solar UV. The hydrogen liberated in this reaction escaped, and the oxygen was absorbed by rocks. Yet there is a possibility that the clouds high above the surface of Venus consist at least partly of ice (see Hunt and Bartlett, 1973; however: Lewis, 1971; Janssen *et al.*, 1973).

The general situation is different for Mars (see Berkner and Marshall, 1967; Glasstone, 1968; Ingersoll and Leovy, 1971; Hammond, 1973; Young, 1973). The atmosphere likewise contains some water, but certainly far less CO_2 than Venus. Molecular oxygen exists on Mars in traces, presumably in a stationary state as a photochemical product (McElroy and Donahue, 1972; Carleton and Traub, 1972). The presence of something of an atmosphere is also indicated by violent dust storms.

Being relatively far from the Sun, and without a greenhouse effect, Mars is cold. According to some authors, the planet may have had, as a consequence of outgassing, an appreciable secondary atmosphere that was lost because of the smaller mass (Urey, 1952 a, 1959). Some conspicuous surface features are interpreted as due to erosion by water, though little water is observed on Mars now. It is also possible that the reddish colour of the surface rocks is due to Fe oxidized by the oxygen that remained after photolysis of water and loss of hydrogen. If Mars has really seen better times, life may gradually have adapted to the present harsh conditions (Ponnamperuma and Klein, 1970).

According to others, there were no Earth-like periods on Mars (Murray and Malin, 1973). Possibly because of the smallness of the mass internal activity may not yet have been sufficient to produce an appreciable atmosphere, and the time for the evolution of life may yet come on Mars (Rasool and DeBergh, 1970). A violent future for Mars may be suggested by the recent discovery of areas of tremendous volcanism and tectonic activity (Murray, 1973).

Many other solar systems in the Universe may contain planets that are, or were, or will be, suitable for the evolution of life (Brown, 1952, 1964; Shapley, 1958; Shklovsky and Sagan, 1966; Ponnamperuma and Klein, 1970; Sagan, 1973b, 1974). This thought has first been expressed by the martyr of science, Giordano Bruno (1584; see Spampanato, 1921; Michel, 1973).

4

ORIGIN OF LIFE

(a) The Concepts of Oparin and Haldane

Largely as a result of the famous experiments of Pasteur, published in 1862, which destroyed any belief in contemporary spontaneous generation, interest developed in the problem how life arose in the distant past. This problem of "biopoiesis" (Pirie, 1953) was discussed, among others, by the experimental physicist Tyndall in 1874. Too little known now are in particular the ideas of Boltzmann, who was not only a theoretical physicist, but also a fervent partisan of Darwin, on the early history of living matter. In 1886 he wrote (all quoted by Broda, 1955):

> Wir machen die Hypothese, es hätten sich Atomkomplexe entwickelt, die imstande waren, sich durch Bildung gleichartiger um sich herum zu vermehren. Von den so entstandenen größeren Massen waren jene am lebensfähigsten, die sich durch Teilung zu vervielfältigen vermochten, dann jene, denen eine Tendenz innewohnte, sich nach Stellen günstiger Lebensbedingungen hin zu bewegen. Dies wurde sehr gefördert durch Empfänglichkeit für äußere Eindrücke, chemische Beschaffenheit und Bewegung des umgebenden Mediums, Licht und Schatten usw.

and in 1904, in a controversy with Ostwald about "Happiness"(!)

> Ob sich während der Jahrmillionen in der enormen Wassermasse auf der Erde das erste Protoplasma 'durch Zufall' im feuchten Schlamme entwickelte, ob Eizellen, Sporen oder sonstige Keime in Staubform oder in Meteoriten eingebettet einmal aus dem Weltenraume auf die Erde gelangt sind, kann uns hier gleich gelten. Höher entwickelte Individuen sind kaum vom Himmel gefallen. Es waren zunächst also nur ganz einfache Individuen, einfache Zellen oder Protoplasmaklümpchen vorhanden. Stete Bewegung, die sogenannte Brownsche Molekularbewegung, ist ja, wie man weiß, allen kleinen Klümpchen eigen; auch ein Anwachsen durch Aufsaugen ähnlicher Bestandteile und eine nachherige Vermehrung durch Teilung ist auf rein mechanischem Wege vollkommen begreiflich. Ebenso begreiflich ist es, daß die raschen Bewegungen durch die Umgebung beeinflußt und modifiziert wurden. Solche Klümpchen, bei denen diese Modifikation in dem Sinne erfolgte, daß sie sich durchschnittlich (mit Vorliebe) dorthin bewegten, wo es besser zum Aufsaugen geeignete Stoffe (bessere Nahrung) gab, gelangten besser zum Wachstum und häufiger zur Fortpflanzung und überwucherten daher bald alle anderen.

But physiological and biochemical knowledge was insufficient for fruitful work, and so the interest waned. The discussion was resumed only after the publication of the pioneer papers by Oparin (1924) and, independently, by Haldane (1929). It gained momentum after Oparin's classic *The Origin of Life* appeared in Russian (1936) and in English (1938).

The essence of the ideas of Oparin and Haldane is a Darwinian approach to events on the early Earth. The new ideas, which have been generally accepted, cannot be treated here in any detail, and for fuller information reference must be made to monographs (Oparin, 1938, 1961, 1964; Bernal, 1967; Kenyon and Steinman, 1969; Rutten, 1971; Orgel, 1973; Miller and Orgel, 1974) and to articles (Sagan, 1961; Wald, 1964; Pattee, 1965; Oparin, 1965b; Hanson, 1966; Lemmon, 1973a). A popular book with beautiful illustrations is that of Ponnamperuma (1972a).

(b) Organic Matter in Space and in Meteorites

The existence of life may be taken as no poor proof that organic substances were available in the secondary atmosphere and in the coexisting hydrosphere. These two spheres led in continuity to the later biosphere. We shall term the production of biomolecules "abiosynthesis" when it takes place without participation of organisms (abiotically). The secular changes in the composition of organic matter in the abiotic hydrosphere through chance and selection are known as "chemical evolution".

Many kinds of small molecules and radicals, often organic, have been observed in space by radioastronomy (Buhl, 1971, 1974; Sagan, 1972; Breuer, 1974; see also Fox, 1971a; Oró, 1972). Typical examples are OH, CO, HCN, HCHO, and even CH:C.CN, cyanoacetylene. Larger organic molecules have been found in a class of meteorites, the "carbonaceous chondrites" (see Du Fresne and Anders, 1962, 1963; Studier *et al.*, 1965; Hayes, 1967; Mueller, 1967; Eglinton and Murphy, 1969; Baker, 1971; Ponnamperuma, 1971; Lawless *et al.*, 1972b; Anders *et al.*, 1973). Famous examples are the Orgueil, the Murray and the Murchison meteorites, fallen in 1863, 1950 and 1969, respectively. The story of the views on the carbonaceous chondrites has been told by Brooks and Shaw (1973). Meteorites of this kind, especially so-called "type I" carbonaceous chondrites, may, apart from lost volatiles, most closely approximate primordial solar system matter (Ringwood, 1959, 1966b; Wood, 1968; Anders, 1971; Anders *et al.*, 1973; Larimer, 1973; Grossman, 1975). The carbonaceous chondrites may really be fairly abundant, compared to others, but they easily disintegrate during and after the fall through the air.

By means of modern methods of analysis, especially gas chromatography, a large number of organic compounds has been identified in carbonaceous chondrites. Among them are many biomolecules, notably aminoacids and bases, including adenine and guanine (Kvenvolden *et al.*, 1970a, 1971; Oró *et al.*, 1971 a, b; Lawless *et al.*, 1971, 1972 a, b; Anders *et al.*, 1973; Lawless, 1973; Kvenvolden, 1974; Lawless and Peterson, 1975). Porphyrins have also been reported (Hodgson and Baker, 1969). Some of the substances are present in tiny amounts only.

In the case of the stones fallen a long time ago, especially of Orgueil (Anders *et al.*, 1964), contamination by terrestrial matter has taken place, but some of the more recently fallen stones have been handled cautiously and efficiently. The best case of a noncontaminated carbonaceous chondrite is probably Murchison. Contaminants have been studied and a history of the search for life in carbonaceous chondrites has been given (Claus, 1968; Claus and Madri, 1972). Strong arguments for lack of contamination are optical inactivity, e.g. of aminoacids, and presence of related molecular species that do not or seldom occur in living matter on Earth, e.g. among aminoacids of the isomers of aminobutyric acid (Kvenvolden *et al.*, 1970a; Oró *et al.*, 1970 a, b). Cell-like structures found in the chondrites are now thought to be abiogenic (see, for example, Rossignol-Strick and Barghoorn, 1971).

Now how was the array of rather complicated organic molecules formed? According to one school (Studier *et al.*, 1965, 1966, 1972; Anders, 1969, 1971; Hayatsu *et al.*, 1972; Anders *et al.*, 1973, 1974; Levy *et al.*, 1973) modified Fischer–Tropsch reactions of hot components of the solar nebula, notably CO, H_2 and NH_3, were catalysed by dust. The products of the reactions were partly trapped by solids in space during cooling. Model experiments were carried out to demonstrate similarity of the products with the mixture found in meteorites (Studier *et al.*, 1966). Photochemical reactions may also have taken place on the surfaces (Breuer, 1974).

In Fischer–Tropsch type reactions chemical equilibria are approached up to a point (see Dayhoff *et al.*, 1964; Eck *et al.*, 1966). According to Anders *et al.* (1973) the composition of the carbonaceous chondrites points to a temperature of production, or rather a temperature where the quasi-equilibria froze on cooling, around 360°K. Equilibria were, however, not fully established; in equilibrium, methane would be an important product and be lost into space (Anders, 1969), and graphite would also be strongly represented (see Urey and Lewis, 1966). It is assumed, moreover, that the resulting mixture was enriched in aromatics by loss of further volatiles at some stage. It is possible that the more complicated organic molecules now found in space are nothing else but volatiles lost; their direct formation in the high vacuum would be improbable.

Molecules from space and from meteorites may have supplied starting materials for terrestrial abiosynthesis (Lederberg and Cowie, 1958; Fowler *et al.*, 1961; Bernal, 1967; Oró, 1969). The trouble with them, if taken up during accretion, is that they found themselves mostly in the interior of the solid Earth. There they were, as a consequence of radioactivity, exposed to heating. So they must have been largely degraded to small molecules before they could, at least in part, reach the surface in degassing. Only organic material from carbonaceous chondrites (Anders, 1970, 1971) or from comets (Oró, 1961, 1965a, 1972) that was subsequently captured by the completed Earth could escape thermal degradation. Concerning the composition of comets, see Delsemme (1973), Rahe (1974) and Whipple (1974).

(c) Abiosynthesis on the Early Earth

Some further organic matter may have been produced in the early Earth in the interaction of metal carbides and nitrides with water (Oparin, 1924, 1964, 1965b), and preserved in cooler parts. Indeed hydrocarbons, carbides and elementary carbon are still present in very ancient rocks. From time to time, eminent authors have, following Mendeleev, suggested that part of the petroleum is of abiotic origin (see Robinson, 1964, 1966; Porfirev, 1971; however, Eisma and Jung, 1969).

However that may be, it is widely held that larger and more complicated molecules have been formed by input of energy into the terrestrial hydrosphere and atmosphere. This idea was explored first by Haldane (1929), who also was the first to propose definitely that life arose in absence of free oxygen. In experiments simulating early Earth conditions, as seen by Urey, mixtures of hydrides (CH_4, H_2O, NH_3, H_2S) were used at first. It is necessary to exclude free oxygen. In an oxidizing atmosphere no organic compounds are formed.

Reactions of hydrides are in most instances endergonic, and at low temperatures always slow. Therefore, energy is introduced. The energizing agent applied in the pioneer experiments was, for reasons of convenience, electric discharges (Miller, 1953). Among the products ("micromolecules") from the mixed reduced gases, many aminoacids have been found (Miller, 1953, 1955, 1957a,b, 1974; Miller and Urey, 1959; Horowitz and Miller, 1962; Miller and Horowitz, 1966; Friedmann *et al.*, 1971; see Lemmon, 1970, 1973b). It is remarkable that aminoacids now found in natural proteins are particularly frequent, e.g. glycine, alanine, aspartic and glutamic acid. Generally, of course, those abiosynthetic methods appear most promising that yield whole families of biomolecules.

The initial dominance of hydrides need not be assumed, as has been pointed out [*3c*]. Supporters of the concept of a more neutral early atmosphere that contained, in addition to H_2, mostly CO_2, CO and N_2, have produced HCN from $CO + N_2 + H_2$ by means of discharges (Abelson, 1966). HCN, being highly reactive, is a suitable starting material for subsequent reactions, as will be seen. A mixture $CO + N_2 + H_2$ is, of course, in the words of Holland, wildly improbable thermodynamically. Incidentally, as early as 1913 Löb had demonstrated the formation of glycine in the application of a discharge to $CO_2 + H_2O + NH_3$.

Being well absorbed, electric discharges (lightning, etc.) may in fact have been the most effective agent for abiosynthesis in Nature (Orgel, 1973). Alternatively, solar UV radiation may have been most important (Haldane, 1929; Urey, 1952b; Miller and Urey, 1959; Lemmon, 1970). UV must have been plentiful (see Sagan, 1965). As long as the atmosphere contained no, or practically no, free oxygen, UV was not captured by ozone and thereby prevented from becoming effective, as it is now. In the laboratory, abiosynthesis with UV radiation has been studied quite early. Groth and Suess (1938) obtained "carbon compounds that were probably the prerequisite for the evolution of organic life" from $CO_2 + H_2O$. Later experiments were by Terenin (1959), Groth and von Weyssenhoff (1960) and Ponnamperuma (1968).

The solar flux of long-wave UV (>200 nm) far exceeds that of short-wave UV (<200 nm). But most likely components of the early atmosphere (H_2, H_2O, CH_4, NH_3, CO, CO_2, N_2) hardly absorb the former. How, then, could they react? It is possible that through the initial (small) action of short-wave UV organic compounds with more complicated molecules and with absorption bands at longer waves were produced. The long waves may also have acted from the beginning through absorption in H_2S, likewise abundant in the early reducing atmosphere; in fact, aminoacids are produced from suitable starting mixtures, containing H_2S, with long-wave UV (Sagan and Khare, 1971; Hong *et al.*, 1974).

Various nuclear (strongly ionizing) radiations (Dose and Rajewsky, 1957; see Fox and Dose, 1972), shock waves (Hochstim, 1963, 1971; Bar-Nun *et al.*, 1970, 1971; Bar-Nun and Tauber, 1972; Bar-Nun, 1975) or magmatic heat (Harada and Fox, 1964; see Fox and Dose, 1972) have also been considered as energy sources for abiosynthesis.

Reaction mechanisms in abiosynthesis may be studied in the mass spectrometer by fragmenting the starting substances and identifying the products (see Lemmon, 1970). Because of their reactivity, radicals are especially important. Whether different energizing agents make the same final products depends on the extent to which the primary products equilibrate.

Among stable intermediates for aminoacid production, aldehydes and nitriles, including HCN, are important (Miller, 1957 a, b; Oró and Kamat, 1961; Oró, 1965a; Abelson, 1966; Ferris *et al.*, 1974; see Kenyon and Steinman, 1969; Lemmon, 1970; Miller and Orgel, 1974). One pathway is the well-known Strecker reaction:

$$R.CHO + NH_3 \rightarrow R.CH = NH \xrightarrow{HCN} R.CH(NH_2).CN \xrightarrow{H_2O} R.CH(NH_2).COOH$$

Likely intermediates further include cyanamide and its dimer (Steinman *et al.*, 1964, 1965; Schimpl *et al.*, 1965; Calvin, 1965), and cyanoacetylene (Sanchez *et al.*, 1966).

Nucleic acid bases—purines and pyrimidines—have likewise been made by chemical reactions in relatively mild conditions, with participation of cyanides (Oró and Kimball, 1962; Oró, 1965a; Sanchez *et al.*, 1967; Ferris *et al.*, 1968; Oró and Sherwood, 1974; see Lemmon, 1970, 1973b; Miller and Orgel, 1974). Nucleic acid bases may also be made by a Fischer–Tropsch type process (Yang and Oró, 1971). Sugars have been obtained by Gabel and Ponnamperuma (1967) from formaldehyde containing systems with energy supply. The spontaneous sugar formation from CH_2O had been observed by Butlerow (1861) and many others (see Oró, 1965a; Fox and Dose, 1972; Lemmon, 1973b). Concerning porphyrins, not easily produced, see Hodgson and Ponnamperuma (1968), Lemmon (1970, 1973b), and Fox and Dose (1972).

Bibliographies on chemical evolution, including lists of abiosynthetically made amino-acids, nucleic acid bases and other biomolecules, have recently been provided by Kenyon and Steinman (1969), Lemmon (1970), West and Ponnamperuma (1970), Fox (1971a), Fox and Dose (1972) and West *et al.* (1972, 1973, 1974, 1975).

(d) The Primeval Soup

The continued supply of external energy by radiation, etc., led to a build-up of substances of high free energy content on the early Earth. The products must have remained dissolved or suspended in the water or taken up from the atmosphere. In addition, asphaltic or tarry matter may have been formed (Grossenbacher and Knight, 1965; Lasaga *et al.*, 1971). On the other hand, decomposition also took place, so that a stationary state was approached.

The aqueous solution and suspension was the medium in which according to Oparin and Haldane life developed—Haldane's "dilute soup". It is, however, quite likely that many of the components were adsorbed by submerged rock, and underwent subsequent reactions there [*4e.*] Covered by water, the substances were now protected from destruction by further UV or electric discharges. At first, of course, there were no enzymes to catalyse reactions. Nevertheless, some kinds of reactions could be fairly rapid (Buvet and Le Port, 1973).

The chemical reactions can be formulated in a general way as

$$aA + bB + \dots = cC + dD + \dots \tag{4.1}$$

To each kind of reaction among solutes, a free energy charge ΔG can be ascribed. Most important were redox reactions. They have later become the general chemical source of all energy for life (see Kalckar, 1941). Redox reactions consist of the transfer of an electron from a reductant to an oxidant. If a proton is transferred along with the electron, i.e. an atom of hydrogen is handed over, one speaks of transhydrogenation (hydrogenation/dehydrogenation). The central role of transhydrogenations in metabolism was early stressed by Wieland (see Wieland, 1922), and specifically for bacteria by Kluyver (see Kluyver, 1952; Kluyver and Van Niel, 1956).

The reaction between pyruvate and hydrogen to give lactate

$$CH_3.CO.COOH + H_2 = CH_3.CHOH.COOH; \; \Delta G_0' = -10.3 \text{ kcal} \tag{4.2}$$

may serve as an arbitrary example. In writing down the equation, we have for simplicity disregarded electrolytic dissociation. So we shall do as a rule in future whenever hydrogen ions are neither set free nor used up.

When a redox potential E can be measured in electrical cells, it is connected with the free energy change by

$$\Delta G = -nFE \tag{4.3}$$

if H_2 appears on the left, as above. F is the Faraday, and n is the number of electrochemical equivalents. In physico-chemical standard conditions (at 298°K; in the other half-cell a normal hydrogen electrode with $p_{H2} = 1$ atm, pH $= 0$), we have

$$\Delta G_0 = -nFE_0. \tag{4.4}$$

When the reaction proceeds in physiological standard conditions (pH 7) the potential difference against the normal hydrogen electrode (E_0') is given (see, for example, Burton, 1957; Dolin, 1961) by (G in kcal, E in volts)

$$E_0' = -(0.0434\Delta G_0/n) - 0.414. \tag{4.5}$$

In our example, with $n = 2$,

$$E_0' = -0.19 \text{ V}. \tag{4.6}$$

(e) Formation of Macromolecules

Another class of important reactions was polymerization, especially of sugars, amino acids and nucleotides, to give polymeric carbohydrates, proteins and nucleic acid—the latter two: informational macromolecules. Straightforward condensations in dilute aqueous solution at ordinary temperature are endergonic. A typical case is that of the condensation of amino-acids to peptides (simpler homologues of proteins), with loss of water. This reaction typically needs, in standard conditions, a free energy of about 1.5 to 4 kcal/mole. Therefore, the reaction of free amino acids in water will normally proceed to a very small extent (a few percent) only.

Various special mechanisms to achieve protein synthesis nevertheless have been invoked (see Kenyon and Steinman, 1969; Fox and Dose, 1972; Lemmon, 1973b). One possibility is that polymers containing aminoacid precursors were produced in one step by heating $HCN + NH_3$ or by sparking $CH_4 + NH_3$ (Matthews and Moser, 1966, 1967; Matthews, 1971, 1975; see Ponnamperuma, 1971). No monomers as tangible intermediates are assumed, but many different aminoacids were obtained in the hydrolysis of the polymer.

Another possibility is that substances with affinity to water abstracted water from the monomers even in an aqueous environment through specific interactions. Cyanate, cyanamide and related compounds have been considered as condensing agents for the synthesis of peptides (Miller and Parris, 1964; Steinman *et al.*, 1964, 1965, 1966; Beck and Orgel, 1965; Calvin, 1965, 1969; Ponnamperuma and Peterson, 1965; Ferris, 1968; Kenyon and Steinman, 1969; Ibanez *et al.*, 1971; Steinman, 1971; Miller and Orgel, 1974). Surprisingly, condensations and also syntheses of monomers may be promoted by UV light (Ponnamperuma *et al.*, 1963; Steinman *et al.*, 1964; Ponnamperuma and Peterson, 1965; Reid *et al.*, 1967; McReynolds *et al.*, 1971; Ponnamperuma, 1971).

Pyrophosphate (or, more generally, polyphosphate) may also act as a condensing agent (Rabinowitz *et al.*, 1968, 1971; Miller and Orgel, 1974). PP was obtained in various moderately plausible abiotic reactions (Ferris, 1968; Miller and Parris, 1971; Österberg and Orgel, 1972; Lohrmann and Orgel, 1973). In this connection, the mobilization of phosphorus from the insoluble phosphates, which must have contained most of the element, presents a problem (Schwartz, 1971, 1972; Schwartz *et al.*, 1975).

Condensations are generally favoured by increased concentration (see Miller and Orgel, 1974). In particular it has been suggested that they proceeded after adsorption by rocks. Clays may have acted as adsorbents and also as catalysts for polymerizations (Bernal, 1951, 1967; Akabori, 1957; Degens and Matheja, 1971; see Kenyon and Steinman, 1969). The most detailed results have been obtained by Katchalsky and associates (Paecht-Horowitz *et al.*, 1970; Paecht-Horowitz, 1971, 1973, 1974; Katchalsky, 1973; Paecht-Horowitz and Katchalsky, 1973). In presence of the clay mineral montmorillonite, rather long polypeptide molecules have been obtained from aqueous solutions of mixed anhydrides of aminoacid and of adenylic acid (AMP; see *6a*). When zeolite was also present, the synthesis can be started with free aminoacid and ATP [*6a*].

The difficulty with processes involving the nucleotides is that these compounds may not have been components of the primeval soup. Model abiosyntheses of adenylate or of other nucleoside monophosphates (Steinman *et al.*, 1964; Ponnamperuma and Mack, 1965; Beck *et al.*, 1967; Rabinowitz *et al.*, 1968; Schwartz and Ponnamperuma, 1968, 1971; Lohrmann and Orgel, 1968; Chang *et al.*, 1970; Miller and Orgel, 1974) have, on the whole, led to modest results only (see Lemmon, 1973b). This is, not surprisingly, true also for the abiosynthesis of nucleoside polyphosphates, e.g. ATP (Ponnamperuma *et al.*, 1963; Sagan, 1965; Waehneldt and Fox, 1967).

This failure is hardly due to faulty experimentation. In the organisms, all informational macromolecules are built up enzymatically with participation of nucleotide triphosphates. But these compounds are structurally so closely defined that their synthesis in reasonable yields requires specific action by catalysts. Indeed now they are always made by enzymes. So the nucleotides could not be made through blind action of energy on the primeval soup. They were no likely products of chemical (abiotic) evolution. Rather, the nucleotides were "designed" at later stages in the evolution of life. A class of forerunners, the existence of which in the primeval soup is not unlikely, are the polyphosphates, including pyrophosphate (see Lemmon, 1970). We have named them as condensing agents already. More will be said about pyrophosphate later [*6c*, *8h*].

Condensing agents can be entirely dispensed with after evaporation of solutions to dryness. Peptide synthesis from mixtures of aminoacids that contain also nonneutral components (glutamic or aspartic acid, lysine) can be forced by application of heat (Fox, 1960, 1965a, b, 1969, 1971a, b; Fox *et al.*, 1970; Fox and Dose, 1972; see, however, Miller and Urey, 1959). A typical temperature is 170°, while in presence of pyrophosphate 50–60° will do, i.e. the condensation may take place in solution. The heat is assumed to have been available from hot spring areas and volcanic magma. The needed temperature is very much lower than that needed for thermal (magmatic) aminoacid synthesis [*4c*].

The "proteinoids" or "protoproteins" obtained by Fox have molecular weights in the thousands. They contain every one of the aminoacids applied, but generally not in the starting proportions or at random. Some selectivity is shown, and certain sequences are preferred; so they are looked at as forerunners of informational macromolecules. The proteinoids have catalytic—e.g. esterase—properties (Rohlfing and Fox, 1969; Fox 1971a; Dose and Zaki, 1971; Fox and Dose, 1972; Dose, 1974).

5

EOBIONTS AND ORGANISMS

(a) From Eobionts to Cells

Somehow "entities", to use an entirely noncommittal term, in the primeval soup must have begun to metabolize, to grow and to assimilate solutes. But they could persist and spread only when they "learned" to divide as soon as their sizes exceeded certain limits. Such processes were the forerunners of the later mechanisms for cell division. Presumably it was an advantage to carry them out in a well-ordered way, and so step by step those entities will have gained the upper hand that had suitable machinery. This way of selection may be looked upon as a primitive early version of Darwin's "struggle for existence".

Entities on the long and surely tortuous way between nonliving and living matter have been called "eobionts" (Pirie, 1953). The term (living) "organism" might be reserved for entities with regular mechanisms for metabolism and proliferation; enzymes are essential for them. Allen (1957) and Hanson (1966) have argued that at an early stage "reflexively catalytic systems" were developed where a final product of a series of reactions catalyses an earlier step of the series. This would be positive feedback or expanded autocatalysis. The process contains the germ for the catalytic reaction cycles now prominent in life [*1e*].

All contemporary organisms form cells, and all cells are derived from cells. "Omnis cellula e cellula" (Virchow). (For the history of this concept, see Florkin and Mason, 1960.) The protoplasm (also used as a noncommittal term) of the cells is surrounded by a membrane. This serves not only to hold the protoplasm together and to separate it from the medium, but it is also itself a most important organelle, capable of many specific functions. These include the transport of solutes into the cell and out of it; such transport may be "active" [*1f*].

This description of an organism obviously excludes viruses. Not forming cells, viruses are distinguished even from the most primitive organisms, i.e. from bacteria (see Lwoff, 1957; Lwoff and Tournier, 1966). In absence of a living host, a clear-cut virus has no metabolism, and it does not proliferate; nor is a virus surrounded by a membrane. Enzyme action, where it exists at all, is very specialized. For example, in some viruses it is limited to the penetration of the host cell. However, there are forms that in various ways suggest a transition to (or better: from) bacteria, e.g. the psittacosis agent (see Lehninger, 1970).

It is now thought that viruses somehow descended from living organisms or from their organelles. However, at least the extreme forms are better not considered as living organisms themselves, though from the point of view of applied biology or of medicine things may look different (see Wildy, 1962; Andrewes, 1965). The origin of the viruses has been discussed by Luria and Darnell (1968), and their evolution by Joklik (1974). In any case, viruses are not relevant to the origin of life or to the evolution of bioenergetic processes in cells.

The organic components of the primeval soup and the eobionts have vanished. If they did not disappear otherwise, they were used up by organisms. This has been pointed out, in a now famous passage, in a letter by Darwin (1871) himself:

> It is often said that all the conditions for the first production of a living organism are now present, which could ever have been present. But if (and oh! what a big if!) we could conceive in some warm little pond, with all sorts of ammonia and phosphoric salts, heat, electricity, etc. present, that a proteine compound was chemically formed ready to undergo still more complex changes, at the present day such matter would be instantly devoured or absorbed, which would not have been the case before living creatures were formed.

We shall now mention current ideas about the path from the prebiotic soup to eobionts and organisms.

(b) Coacervates and Microspheres

Oparin has long suggested (see Oparin, 1938, 1961, 1964, 1965a,b, 1968, 1971) that droplets containing macromolecules separated in the primeval ocean in form of "coacervates". Such microscopic droplets had been described by Bungenberg de Jong (see Booij and Bungenberg de Jong, 1956). Typically, they are produced by mixing solutes of opposite electric charges. Only those droplets survived that were adjusted to the conditions. They may have sunk towards the bottom, and been protected there from destructive UV radiation.

While Bungenberg's coacervates are static systems, in the primeval soup gradually "dynamic" droplets may have developed which had increased stability as a consequence of a balanced intake and loss of components. Inside, concentrations of solutes, e.g. of amino-acids, could be far higher than in the supernatant, and reactions could therefore be relatively rapid. The reactions could also be more specific than in dilute solutions; some droplets may have included catalysts, i.e. precursors of enzymes. Later, some droplets learned to respond to external disturbance by compensating changes. For the dynamic state and for regulation, a source of free energy was needed.

Much work on artificial coacervates has been carried out by Oparin and his colleagues, and the behaviour of coacervates in different conditions demonstrated. In striking experiments (see Oparin, 1971) droplets holding phosphorylase synthesized starch from glucose-1-phosphate in the medium, as this phosphate diffused in. If amylase was also included, the starch was hydrolysed to maltose, and this diffused out.

The coacervates studied by Oparin are based on biogenic macromolecules. This is why Bernal (1967) wanted to postpone formation of coacervates to a later stage in evolution. However, Herrara (1942) produced microscopic droplets with a fabric of non-biogenic macromolecules by incubation of solutions of ammonium thiocyanate and of formaldehyde, and postulated the existence of similar droplets in the primeval ocean (see Kenyon and Steinman, 1969).

Fox has made "microspheres" from his artificial proteinoids by treatment with water or salt solutions. Suitably prepared microspheres are stable and uniform, and have ultrastructure. In some cases, they form a double-layered boundary and selectively take up solutes by diffusion. They shrink or swell in hyper- or hypotonic solutions, respectively. The microspheres may grow by accretion and proliferate through budding or similar processes, in a manner partly reminiscent of microorganisms. At some stage, they may be looked at as "protocells" (Fox *et al.*, 1970; Fox, 1971a,b, 1973; Fox and Dose, 1972).

In the original ideas about coacervates and microspheres nucleic acids do not occur. At their stage, proteins are supposed by the authors to have been the only informational macromolecules. If so, the proteins have lost this exclusive role afterwards. But it is, of course, possible to arrange for participation of nucleic acids in coacervates and microspheres, especially in microspheres made from basic proteinoids. According to Fox (1973), mechanisms involving nucleic acids arose as an "evolutionary refinement", and at some stage information may have begun to flow in either direction between the two kinds of macromolecules. Some results on the synthesis of oligonucleotides by proteinoid microspheres acting on ATP have been reported (Jungck and Fox, 1973; Fox, 1974; Fox *et al.*, 1974).

Attention has been drawn from the evolutionary point of view by Lipmann (1971) to contemporary enzyme systems that make polypeptides from energy-rich derivatives of aminoacids without direct participation of nucleic acids. Such systems exist, e.g. for the biosynthesis of the bacterial antibiotics gramicidin S and tyrocidin. Both can be considered as simple informational "macro"-molecules, each consisting of ten aminoacid residues. However, the system requires catalysis by enzymes, and at least in the present world all enzymes are made on the basis of information carried by nucleic acids. The specific qualities of proteins and of nucleic acids have been mentioned already [*2b*].

(c) The Origin of the Genetic Code

In contrast to Fox, Orgel (see Orgel, 1973; Miller and Orgel, 1974) considers that proteins as such could never replicate, and he gives precedence to nucleic acids (polynucleotides). Nonenzymatic, therefore protein-free, formation of polynucleotides has been demonstrated, e.g. by condensation of suitable derivatives of bases on a template made of artificial polynucleotide (Sulston *et al.*, 1968; Orgel and Sulston, 1971). Processes of this type may have been early, and may later have been promoted by interaction with "noninformed" polypeptides or proteins. Surfaces may have acted as catalysts. Semipermeable membranes may have enveloped the entities only after the beginning of enzymatically catalysed biosynthesis, when its products had to be retained and protected.

A few words about the contemporary nucleic acid-directed synthesis of proteins have been said before [*2b*]. It is well known that triplets of nucleotides "code" for single aminoacids. For instance, the triplet UCG (uracil, cytosine, guanine) in messenger RNA ensures the incorporation of a molecule of serine into a protein chain. In recent years, the genetic code has been fully elucidated, i.e. the coding action of every one of the sixty-four possible triplets has been defined (see, for example, Yčas, 1969; Caskey, 1970; Miller and Orgel, 1974). As far as is known, the code is universal for all organisms.

The question arises whether the relationship between each triplet and its aminoacid, which is in each case mediated by a rather complicated system involving enzymes and transfer RNA, is based on specific preference in affinity or ease of reaction (chemically determined), or whether the relationship is one of chance (a frozen accident). In the latter case, a particular genetic system, connecting a particular triplet of bases with a particular aminoacid, has survived because the system happened to be efficient on other grounds. The question has not been decided as yet. Ideas about the origin of the genetic code have been put forward by Rich (1962), Woese (1967, 1973 a,b), Crick (1968), Orgel (1968, 1973), Fox (1974), Jukes (1974), Miller and Orgel (1974) and Dillon (1974). According to Orgel (see, however, Rich, 1962) the code must always have been a triplet code; a later transition from another type of code to a triplet code would have created a prohibitive upheaval.

After synthesis, the informational macromolecules must take up the right three-dimensional configuration to become effective. For instance, enzyme molecules act as catalysts only in the correct "tertiary" structure. Moreover, different kinds of molecules must combine spatially to form "quaternary" structures (many enzymes) or supermolecular organelles, e.g. membranes or ribosomes. For some enzymes and organelles (ribosomes) it has been shown that in the correct conditions this assembly occurs spontaneously. Then no additional information for the purpose need be carried in the genome ("self-assembly"). However, on general grounds it does not appear unlikely at all that this may not always apply (see Kushner, 1969; Florkin, 1974).

(d) Considerations about Self-organization

A rather different approach to the problems of eobiont formation has been used by Eigen (1971a, b; see Schuster, 1972). The aim is a physico-chemical theory of selection among a population of macromolecules, and the elucidation of kinetic principles for self-organization. Eigen's considerations can be put into a form where they are consistent with the principles of irreversible thermodynamics, as developed by Prigogine and Glansdorff (1971).

At first a most highly simplified and purely hypothetical system is looked at in which polymers can grow by replication at the expense of monomers. It is assumed that the medium can maintain the growth. In particular, the energy supply is assured. On the other hand, the molecules depolymerize again with a given probability per unit time.

When several kinds of monomers, e.g. several nucleotides, are present, the composition of the individual chains represents information ("informational polymer"). The composition is given by the sequential order of the monomers. Now it may be assumed that the chains with different composition form, and decompose again, with different speeds. Therefore when competition for monomers takes place, the fittest kinds of chains survive best. But the chains that are fittest among those existing may still be overtaken by even fitter chains arising in chance mutations, i.e. in faulty replication. In the end, information has been created through "self-organization". In ordinary information theory, information is only transmitted, and its amount is, at best, maintained.

Values may in this connection be ascribed to the particular compositions. The concept of value is a characteristic feature of these considerations. Value is defined by the difference (rate of production times degree of correctness) minus rate of destruction.

The usefulness of a combined system, consisting both of nucleic acids and of proteins, is quite plausible, at least *a posteriori*. The calculations of the Eigen type may be extended to the case of mutual promotion of fitness in systems containing both kinds of known informational macromolecules. The presence of RNA as well as DNA in our real world is, of course, a further complication. Apparently nobody has gone to the trouble of computing or trying out possible alternative informational systems, which might consist of macromolecules of quite different structures!

While many experiments, with a lot of empiricism, have been done with coacervates and microspheres, the Eigen-type approach has been largely theoretical up to now. Yet the independent experiments of Spiegelman are relevant (see Spiegelman, 1971; Mills *et al.*, 1973). Proliferation of RNA of viral origin, which does not need DNA, in a soluble system that contains nucleotides as building units, the enzyme replicase, for which this RNA naturally codes in *E. coli*, and a few other defined components has been studied. Selection of greatly modified RNA molecules, apparently particularly fit in the given environment, has been demonstrated.

Oparin (1965b) maintains that the concept of evolution should not be applied to single molecules, even if they are genes, as was done in extreme form by Muller (1929, 1955; see also Sagan, 1965). The concept should rather be applied to the system as a whole, namely, informational macromolecules and the medium in direct interaction with them. The importance of replication compared to metabolism ought not to be overrated, as was done by Muller. Similarly, Lipmann (1965) and Hanson (1966) have warned against a one-sided concentration of attention on the development of the genetic transfer system. Here also the interesting, but aberrant views of Hinshelwood (1961; Dean and Hinshelwood, 1966) may be mentioned who argued against the excessive attention given to properties of particular molecules (in bacterial systems), and pleaded for interpretation through a kind of system analysis on the basis of traditional chemical physics.

It is quite likely that at some stage the ideas of Oparin and Fox, on the one hand, and of Eigen and Spiegelman, on the other hand, will turn out to be compatible. As Orgel (1973) says, coacervates and organic catalysts, including noninformed polypeptides (see also Woese, 1973b), may have facilitated the formation or replication of the most primitive nucleic acids. Further speculations about the self-organization of molecular systems and the evolution of the genetical machinery in very concrete terms, with reference to mechanisms existing in Nature now, have been offered by Kuhn (1972).

(e) Acquisition of Enzymes

Acquisition of new enzymes by bacteria has often been observed. An instructive recent example of "evolution in action" has been given by Betz *et al.* (1974), who introduced novel substrates (various substituted aliphatic amides) into the medium of *Pseudomonas aeruginosa* and selected strains that had mutated to use these substrates (see also Clarke, 1974; Hartley, 1974; Rigby *et al.*, 1974).

In a rather similar way, early organisms must gradually have added to their armoury of enzymes as the organic substrates in the primeval soup approached exhaustion. Horowitz (1945, 1965) has put forward the concept of "retrograde evolution", in particular in respect to biosynthetic (anabolic) enzymes. It is thought that originally the organisms had available in solution all the needed building blocks. But after one of the essential substrates ("A") had been used up, only organisms could survive that acquired an enzyme to convert a precursor ("B") into the now missing substances A. This process was repeated when precursor B itself was exhausted, etc. It is the great merit of this concept that only the capacity to make one particular substance has to be gained at a time, i.e. that the enzymes for the whole biosynthetic pathway could be acquired stepwise. The process occurred independently in all the synthetic pathways, used by the given organism. In this way, a whole series of sequential enzymes could be made by the organism "from the top down".

Lewis (1951), Horowitz (1965), Watts and Watts (1968), Ohno (1970) and Koch (1972) have analysed the neoformation of biosynthetic enzymes in more detail (see also Dayhoff and Eck, 1969). The basis is that a gene, as a consequence of mutation, duplicates. It can then make two molecules of enzyme instead of one. One of these half-systems becomes available for gradual modification so that in the end a new kind of enzyme is there. Experimental evidence exists (Hartley, 1974; Rigby *et al.*, 1974), but this "tandem gene duplication" mechanism is also criticized (Wu *et al.*, 1968), and its various features are still under discussion (Hegeman and Rosenberg, 1970; McFadden, 1973; Clarke, 1974; Hartley, 1974).

Metabolic pathways growing through retrograde evolution may well acquire further enzymes, as required, by modifying ("recruiting") preexisting enzymes from other pathways; the original functions of these enzymes presumably get lost in the process. Fusion of genes has been assumed to occur in Nature (Bonner *et al.*, 1965; Cassio and Waller, 1971; Veron *et al.*, 1972; Jacq and Lederer, 1974), and has been provoked artificially (Yourno *et al.*, 1970; Rechler and Bruni, 1971; Kohno and Yourno, 1971).

(It is possible, though in some ways problematical, to imagine a similar chain of events, as envisaged by Horowitz, in respect to the catabolic pathways that serve energy metabolism (Hegeman and Rosenberg, 1970; Jeffcoat and Dagley, 1973). Here the gaps are filled "from the bottom up" (see Ornston, 1971).)

Gradually the organisms improved their treasure of enzymes, and they learned to live on poorer and poorer media. Using nothing but CO_2 as source of carbon, the autotrophs [*11, 12, 15*] are distinguished by greatest synthetic power. In contrast to what was often believed, the autotrophs must therefore have been latecomers (see Kluyver and Van Niel, 1956; Sorokin, 1957). This is, of course, also implied in the hypotheses of Oparin and Haldane. In Van Niel's words: "Physiological evolution, insofar as it pertains to the progressive development instead of the loss of function, can thus be paraphrased as the stepwise acquisition of complete independence from external reducing substances."

Autotrophs show a tendency to change back into heterotrophs when supplied with suitable nutrients. Quite generally, readiness for loss is a property of cells that has been well investigated with simple systems.

(f) Loss of Functions

The fruitful hypothesis of Horowitz starts from the obvious fact that in a depleted medium a prototrophic strain (itself capable of making a needed metabolite) is superior to an auxotrophic strain (dependent on the ready-made metabolite). In contrast, it has been shown (e.g. Zamenhof and Eichhorn, 1967) that in complete media auxotrophs grow better. Presumably it is their advantage that they need not make so many metabolites, and can concentrate on their most important tasks.

(At higher stages, unnecessary functions could be reversibly suspended by control. Among the contemporary mechanisms for control on a molecular level, inhibition by end products or other substances (allostery) and repression of enzyme synthesis are best known. A search for the mechanism of the evolution of control and regulation (see Davis, 1961) has become urgent. Stanier *et al.* (1970) suggest that the capacity to regulate catabolic pathways has been acquired late in their evolution.)

Thus unneeded genes and enzymes tend to be lost. This is of tremendous importance in evolution, and particularly so in microbial evolution. The loss of functions has been emphasized by Lwoff in an important and amazingly one-sided book, published in dark Paris in 1944, which grandly neglects progress. It says:

> Le lecteur pensera peut-être que nous avons complaisamment insisté sur l'aspect pathologique de l'évolution physiologique, sur son côté regressif. Ce n'est pas que nous n'avons pas cherché des documents ou des faits biochimiques ou physiologiques permettant d'affirmer ou même de soupçonner l'existence d'une évolution physiologique progressive. Ces documents, nous ne les avons pas trouvés . . .
>
> Pour tout une catégorie de penseurs résolument optimistes, les êtres vivants sont en voie de perfectionnement constant et tendent, du fait de la direction efficace d'un principe bienveillant, vers un

état idéal et parfait . . . Mais peu (de biologistes) accepteront de gaité de coeur l'idée que la structure physico-chimique même des organismes puisse être la cause d'une évolution qui conduit les espèces vers l'extinction dont la dégradation physiologique est le prélude . . . De nombreuses espèces montrent les signes indéniables d'une dégradation physiologique évolutive. Des groupes entiers sont acheminés vers l'anéantissement, en vertu d'un déterminisme rigoureux et implacable. La dégradation et la mort semblent les états les plus probables des espèces vivantes aussi bien que des individus.

(g) The Origin of Optical Activity

A problem of great fascination is that of the origin of optical activity (chirality). Molecules with an "asymmetric" carbon atom appear in two chiral forms, one the mirror image of the other. They show opposite optical activity. Pasteur established that those biogenic molecules that on structural grounds are capable of optical activity usually indeed are active. For example, all aminoacids in proteins (except glycine, which has no chiral centre) belong, in the terminology of E. Fischer, to the L series. Now the chances for optical isomers (L and D compounds) to be formed from fully symmetric reactants in a fully symmetric environment are strictly identical. This applied to original abiosynthesis, too. So how could optical activity arise in Nature at all?

Asymmetric action has been ascribed to various physical agencies (for references, see Ageno, 1972; Elias, 1972; Thiemann, 1974). If one of these "blind" processes were responsible for asymmetry, optical activity would be an incidental fact without deeper significance. Selective adsorption on optically active quartz and selective photochemistry by polarized light have been invoked; daylight shows very faint circular polarization. After the downfall of mirror symmetry in radioactive beta-decay, an influence of this process on abiosynthesis has been proposed. But obviously these agencies, even if they have any slight influence at all, cannot account for the truly massive fact of natural optical activity.

The curious feature is that a fully satisfactory explanation of the optical activity of biomolecules has been given quite a time ago by Wald (1957). The origin of optical activity is attributed to evolving life itself:

Only the fact that chemistry is learned from the plane surfaces of paper and blackboard makes such selectivity seem strange. We tend to think of optical isomers as very much alike, but in fact they represent profound differences in shape; and, in the type of reaction upon which life depends, involving the ceaseless, intimate fitting together of molecular surfaces, shape is all-important. Organisms made the choice between optical antipodes long ago. To tamper with that choice now would be like trying to draw a left glove on a right hand . . . Optical activity appeared as a consequence of intrinsic structural demands of key molecules of which organisms were eventually composed, through the selection of optical isomers from racemic mixtures.

It is admitted that the optical isomers are chemically identical. But it is well known that the identity may be lost when two such asymmetric molecules combine. For instance, when an L compound reacts with an otherwise identical partner of L or of D configuration, the products (diastereomers) differ chemically. For example, the two optical isomers of glyceroaldehyde (asymmetric C atom starred)

$$\begin{array}{c} CHO \\ | \\ H^{*}COH \\ | \\ CH_2OH \end{array} \quad \text{and} \quad \begin{array}{c} CHO \\ | \\ HO^{*}CH \\ | \\ CH_2OH \end{array}$$

are chemically identical. But the four possible esters (at the terminal OH groups) with the two optical isomers of lactic acid, $CH_3.H^{*}COH.COOH$ and $CH_3.HO^{*}CH.COOH$, are only pairwise identical, i.e. we have two pairs of image and mirror image. The L glyceraldehyde ester of D lactic acid is identical in its chemical and physical properties with the D glyceraldehyde ester of L lactic acid, but not with the L glyceraldehyde ester of L lactic acid.

Consequently, organisms would be identical in all chemical and therefore biological properties with existing organisms if all optically active molecules without exception were replaced by their optical antipodes. But this would not be true for organisms with only partial replacement. Therefore competition and selection would tend to promote the survival of organisms in whose metabolism an optimal mixture of biomolecules, in respect to chirality, existed.

More particularly, Wald has reported that in the composition of informational macromolecules that are at least partly helical (proteins and nucleic acids) only building units of one and the same series (D or L) can be tolerated easily. Otherwise, serious instabilities would occur. The position may be compared to that in the building of a staircase. *A priori*, right-hand and left-hand staircases are surely equally good. But not a mixture. This does not mean, however, that in the entire organisms no optical antipodes can coexist. Quite often D aminoacids have been found as components of peptides, for instance.

Now how did Nature decide in favour of L amino acids for proteins, etc., although a world of antipodes would have been just as good? It is quite likely that initially chirality was not strictly defined, and that eobionts with (partly or fully) opposite symmetries of their helical molecules coexisted. But in the end, unified chirality prevailed in each group of eobionts that was genetically linked. Among the two groups that formed in this way that one survived that had better fitness on other grounds. Thus the ultimate selection was due to chance, but a clear choice had to be made. On the planets of other solar systems, the opposite choice will have been made in about the half of the cases.

6

ENERGY-RICH COMPOUNDS

(a) The Adenylic Acid System

It will be seen that the simplest processes in energy metabolism are the fermentation processes. Each of them consists of a series of exergonic, anaerobic and nonphotosynthetic reactions. The fruitfulness of the thermodynamical approach to these processes was well demonstrated already in the earlier literature. In particular, the fine books of Meyerhof (1930) and Stern (1933) took the reader as far as was possible before the role of the "energy-rich" compounds was grasped.

Experience showed, however, that the energy released in fermentation and also in other processes of energy metabolism becomes available to the cells mainly by being used to build up energy-rich compounds. This concept, central for bioenergetics (see Atkinson and Morton, 1960; Huennekens and Whiteley, 1960; Kalckar, 1966b, 1969; Todd, 1968; Lehninger, 1970, 1971) has been formulated by Lipmann (1941) in an epoch-making paper in the Meyerhof tradition. Lipmann (1971a) himself has told the story. From time to time, the concept of the energy-rich compound has been criticized, but even the most recent attacks (Banks, 1969; Banks and Vernon, 1970) have, in the view of the present author, been successfully repulsed (Huxley, 1970; Pauling, 1970; Wilkie, 1970a).

Energy-rich compounds are distinguished by the high exergonicity of at least one of their enzyme-catalysed reactions for the transfer of a group. In another expression of Lipmann, the relevant "group potential" is high, and in the transfer it decreases. In standard conditions, the difference of free energy is $\Delta G_0'$. In transfer, the bond initially joining the group to the rest of the molecule is severed. This is the "high-energy bond". The term must, however, not be taken literally. Not more is implied than the high group potential. Especially, it must be kept in mind that G always expresses free energy (more exactly: free enthalpy) and not energy (more exactly: enthalpy).

The paragons of energy-rich compounds are nucleotides of the adenylic acid system† (Fig. 6.1). They transfer phosphoryl groups. The energy-rich character is shown most obviously in hydrolysis, when the phosphoryl group is transferred to the standard acceptor, water. For instance, for ATP:

$$\text{ATP} = \text{ADP} + \text{P}; \Delta G_0' = -\ 7.3 \text{ kcal.} \tag{6.1}$$

(To make sense to organisms, in bioenergetic processes this reaction is coupled to endergonic reactions which are to be driven; see *6b*.) Within the cell, where the concentration of P is low, ΔG far exceeds $\Delta G_0'$ in absolute value [see *1d*], and may be as much as −12 kcal/

†ATP=adenosine triphosphate; ADP=adenosine diphosphate; AMP=adenosine monophosphate= adenylic acid; P=orthophosphate; PP=pyrophosphate. All in solution.

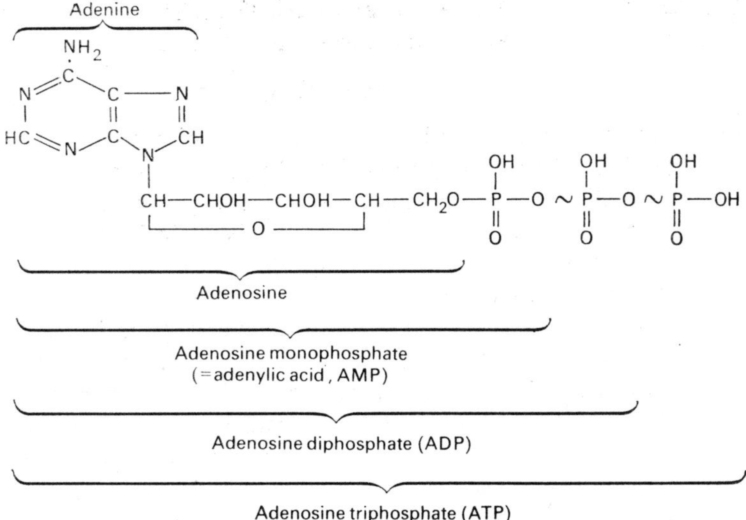

FIG. 6.1. Chemical formulae of the adenine nucleotides. Electrolytic dissociation not shown.

mole. Recent results in the thermodynamics of ATP have been reviewed by Alberty (1969), and the influence of pH and Mg concentration discussed.

Another possible notation for ATP is ARP ~ P ~ P, where A indicates adenine, AR adenosine, R ribose and the squiggle the "high-energy bond"; the squiggle is another of Lipmann's inventions (see Kalckar, 1966a). The bond between R and P is not a high-energy bond; adenylic acid is not energy-rich. In Table 6.1 the phosphoryl group potentials of some important intermediates in energy metabolism are listed. Those above ATP donate phosphoryl groups to ATP, those below accept them in standard conditions.

TABLE 6.1. SOME PHOSPHORYL GROUP POTENTIALS

(Kcal/mole. Negative values)

(Lehninger, 1971)

Phosphoenolpyruvate	14.8
1,3-Diphosphoglycerate	11.8
Phosphocreatine	10.3
Acetyl phosphate	10.1
ATP	7.3
Glucose 1-phosphate	5.0
Fructose 6-phosphate	3.8
Glucose 6-phosphate	3.3
3-Phosphoglycerate	2.4
Glycerol 3-phosphate	2.2

Reasons for the high or low group potentials of different compounds (energy-richness of bonds) have been discussed many times (see Coryell, 1941; Hill and Morales, 1951; Oesper, 1951; Grabe, 1959; George and Rutman, 1960; Pullman and Pullman, 1960, 1962, 1963, 1967; Fernandez-Alonso, 1964; Phillips *et al.*, 1969; Boyd and Lipscomb, 1969; Fischer-

Hjalmars, 1969; George *et al.*, 1970; Alving and Laki, 1972). These reasons differ between different classes of compounds. Above all according to eqn. (1.3)

$$\Delta G = \Delta H - T\Delta S, \qquad (1.3 = 6.2)$$

a high negative value of ΔG may either be due to a high negative value of ΔH or to a high positive value of $T\Delta S$. The former may be attributed to improved resonance stabilization or to improved distribution of electric charge in the transition from the reactants to the products. A good negative value of ΔH is, for instance, the main factor for the energy-rich nature of the constituent of muscle, creatine phosphate: the products of its hydrolysis are better stabilized by resonance than the reactant.

On the other hand, a high positive value of $T\Delta S$ may arise from increased electrolytic dissociation of products. As the newly formed hydrogen ions must in neutral solution be brought to the concentration 10^{-7}, a large term for dilution entropy comes in. Also, the products may surpass the reactants in freedom of internal movement. For example, internal rotation is easier in pyruvate than in the enolpyruvate moiety of phosphoenol pyruvate (PEP), a key compound in glycolysis.

Among the energy-rich compounds, some, including the energy-rich nucleotides, are independent of the redox state of a system, as emphasized by Krebs and Kornberg (1957). Therefore the nucleotides can transfer energy between compounds at different redox states, e.g. between the components of the respiratory chain [*13d*]. A further advantage of the energy-rich nucleotides (compared, for example, with acetyl phosphate; see below) is their relative resistance to non-enzymatic hydrolysis (Lipmann, 1951).

The energy-rich nucleotides are pre-eminent among the energy-rich compounds of the present world, and among them ATP has the first place. Mostly ATP is the primary energy-rich compound formed in energy metabolism. On the side of utilization, ATP is needed for the most crucial biosynthetic reactions, including the syntheses of proteins and nucleic acids. ATP is also the only source of energy in fibre, including muscle, contraction, and at least in some cases it has been shown to provide energy for osmotic work (active transport).

It is amusing to work out the amount of ATP that is generated per day and per unit weight of living matter. Respirometry by the Warburg method shows that among fermenting bacteria a $Q_{CO_2}^{N_2}$ value of 300 (300 ml CO_2 set free per g dry weight and hour under nitrogen) is not unusual (among lactobacilli; see Stephenson, 1949; Gibbs, 1962). As 1 molecule of acid releases 1 molecule of CO_2 from bicarbonate, this CO_2 production corresponds to 1.3×10^{-2} moles of lactic acid. (In yeast, $Q_{CO_2}^{N_2}$ is 250, and in the green alga, Chlorella, only about 1.) Assuming [*7c*] that with the bacteria 1 mole of lactic acid corresponds to 1 mole of ATP, one works out that 1 g of dry bacterial matter produces, and decomposes again, 0.31 mole or 180 g of ATP (molecular weight 507) per day. With a typical content of 2 μmoles ATP/g bacterial matter it is calculated that the ATP is turned over in bacteria 300,000 times per day, and that the mean life of the ATP molecules is one-third of a second only.

In respiring bacteria, ATP production is far larger. Q_{O_2} may reach a value of the order of 2000, i.e. 2000 ml of CO_2 are produced per g dry weight and hour (*Azotobacter*; see Schlegel, 1972). (In yeast, Q_{O_2} is only of the order of 100, in mammalian kidney or liver of 10–20, and in organs of higher plants of 0.5–4.) Recalling that in respiration 1 mole of CO_2 produced (or of O_2 consumed) corresponds to one-sixth of a mole of standard nutrient (hexose), one finds that 1 g bacteria makes 7000 g ATP per day. Similarly, a human person of 70 kg weight makes from carbohydrates worth 3000 kcal (700 g), a reasonable amount to be consumed per day on a pure carbohydrate diet in a short-term experiment, 75 kg (!) ATP in a day. This corresponds, however, only to about 3 g ATP/g dry weight of man, and to a daily turnover number of the order of 1000 only. Incidentally, the market value of 75 kg ATP, in technical quality is, about $150,000.

To many it may be surprising that the energy production of organisms per unit weight very much exceeds that of the Sun (2 erg/g sec). Thus a man of 70 kg, at 3000 kcal/day, produces 2×10^4 erg/g sec, and respiring *Azotobacter*, just mentioned, reaches 10^8 erg/g sec.

Again and again it has been shown that catabolic and anabolic enzyme systems are influenced in their catalytic (!) action antagonistically by adenosine phosphates that are, or are not, energy-rich. This influence, then, has not a thermodynamic, but a kinetic basis: through interaction with "modulators" or "effectors" (here: ATP, ADP, AMP) the activity of enzymes may be enhanced or depressed. This is an example of the widespread phenomenon of allostery [*5f*]. The potential usefulness of such modulation of metabolism is obvious. It has been found practically convenient (Atkinson, 1969, 1971) to express the influence of the adenosine phosphates through the "energy charge", defined as

$$\frac{[\mathrm{ATP}] + 1/2\ [\mathrm{ADP}]}{[\mathrm{ATP}] + [\mathrm{ADP}] + [\mathrm{AMP}]}.$$

The values in brackets give the concentrations. The energy charge indicates the extent to which the joint pool of the adenosine phosphates is "filled" with free energy, i.e. the energy status of the cell.

(b) Energetic Coupling

We are returning to a consideration of thermodynamical data. Some compounds have even higher phosphoryl potentials than ATP, so that their phosphoryl group can be transmitted to ADP. Phosphoenol pyruvate (PEP) is an example. On the other hand, the phosphoryl potential of glucose-6-phosphate is much less; therefore, ATP easily phosphorylates glucose in the "hexokinase" reaction. Obviously, such transphosphorylations are exergonic.

Long ago, Kluyver, influenced by Rubner, pointed out that in biosystems endergonic reactions can proceed only when suitably coupled with exergonic reactions, and raised the problem of the mechanism. Lipmann then showed that this coupling involves the participation of an energy-rich compound as a common intermediate in the two reactions (see Lehninger, 1971). In this way, transfer of metabolic energy is based on group transfer.

The final product may still be energy-rich. An example is the build-up of ATP at the expense of acetyl coenzyme A (AcCoA; likewise discovered by Lipmann). The thioesters of coenzyme A (CoA), including AcCoA, are energy-rich with $\Delta G_0'$ for hydrolysis around -8 kcal/mole. For instance, in *Clostridium kluyveri* the following two reactions are coupled:

$$\mathrm{AcCoA} + \mathrm{P} = \text{acetyl phosphate (AcP)} + \mathrm{CoA};\ \Delta G_0' = 2.5\ \text{kcal} \quad (6.3)$$

$$\mathrm{AcP} + \mathrm{ADP} = \text{acetate} + \mathrm{ATP};\ \Delta G_0' = -2.5\ \text{kcal} \quad (6.4)$$

$$\mathrm{AcCoA} + \mathrm{ADP} + \mathrm{P} = \text{acetate} + \mathrm{CoA} + \mathrm{ATP};\ \Delta G_0' = 0 \quad (6.5)$$

In this case, energy-rich AcP is the common intermediate. Incidentally, AcP may serve not only as a phosphorylating, but also as an acetylating agent (see Stadtman, 1966).

(c) The Origin of Energy-rich Compounds

Weight by weight, redox systems contain more energy than energy-rich compounds. For instance, the total free energy change in the lactic acid fermentation of glucose [*7c*] is -47

kcal/mole, or —260 kcal/kg. The corresponding values for the utilization of glucose by respiration [*13a*] are —686 and 3840 kcal/mole, respectively. In contrast, ATP (ΔG in the cell = —12 kcal/mole) contains only —24 kcal/kg. Consequently, the bulk of the energy reserves of all cells is stored in the form of fermentable (or, later, respirable) substrates. On the other hand, the energy-rich compounds, temporarily formed in the fermentation or respiration of the substrates, are effective through rapid turnover. They are better suited to the rapid and specific mobilization of energy. They may be looked at as energy transfer agents (energy currency).

Metabolic processes in eobionts may already have involved energy-rich compounds. Lipmann (1965) remarked that the application of the energy of redox reactions to increase group potential was the "first event on the way to life". The build-up of the structural elements for resistant coacervate droplets and of early nucleic acids as well as mechanisms for proliferation may have required energy-rich compounds. However, we know nothing about them.

Nucleotides can hardly have been the first energy-rich compounds in the primeval soup, as they are rather complicated substances, difficult to produce [*4e*]. Moreover, it is uncertain whether among the energy-rich nucleotides, when at last they arrived, ATP came first, although adenine, owing to strong resonance, is more stable than other bases in contemporary nucleotides (Pullman and Pullman, 1962). Many of the analogs of ATP are energy-rich in a similar sense. Some of them, in which adenine is replaced by another base, e.g. guanine, function now in limited fields. Compounds containing another sugar instead of ribose are also worth consideration.

Miller and Parris (1964) and Lipmann (1965) have discussed inorganic pyrophosphate, PP, as a forerunner of ATP in evolution, after Calvin (1957, 1963) and Jones and Lipmann (1960) had first suggested PP in this connection. PP has been mentioned as a possible primeval condensing agent [*4e*]. Inorganic pyrophosphate is found in recent bacteria of many kinds (see Harold, 1966; Dawes and Senior, 1973). The idea of PP as a forerunner of ATP has been strongly supported by H. Baltscheffsky (1967a, 1969a, 1971) on the basis of his results with photosynthetic bacteria [*8h*], where PP can, to some extent, replace ATP; it has been emphasized that in these bacteria the reactions of PP do not involve ATP, i.e. PP does not presuppose ATP. Higher polyphosphate may also have a role in energy metabolism (Hoffmann-Ostenhof and Weigert, 1952; Cole and Hughes, 1965; Kulaev, 1971; Kulaev and Bobyk, 1971).

Whatever the remaining role of PP may be, the ATP system has become universal. Growth of bacterial or fungal biomass is generally correlated with ATP production (Monod, 1942; Bauchop and Elsden, 1960; Gunsalus and Shuster, 1961; Senez, 1962; see Van Niel, 1966; Rose, 1968; Payne, 1970; Decker *et al.*, 1970; Haukeli and Lie, 1971; Forrest, 1969; Forrest and Walker, 1971; Stouthamer, 1969; Stouthamer and Bettenhaussen, 1973). In the evolution of the microorganisms new methods for the production of ATP were developed, and in the transition to higher organisms, with division of labour between cells, new methods for the utilization of ATP.

7

FERMENTATION

(a) The Simplest Extant Organisms

As we have seen, it is the common bioenergetic goal of all organisms in the contemporary world, of the most ancient and the most recent, to make enough ATP. But what are the most ancient organisms still with us? Clearly, no living organisms are unchanged descendants of the original beings that arose from eobionts. What matters here in selecting the organisms to be considered as the most ancient, is greatest similarity, in respect to bioenergetics, with the most distant ancestors.

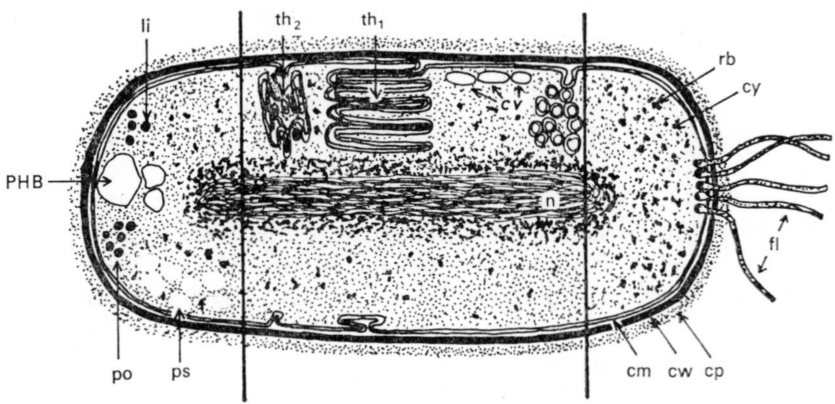

FIG. 7.1. Combined scheme of a bacterial (prokaryotic) cell (simplified after Schlegel, 1972). On the right, the fundamental constituents are shown, on the left, storage materials, and in the middle, structures for photosynthesis. cm = cell membrane, cp = capsule, cv = chlorobium vesicles, cw = cell wall, cy = cytoplasm, fl = flagella, li = lipids, n = nucleus, PHB = polyhydroxybutyrate, po = polyphosphate, ps = polysaccharide, rb = ribosomes, th_1 = lamellar thylakoids, th_2 = tubular thylakoids.

There is no doubt that bacteria are the most primitive self-supporting organisms in the present world. In the words of Pallade (1964), the bacterial cell apparently represents the minimal but sufficient formula for the cellular level in the hierarchy of patterns of living matter. Its emergence was a crucial event in the history of life. Figure 7.1 pictures, after Schlegel (1972), the essential features of bacterial cells.

Every bacterial (or other) cell is surrounded by a cell membrane, which contains proteins and lipids. The membrane acts as a barrier, but it is also the place of the mechanisms for transport into the cell and out of the cell, including the pumps for active transport. Moreover, bacterial photosynthesis and respiration occur along the membranes or their intracellular extensions. Viruses are not surrounded by membranes. Hypotheses about the structure and functions of the cell membranes abound (see, for example, Salton, 1960, 1967; Robertson, 1967; Stein, 1967; Rogers and Perkins, 1968; Korn, 1969). The cell "envelope" may, but need not, contain, in addition to the cell membrane and outside of it, a cell wall (Salton, 1964, 1967; Rogers and Perkins, 1968). Principal functions of the wall are protection of the bacterium against mechanical attack and osmotic stress.

The bacteria present us with a tremendous variety of processes in bioenergetics and otherwise. The genetic and therefore phenotypic diversity among families of bacteria may be greater than among classes of vertebrates (Wilson and Kaplan, 1964; De Ley, 1962, 1971). It is as if Nature had at the beginning of evolution wildly experimented before selecting the best mechanisms even in fundamentals.

The bacteria also are quite diverse in respect to the rates of their life processes, including their bioenergetic processes. However important and interesting this aspect, it will be left out from the proposed treatment of bacterial (and other) energy metabolism.

Most ancient, then, may be those bacteria that can make ATP in a solution of organic compounds anaerobically without direct participation of light. This was presumably the way of life of the early organisms. Fortunately, although the layman and the medical man may not always think so, such organisms are still with us.

Thus we exclude organisms with additional bioenergetic mechanisms, namely, for the utilization of light, for protection against free oxygen or for respiration. The creatures must then make ATP by fermentation. No disproof of the possibility exists that the primary non-photosynthetic anaerobic heterotrophs (fermenters) have all died out as such, and that all extant fermenters are simplified descendants of photosynthetizers or respirers. This does not appear as an attractive idea, however.

Pasteur recognized fermentation as "life without air" (see Lipmann, 1946). His original words (1875) were: "La fermentation est la conséquence de la vie sans gaz oxygène libre . . ." From a more modern point of view, fermentation may be broadly defined as a sequence of anaerobic reactions that leads to a net synthesis of ATP. Participation of O_2 in energy production is excluded; reaction sequences with utilization of O_2 for energy production constitute, by definition, respiration.

(It will later be seen that the broad definition of fermentation is not sharp enough. On the one hand, O_2 may remove products of fermentation and thereby drive the reactions that yield ATP without, however, itself taking a direct part in ATP production [*14c*]. On the other hand, some alternative inorganic electron acceptors, notably nitrate, can substitute for O_2 in respiration, and respiration then occurs anaerobically [*16*]. From the most modern point of view the best criterion is that fermentation proceeds in solution while respiration needs compartmentation by membranes [*13b*] and involves electron flow [*13d*]. This applies, as experience shows, to all ATP generating processes that involve oxygen and to related processes with substitutes of oxygen (sulphate, nitrate). In this way the intuitive distinction between fermentation and respiration emerges as less arbitrary than might have been thought.)

It will then be best to consider first the extant, non-photosynthetic, strictly anaerobic, bacteria, the fermenters. This has been the line of most previous workers (e.g. Oparin, 1938, 1968; Krebs and Kronberg, 1957; Arnon *et al.*, 1958; Decker *et al.*, 1970). Of course, the contemporary fermenters must differ from the earliest organisms in their metabolism if only because the former now depend on products of higher organisms rather than on components of the prebiotic soup.

Mechanistically, the way of formation of ATP in fermentation is called "substrate-level phosphorylation" (see Racker, 1965). This name indicates that some of the intermediates in the degradation of the substrate of fermentation are themselves phosphorylated. Substrate-level phosphorylation is likely to be ancient, as it occurs simply in solution, and requires neither light nor free oxygen. The energy is ultimately derived from redox reactions. Substrate-level phosphorylation differs from photosynthetic phosphorylation [*8e*] and from oxidative phosphorylation [*13b*], both of which need compartmentation, and which are characteristic for photosynthesis and for respiration, respectively.

We may introduce the terms "prokaryotes" and "eukaryotes" already here, so that we need not be afraid of them, although we shall have to return to these concepts in much more detail later [*18*]. The prokaryotes are the more primitive organisms, namely, the bacteria and the blue–green algae—not only fermenters! All other organisms, more advanced, are eukaryotes: the plants except the blue–green algae, all animals and all fungi. In contrast to the prokaryotes, the eukaryotes have well-defined nuclei, nearly always mitochondria, and often chloroplasts. These organelles are far more elaborate than anything contained in prokaryotes. Yet photosynthesis and respiration are observed not only among eukaryotes, but also among prokaryotes.

(b) Clostridia

The present fermenters use a large variety of multi-stage processes for energy production. To establish the mechanism of fermentative or of other processes, it is in general essential (a) to identify the products, (b) to detect the required enzymes, and (c) to follow, with isotopes, the kinetics of the process. Unfortunately, the classification of the many pathways of fermentation is still largely empirical and mainly serves practical taxonomy. Only beginnings have been made to group them naturally, and also to assess the survival value of each of them in one or another environment. Excellent critical surveys have been given by Wood (1961), DeLey (1962), Rose (1968), Stanier *et al.* (1970), Decker *et al.* (1970), Schlegel (1972) and Sneath (1974).

Among many groups of (non-photosynthetic) bacteria strictly anaerobic species (obligate fermenters) are found, e.g. among the spirochaetes (see Canale-Parola *et al.*, 1968) and among the poorly known Gram negative organisms that inhabit the animal body, the rumen, the saliva, etc. (see Stanier *et al.*, 1970). It is not clear how far they should be considered as parasites or as symbionts. They may be degenerate rather than primitive.

The strictly anaerobic way of life is consistently adhered to by two other (non-photosynthetic) groups that are clearly primitive, the clostridia and the methane formers, both Gram positive. We shall assume that all ancestors of the recent clostridia and methane formers were non-photosynthetic anaerobes. The methane formers, though very important and much investigated, still present great theoretical problems. Also their relationship to the clostridia is not clear. So only some words will be said about them [*7e*]. We shall mostly concentrate on the clostridia.

(c) Glycolysis

Among the energy-yielding processes, glycolysis (the EMP pathway named after Embden, Meyerhof and Parnas) may be selected for illustration. Lipmann (1946) has called it the simplest kind of fermentation. Certainly, it is the best known; one reason is its great economic importance.

Glycolysis leads from carbohydrates to pyruvic acid. Carbohydrates can after phosphorylation enter the pathway. Polymeric carbohydrate is a common storage material, e.g. in contemporary clostridia. For simplicity we shall take glucose as starting-point. The path from glucose to pyruvate has nine steps, each catalysed by a specific soluble enzyme. Pyruvate is a less reduced compound than glucose, and the lost ("active") hydrogen, shown as [H], reappears in NADH.† As far as the substrate is concerned, the overall equation (see Fig. 7.2) is:

$$C_6H_{12}O_6 = 2CH_3.CO.COOH + 4[H] \qquad (7.1)$$

Moreover, there is a net gain of 2ATP from 2ADP + 2P.

In glycolysis, phosphorylated intermediates include diphosphoglycerate and phosphoenol pyruvate. Both these substances transfer phosphoryl groups to ADP (for mechanisms, see Racker, 1965; Lehninger, 1971), and in this way ATP is made.

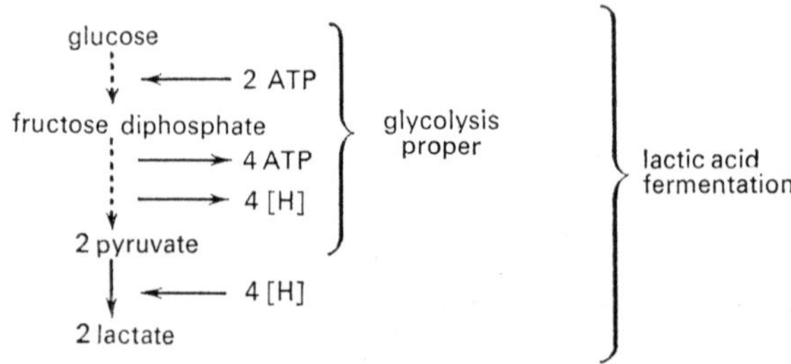

FIG. 7.2. Abbreviated scheme of glycolysis, merely from the point of view of bioenergetics. Here and in future ADP + P will, for typographical clarity, not be shown explicitly as reactants or products. [H] always indicates NAD(P)H unless specified otherwise.

The term "glycolysis" is generally used somewhat loosely. We shall restrict it to the steps leading from carbohydrate to pyruvic acid, and shall not use it as a synonym of lactic acid fermentation or of alcoholic fermentation. These are terms that include not only glycolysis proper, but also the two common variants for the disposal of the active hydrogen, [H], that appears together with pyruvic acid in eqn. (7.1).

In lactic acid fermentation, [H] is transferred to pyruvate itself. The last step is then

$$CH_3.CO.COOH + NADH + H^+ = CH_3.CHOH.COOH + NAD^+. \qquad (7.2)$$

In this "homolactic" process, with lactic acid as the only product, glucose may as a final result be considered as "split":

† Concerning names of pyridine nucleotides: NAD^+ is nicotinamide adenine dinucleotide (formerly DPN^+), $NADP^+$ is nicotinamide adenine dinucleotide phosphate (formerly TPN^+). NADH and NADPH are the reduced forms (DPNH or TPNH); on reduction, 1 H^+ also appears in each case. For typographic simplicity, the + sign will, as a rule, be left out from now. When the compound is not specified, we shall write NAD(P) and NAD(P)H; the redox potentials of NAD and NADP are about the same.

$$C_6H_{12}O_6 = 2CH_3.CHOH.COOH; \Delta G_0' = -47 \text{ kcal.} \tag{7.3}$$

Not only the reduction of pyruvate, but lactic acid fermentation as a whole may be considered as a transhydrogenation reaction in so far as in the splitting of glucose two hydrogen atoms are transferred from the carbon atoms 3 and 4 to atoms 1 and 6 of the glucose. This transfer has been confirmed with isotopic tracers. Intramolecular transhydrogenation is still more obvious in the overall equation for alcoholic fermentation. Here pyruvate is first decarboxylated before reduction. The final result, again omitting ATP synthesis, is

$$C_6H_{12}O_6 = 2C_2H_5OH + 2CO_2; \Delta G_0' = -56 \text{ kcal.} \tag{7.4}$$

$\Delta G_0'$ for the synthesis of 1 ATP was taken by Burton as about −9, while now it is rather thought to be −7.3 kcal [*6a*]; hence the overall value of $\Delta G_0'$ for lactic acid fermentation with coupled ATP synthesis in standard conditions—sticking, for consistency, to Burton—is −29 kcal. For alcoholic fermentation, it is −38 kcal.

The enzymes of glycolysis have been found in virtually all bacteria investigated (see Wood, 1961), though glycolysis may have assumed its present importance only after plants began to make carbohydrates in quantity (Gaffron, 1965); carbohydrates may not have been so common in the prebiotic soup. From this point of view, it is interesting that other kinds of organic compounds, notably alcohols, carboxylic acids and aminoacids, are also used as substrates for fermentation by clostridia. The width of the range of fermentable substrates has been taken as a further argument for the antiquity of the clostridia (Decker *et al.*, 1970).

The interesting similarity of the reaction pathways in an alkaline, non-enzymatic, degradation of carbohydrates with those in glycolysis, and the possibility of an evolution of the latter from the former in primeval, more alkaline, waters (see Abelson, 1966; Holland, 1972) has been pointed out (Degani and Halmann, 1967).

Glycolysis, developed by the most primitive bacteria, has been retained in further evolution. Probably all plants and animals are capable of it. Pyruvic acid has become the starting-point of the citric acid cycle and therefore of the form of respiration that is practised by most organisms.

Lactic acid fermentation on the basis of glycolysis is observed among many kinds of bacteria. The homolactic process probably arrived relatively late. It can best be studied with the lactic acid bacteria proper (see DeLey, 1962), to which further reference will be made [*14f*]. A few words about the important role of lactic acid fermentation in vertebrates will be said later [*22f*].

(d) Hydrogen Fermentations

In clostridial fermentations, pyruvate can be used up in other directions than production of lactic acid or ethanol. The other products also arise from the need to remove the active hydrogen from the NADH formed in the glycolytic production of pyruvate (see Stanier *et al.*, 1970; Forrest and Walker, 1971; Gest, 1972; Schlegel, 1972). Further NADH appears in the synthesis of many building units of cells, e.g. aminoacids, from substrates on the oxidation level of carbohydrates (see Krebs, 1972).

Among products of fermentation, we may mention higher alcohols (butanediol) and ketones (acetone). The relative or complete neutrality of the products may be an advantage to the bacteria. But various acids are also made, including butyric and succinic acid. The succinate is made by means of enzymes that are entirely different from the succinate dehydrogenase sinvolved in the respiratory chains of aerobes (Singer, 1971; *13c*).

Ferredoxins [*2d*] are important transfer agents for electrons in fermentations. They are proteins with strongly negative standard potentials, similar to that of hydrogen (Mortenson *et al.*, 1962; Mortenson, 1963; Valentine, 1964; Buchanan, 1966; Malkin and Rabinowitz, 1967; Hall and Evans, 1969; Arnon, 1969b; Buchanan and Arnon, 1970; Benemann and Valentine, 1972; Yoch and Valentine, 1972; Hall *et al.*, 1973 a, b, 1974, 1975; Lovenberg, 1973). In the ferredoxins ["*fd*"] the Fe, which accepts or donates electrons, is, in contrast to the cytochromes [*2d, 8c*], not present in the form of haem. The ferredoxins also contain inorganic sulphur. Ferredoxins from different sources, though differing in composition, can often replace each other mutually. Ferredoxin is involved in the release of free hydrogen (and CO_2) by clostridia. Many Fe-S proteins with higher standard potentials also exist.

(In anaerobic conditions, the much more advanced facultative aerobes, including *E. coli*, likewise produce—among other products—$H_2 + CO_2$ by fermentation, but the pathways differ from those in the strict anaerobes [see Gray and Gest, 1965].)

Alternatively, reduced ferredoxin may be used for the reduction of atmospheric nitrogen (see Winter and Arnon, 1969; Yoch and Arnon, 1969; Postgate, 1970). The evolution of nitrogen fixation will not be discussed here (see Burris, 1961, 1966; Imshenetskii, 1961; Takahashi *et al.*, 1963; Postgate, 1968a, 1971, 1974; Benemann and Valentine, 1972; Silver and Postgate, 1973; Streicher and Valentine, 1973). It may merely be pointed out that the necessity for the process surely arose in an early period, and indeed the strictly anaerobic clostridia already fix nitrogen. So do the (later) photosynthetic bacteria, the blue–green algae, and also some aerobic bacteria. The eukaryotes do not fix nitrogen (Millbank, 1969).

(e) Methane Formers

A further strictly anaerobic group are the methane formers, methanobacteria, briefly mentioned before (Söhngen, 1910; see Kluyver and Van Niel, 1956; Wood, 1961; Barker, 1956, 1967; Stadtman, 1967; Wolfe, 1971; Stanier *et al.*, 1971).

Typically, the methane formers obtain energy by reduction of CO_2 to CH_4 with H_2:

$$4H_2 + CO_2 = 2\,H_2O + CH_4;\ \Delta G_0' = -33\ \text{kcal.} \tag{7.5}$$

It has been shown with isotopes that in the species tested all the carbon of the methane is derived from CO_2 (Stadtman and Barker, 1951; Stadtman, 1967). Normally, however, they do not appear to be autotrophs.

For a long time, the dismutation of simple organic compounds (alcohols, carboxylic acids) by methane formers has also been observed. For instance (see Quayle, 1972):

$$4CH_3OH = CO_2 + 2H_2O + 3CH_4;\ \Delta G_0' = -74\ \text{kcal.} \tag{7.6}$$

$$3HCOOH = 3CO_2 + 2H_2O + CH_4;\ \Delta G_0' = -62\ \text{kcal.} \tag{7.7}$$

A thought-provoking recent result (Bryant *et al.*, 1967; Wolfe, 1971; Reddy *et al.*, 1972) with *Methanobacterium omelianskii* is that this "organism" is really a (symbiotic?) association of two different bacteria, one responsible for the reaction

$$CH_3CH_2OH + H_2O = CH_3COOH + H_2; \Delta G_0' = 11 \text{ kcal} \quad (7.8)$$

and the other for the reaction

$$CO_2 + 4H_2 = CH_4 + 2H_2O; \Delta G_0' = -33 \text{ kcal.} \quad (7.5 = 7.9)$$

Only the latter species should therefore be considered as a methanobacterium. Possibly similar results will be obtained with other "species", and maybe the true methane bacteria always use H_2 as a reductant for CO_2. Perhaps the principle of interspecies hydrogen transfer is of more general significance (Peck, 1974).

CO_2 is reduced stepwise by the methane bacteria, and other C_1 compounds (in bound form) have been detected as intermediates. The last step is the production of CH_4 from a methylcobalamin, i.e. a substance with a corrinoid ring system and related to (Co containing) vitamin B_{12}. At least this applies in artificial conditions. But on the whole, in spite of much work done and of their great interest, not much is known as yet about the mechanisms for ATP production in the methane bacteria (see Barker, 1956; Stadtman, 1967; Forrest, 1969; Ljungdahl and Wood, 1969; Wolfe, 1971; Quayle, 1972).

Incidentally, CO_2 is reduced to acetate with organic compounds or H_2 in energy-yielding reactions by some clostridia (see Kluyver and Van Niel, 1956; Wood and Stjernholm, 1961; Ljungdahl and Wood, 1969; Quayle, 1972), e.g. *C. kluyveri* [*10b*], *C. aceticum* (Wieringa, 1940; El Ghazzawi, 1967; Linke, 1969; Andreesen and Gottschalk, 1969; Andreesen *et al.*, 1970), *C. thermoaceticum* (Andreesen *et al.*, 1973), and *C. formicoaceticum* (Andreesen *et al.*, 1970, 1974). The mechanisms show some similarity to those in methane bacteria, pointing to phylogenetic relationship, but they and their relationships need further elucidation (Andreesen *et al.*, 1969; Schulman *et al.*, 1972, 1973).

According to Schlegel (1972), the methane formers ecologically depend on other fermenters in the sense that they exploit the end products of the latters' energy metabolism. This applies, for example, to the methane formers in the rumen or in anoxygenic water bodies (swamps). The concept is supported by the results on "*M. omelianskii*", mentioned before. As far as the methane formers cover all their energy requirements with the electron donor H_2, they deserve to be called anaerobic chemolithotrophs [*10c*].

(f) The Pentose Phosphate Pathway

The phylogenetic role of the "pentose phosphate pathway" or "hexose monophosphate shunt" (see, for example, Krebs and Kornberg, 1957; Harrison, 1960; Wood, 1961; Horecker, 1962; DeLey, 1962; Pon, 1964; Axelrod, 1967; Lehninger, 1970) has been discussed by Horecker (1961, 1965). Horecker has also dealt with the evolutionary significance of further pathways for carbohydrate utilization, e.g. the Entner–Doudoroff (ED) pathway, which are found in many microorganisms, including clostridia (see also DeLey, 1962, 1968a; Axelrod, 1967; Andreesen and Gottschalk, 1969; Stanier *et al.*, 1970; Schlegel, 1972).

One role of the pentose phosphate pathway is production of pentoses, which are constituents of nucleotides, from hexoses. Pentoses do not arise in glycolysis, and therefore a separate pathway for hexose degradation to pentose is expected to exist. The second role of the pentose phosphate pathway is the production of metabolic (active) hydrogen, [H] in the form of NAD(P)H, as shown in Fig. 7.3.

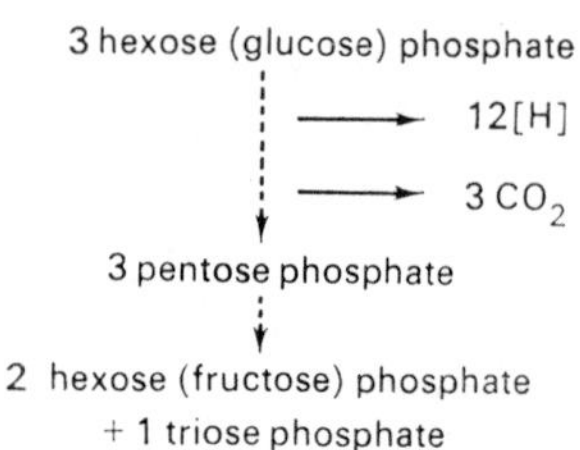

Fig. 7.3. Abbreviated scheme of the pentose phosphate pathway, merely from the point of view of bioenergetics. The overall reaction shown requires three cycles (turns of the wheel).

This metabolic hydrogen is needed for many kinds of reductive syntheses, conveniently listed by Horecker (1965). They include those of glutamate, and therefore of other amino-acids, from ketoglutarate, of higher fatty acids from acetic acid, and of deoxyribonucleotides from ribonucleotides. According to Horecker

> each of the two major pathways of carbohydrate metabolism has a role in providing specific building blocks for cell synthesis . . . The Embden–Meyerhof pathway and the citric acid cycle not only furnish energy as ATP and reduced diphosphopyridine nucleotide (NADH), but also provide precursors of alanine, glutamic acid and the amino acids derived from it, and aspartic acid and its many metabolic products, including the purines and pyrimidines. The pentose phosphate pathway provides energy as reduced triphosphopyridine nucleotide (NADPH) as well as precursors for a number of aminoacids such as histidine, and aromatic amino acids. It also provides ribose and deoxyribose.

Many enzymes of the pentose phosphate pathway are soluble. Because of their presence in anaerobes, e.g. clostridia, the reactions of the pathway may be assumed to be very ancient. According to Horecker (1961), this applies especially to reactions leading from hexoses to pentoses without dehydrogenation, which involve transketolase and transaldolase. These enzymes seem to occur universally. A further argument for the antiquity of the pathway is that some of the reactions, in the reverse direction, are part of the Calvin cycle for the assimilation of CO_2 by photosynthetic cells. Though this cycle is not found among fermenters, it is common among the photosynthetic bacteria [*10e*].

Anaerobically, NADH rather than NADPH is often found to serve as hydrogen donor. According to Horecker (1965) this is probably a relic from early conditions of life in a reducing atmosphere. Loss of reducing power as NADH through the respiratory chain [*13d*] naturally cannot occur in the fermenters. Likewise, in (anaerobic) photosynthetic bacteria the crucial reduction step—phosphoglycerate to glyceraldehyde phosphate—is still generally carried out by means of NADH. But in oxygenic conditions, where the electrons of NADH flow into the respiratory chain and ultimately to free oxygen, NADH cannot, in general, provide reducing power for synthesis. Therefore in aerobic organisms the problem is how to obtain useful reducing power nevertheless. As a consequence, the tendency appears to link the reactions of the pentose phosphate pathway to NADP. Equally, phosphoglycerate is reduced by NADPH in the (aerobic) plants. NADPH is mostly insensitive to the enzymes of the respiratory chain, and thus remains available (see, however, below).

ATP can be built up in fermentative reactions of glyceraldehyde phosphate, a product of the pentose phosphate pathway. This product gives rise to lactate or ethanol, as in the EMP pathway.

Has the pentose phosphate cycle then, in the later oxygenic conditions, a bioenergetic role as an integrated cycle? As shown in the scheme given before, one molecule of hexose could be wholly degraded to CO_2 if no pentose were drained off and the wheel turned six times with intermediate condensation of triose phosphate to hexose diphosphate:

$$\text{glucose-6-phosphate} + 12\text{NADP} = 6CO_2 + 12\text{NADPH} + P + 12H^+. \quad (7.10)$$

Some authors doubt the reality of an integrated pentose phosphate cycle (Horecker, 1965; see also Krebs and Kornberg, 1957). The replacement of NAD by NADP in evolution has been taken as evidence against the cycle concept. At least in mitochondria, i.e. in eukaryotes, NADPH as such cannot enter the respiratory chain and therefore be utilized for ATP production. It is significant that (in rat liver) one of the pyridine nucleotides (NAD) has been found mostly in the oxidized, the other (NADPH) in the reduced form (Glock and McLean, 1955; Jacobson and Kaplan, 1957; see Krebs and Veech, 1970). According to Lehninger (1970), the pentose phosphate pathway ought to be considered as a branching pathway with diffuse ends, capable of great metabolic flexibility, but therefore not predominantly serving production of ATP. The regulation of the pathway has been investigated by Eggleston and Krebs (1974).

Other authors think in terms of an integrated cycle for energy production (Wood *et al.*, 1963). Especially the blue–green algae, which were probably the first respirers, probably have no complete citric acid cycle as a source of metabolic hydrogen (electrons) for the respiratory chain; this pathway may have been lost. Their respiration may be based on NADPH from the pentose phosphate pathway (Biggins, 1969; Peschek, 1975).

(g) Storage Materials

When supplied with enough nutrient, bacteria may build up storage materials, for carbon and energy, from substrates. Common are polymeric carbohydrates which are found, for example, in clostridia, lactic acid bacteria and photosynthetic bacteria. Lipids, notably polymeric hydroxybutyric acid (PHB; Lemoigne, 1925), are also important. PHB-free mutants of bacteria (*Hydrogenomonas*) that normally make PHB are known (Schlegel *et al.*, 1970). Clearly necessary enzymes are missing. Both PHB and polymeric carbohydrates are contained within the cells as solids, and therefore are not active osmotically. When needed, they are used as substrates for energy-yielding catabolism. Some bacteria have several kinds of storage material in parallel.

For the biosynthesis of PHB, energy-rich acetyl coenzyme A is condensed to acetoacetyl coenzyme A, and this is reduced with NAD(P)H to hydroxybutyrate, and polymerized. The reports and review articles on PHB refer partly to the photosynthetic bacteria, which will be treated later (Gaffron, 1933, 1935, 1946; Doudoroff and Stanier, 1959; Stanier, 1961; Ormerod and Gest, 1962; Schlegel and Gottschalk, 1962; Dawes and Ribbons, 1964; Gray and Gest, 1965; Doudoroff, 1966; Gest, 1966; Van Gemerden, 1968; Bosshard-Heer and Bachofen, 1969; Dawes and Senior, 1973).

It is reported that some bacteria do not build up specific storage materials, e.g. *Pseudomonas aeruginosa* (Campbell *et al.*, 1963). Such organisms are thought to degrade informational macromolecules for energy, when needed. Polyphosphate [*6c*] is considered by some as a storage material for energy and/or phosphorus.

8

BACTERIAL PHOTOSYNTHESIS IN GENERAL

(a) A Critical Stage for Life

If our assumptions are right, the only free energy on which eobionts and early organisms (fermenters) could draw was chemical energy of solutes. As biomass increased, and the cells produced more and more effective enzymes for the utilization of the components of the medium, sooner or later critical conditions arose. The production rate by physical agents (radiations, etc.) of compounds with high free energy contents fell behind the rate of their consumption in life processes. The production rate was more or less constant while the needs of the growing populations increased exponentially—a super-Malthusian situation. Data about growth rates of recent microorganisms in Nature have been collected by Brock (1973b).

Yeast in a barrel of sugar solution would be a modern model. In alcoholic fermentation part (35 % in plausible conditions) of the free energy change in the catabolism of glucose is recovered through synthesis of ATP, as has been seen. The rest of the free energy is degraded to heat. The free energy at first laid down in ATP is used to support, in one way or other, the life activities of the cells, and some of it is temporarily stored as chemical energy of biomolecules. But this part is also in the end, when the cells and their components perish, converted into heat. Thus living matter, through enzyme-catalysed reactions, acts as a kind of machine to destroy free energy, to produce entropy, and to exhaust the medium.

In this way, the position of life must gradually have become precarious. Shortage of free energy must have increased selection pressure on the existing bacteria (still anaerobic, non-photosynthetic). Maybe the point was not so far off where life might have come to a stop for lack of free energy. A kind of "heat death" ("Wärmetod" of Rudolph Clausius) threatened our ancestors; the free energy of the solutes in the medium approached a minimum, and their entropy a maximum.

At this critical stage, organisms must have tried to exploit additional sources of free energy. In retrospect, it appears that the only feasible additional source was light; we can hardly expect bacteria to exploit nuclear energy! It was J. R. Mayer, the discoverer of the First Law of thermodynamics, who wrote in 1845: "Die Pflanzen nehmen eine Kraft, das Licht, auf, und bringen eine Kraft hervor: Die chemische Differenz." (At that time, the term Kraft (force) was used for what is now called energy.) The new source of energy is inexhaustible. It has been sufficient to support not only the photosynthetic anaerobes themselves, but also the coexisting non-photosynthetic anaerobes and (later) the aerobes; in effect, all living things.

The light effective in building up solutes was ultraviolet. At the stage in question, UV light still reached the surface of the Earth. Nevertheless, all photosynthetic organisms known exploit only visible and infrared light. Only occasionally the suggestion has been made (Oparin, quoted by Pavlovskaya, 1971) that UV light was used directly by early photosynthetic bacteria.

Short-wave UV light, precisely because it is violently active photochemically, is lethal (see Giese, 1964, etc.). In contrast, visible or infrared light, when taken up by a well-constructed system, is highly suitable for the performance of orderly reactions, as needed by life. Organic molecules are excited photochemically, but normally not to an extent that they decompose violently and irregularly. Rather they will, in a specific way, rearrange or react with the "cold" (unaffected) environment (Gaffron, 1965). Therefore quantum yields of useful reactions may be high, and side reactions rare. Also in respect to total energy flow, the bargain was good. There is far more energy in the visible + infrared than in the ultraviolet part of the solar spectrum (see Miller and Urey, 1959; Gates, 1966; Fig. 8.1).

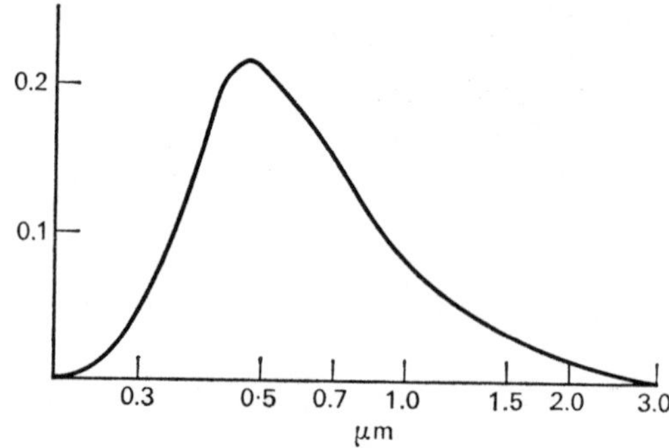

FIG. 8.1. Spectral irradiance (watts cm^{-2} μm^{-1}) outside the atmosphere (after Robinson, 1966). The boundary UV/visible is at about 0.38 μm = 380 nm, the boundary visible/infrared at about 0.74 μm.

It seems in order now to introduce some more terms referring to energy metabolism (*Cold Spring Harb. Symp. Quant. Biol.* **11**, 302 [1946]). The reliance of the fermenters, in respect to reducing agents (electron donors), on external organic substances is expressed by the term "organotrophy". In contrast, organisms that use inorganic reductants, to be introduced later, are called "lithotrophs". This division seems to fit bioenergetics better than that between "heterotrophs" and "autotrophs", where attention is focused on synthetic capabilities, and which is beset with difficulties (see Butlin and Postgate, 1954; Woods and Lascelles, 1954; Umbreit, 1962; Lees, 1962; Rittenberg, 1969; Kelly, 1971). Thus a strict autotroph must be a lithotroph, but the reverse is not true; e.g. a lithotroph may require the external supply of an organic growth factor. Further, as far as needed, photosynthetic organisms will be indicated by the prefix photo-, and organisms that use only "dark" reactions by the prefix chemo-. A new term, phytotrophy, will be suggested for the kind of photolithotrophy that is practiced by plants [*12*].

(b) The Overall Role of Phototrophy

The simplest extant phototrophs, the photosynthetic bacteria, were discovered by Engelmann in 1883. Their practical importance in the present biosphere is limited (see Stanier and Cohen-Bazire, 1957). But they carry decisive evidence about steps in evolution, and we are fortunate that a few species at least have (surely in modified forms) survived.

Up to a point, the photosynthetic bacteria can obtain energy from various fermentations, especially in absence of light (Nakamura, 1939; Gest, 1951, 1972; Kohlmiller and Gest, 1951; Hendley, 1955; Gest and Kamen, 1960; Bennett *et al.*, 1964; Gray and Gest, 1965; Schön and Drews, 1966). For instance, exogenous pyruvate may be converted by photosynthetic bacteria, as by clostridia, to H_2, CO_2 and fatty acids. Endogenous fermentation is obviously needed to tide the anaerobic photosynthetic bacteria over dark periods.

Wherever exogenous fermentation does not suffice for the growth of photosynthetic bacteria (see Gest, 1951; Stanier, 1961; Sojka *et al.*, 1967), loss of necessary enzymes or loss of transport mechanisms in the cell membranes may have occurred. The importance of loss is shown by the fact that some photolithotrophs (and chemolithotrophs; *15a*) are even strict autotrophs (see Umbreit, 1962; Smillie and Evans, 1963; Kelly, 1967, 1971; Smith *et al.*, 1967; Peck, 1968; Rittenberg, 1969), although they surely descended from heterotrophs. In these circumstances, the discovery (Uffen *et al.*, 1971; Uffen, 1973) of long-term growth of *Rhodospirillum rubrum* and other non-sulphur purple bacteria by fermentation is important.

In particular, photosynthetic bacteria do not, in spite of the presence of the glycolytic enzymes, deal well with carbohydrates (Gaffron, 1962b; see Gest, 1951; Wood, 1961). Carbohydrates may not have been so abundant in early times, as pointed out [*7c*]. Perhaps they were uncommon when the photosynthetic bacteria evolved from the fermenters. But clostridia do have the enzymes for glycolysis [*7c*].

In photosynthesis at least part of the energy is always gained in the form of ATP. Indeed some of the principal results of recent research are (a) that in all photosynthetic processes ATP is obtained, and (b) that NAD(P)H may also be a product; at least in most cases it is required for the assimilation of CO_2.

In the coming sections, we shall often have to use results obtained with plants. While photosynthetic bacteria are more ancient and simpler, plants have been better investigated.

(c) The Role of Chlorophyll

The chlorophylls are the key compounds in photosynthesis. They are a group of closely related derivatives of porphyrins, compounds with four pyrrol rings, and they invariably also contain Mg. Most authors think that early photosensitizers likewise consisted of pyrrol compounds. The extraordinary stability of these substances has been pointed out (Gaffron, 1962a, 1965), and has been attributed to resonance (Pullman and Pullman, 1962, 1963). The chlorophylls always act in conjunction with proteins.

The chlorophylls have been called photocatalysts. This is in any case formally correct, as the chlorophyll is not permanently transformed as a consequence of the photoreaction. Most probably, however, the molecule suffers a transient chemical change. The suggestion is due to Katz (1949) and Levitt (1953) that the essential feature of the photoreaction consists in the donation of an electron by a chlorophyll to an acceptor. In the terminology of Lewis and Lipkin (1942), the acceptor undergoes photoreduction, while chlorophyll is photooxidized. The basis of this idea has found general agreement (see Arnon, 1959; Kamen, 1963; Clayton, 1972, 1973; Katz and Morris, 1973; Parson and Codgell, 1975).

The excited chlorophyll acts as a strong reductant, and therefore the electron may be forced even upon a very weak oxidant. This acceptor thereby becomes a strong reductant. The hole in the chlorophyll is subsequently filled by an electron from another donor in a dark reaction. Relevant reactions of chlorophyll solutions will be mentioned later [*8h*].

In plants, the photoreductant is always chlorophyll a, in photosynthetic bacteria always bacteriochlorophyll a, of which forms exist that are esterified with different alcohols (Brockmann and Knobloch, 1972; Künzler and Pfennig, 1973; Gloe and Pfennig, 1974). The formulae can be taken from Fig. 8.2. The main absorption band of bacteriochlorophyll a lies between 800 and 900 nm rather than between 600 and 700 nm as in chlorophyll a. These data refer to absorption within the cell, not in solution.

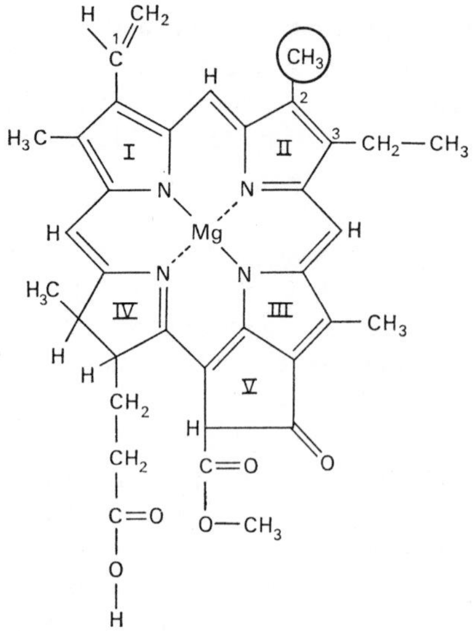

FIG. 8.2. Chemical formula of chlorophyll a (partly after Rabinowitch and Govindjee, 1969). All corners except those marked N are occupied by C atoms. The carboxyl group is esterified with phytol, $C_{20}H_{39}OH$. In chlorophyll b, the circled methyl group is replaced by the aldehyde group, CHO. In bacteriochlorophyll a, the vinyl group "1" is substituted by an acetyl group, and between the positions 2 and 3 there is no double bond, i.e. the ring II is reduced.

Chlorobium chlorophyll in the green photosynthetic bacteria, and chlorophyll b in most eukaryotic plants, including all higher plants, are photosensitizers. They transfer captured light energy from parts of the spectrum, where bacteriochlorophyll a or chlorophyll a, respectively, do not absorb well, to these substances. A similar role is played in prokaryotic plants (blue–green algae) and in some of the eukaryotic algae, the red algae, by phycobilins, open-ended tetrapyrrols.

Carotenoids are also associated with chlorophyll. They are ascribed a dual function. They serve as additional photosensitizers and also (jointly with other substances) as antioxidants (Engelmann, 1884, 1887; Duysens, 1952; Blinks, 1954, 1964; Griffiths *et al.*, 1955; Sistrom *et al.*, 1956; Stanier and Cohen-Bazire, 1957; see Rabinowitch, 1945; Duysens, 1964; Cohen-Bazire and Stanier, 1958; Goedheer, 1959, 1969; Batra, 1968; Krinsky, 1967, 1968; Liaaen-Jensen and Andrewes, 1972).

In all photosynthetic organisms the redox compounds include cytochromes as donors (not necessarily as direct donors) of electrons to chlorophyll. Cytochromes ("cyt"), mentioned in a different context [*2d*], are haemoproteins in which the iron (as in the non-haem proteins, the ferredoxins) may carry either two or three charges; different cytochromes vary greatly in their redox potentials (see Hill, 1965; Keilin, 1966; San Pietro, 1967; Bartsch, 1968; Bendall and Hill, 1968; Horio and Kamen, 1970; Kamen and Horio, 1970; Lemberg and Barrett, 1973). After transferring (probably by way of an intermediate) an electron to chlorophyll, the cytochrome, now oxidized, obtains an electron from a further redox compound, etc. In photosynthetic bacteria, cytochromes were first found by Elsden *et al.* (1953). The change in the cytochrome spectrum on illumination of photosynthetic bacteria was observed by Duysens (1953).

Photochemical systems different from the chlorophyll system arose in organisms at various stages (see Giese, 1964, etc.). For instance, they control phototaxis, phototropism and (the phytochrome system) differentiation. The light sensitive substances are in some cases related to chlorophyll, e.g. the open-chain phycobilins in phytochrome. In many other cases the substances have not yet been identified. In all these systems, in contrast to the chlorophyll system, light serves as a signal rather than a major source of energy.

But a striking exception has been found. In *Halobacterium halobium* (Blaurock and Stoeckenius, 1971; Oesterhelt and Stoeckenius, 1971, 1973; Racker and Stoeckenius, 1974; Danon and Stoeckenius, 1974; Kayushin and Skulachev, 1974) a chromoprotein containing the carotenoid retinal and called bacteriorhodopsin captures light energy to build up ATP. The organism is an obligate aerobe, and in the dark it makes ATP mainly through respiration. As oxygen is needed for the production of retinal, the bacterium can, in its present form, have arisen only after the advent of oxygen in the atmosphere, but the former existence of an anaerobic pathway for the biosynthesis of retinal cannot be excluded. Because of its astonishing simplicity, the *halobacterium* system has tremendous value for fundamental research in photosynthesis and more generally in bioenergetics. In particular, the system is well suited for tests of Mitchell's chemiosmotic hypothesis for the mechanism of energy conservation by means of membranes. Reference to this hypothesis will be made later [*8h*].

The visual systems of the different phyla of metazoa also operate, as far as is known, through rhodopsins, i.e. chromoproteins containing retinal. In fact, the term rhodopsin was coined by Wald for the light sensitive substance in the rods of the mammalian retina. If ATP production by these systems ever existed, it has been generally abandoned in favour of the signalling function. But the question may be posed whether the different rhodopsins (which are, of course, distinguished by their protein components, the opsins) arose polyphyletically, i.e. by convergence. Since the discovery of bacteriorhodopsin this question has become even more acute.

(d) The Photosynthetic Unit

Through studies begun by Emerson and Arnold (1932) and interpreted by Gaffron and Wohl (1936) it has become clear that both in photosynthetic bacteria and in plants always many molecules of chlorophyll collaborate in a "photosynthetic unit"—in plants typically consisting of several hundred, in purple bacteria maybe of 40 molecules (Duysens, 1967; Kok, 1967). At least in higher plants the size of the photosynthetic unit is not uniform in one and the same plant, the size differs between species, and it is also influenced by environmental conditions (Schmid and Gaffron, 1966, 1969, 1971).

All the chlorophyll molecules are capable of light absorption and therefore of excitation, but the energy of the quantum then migrates towards a particular "reaction centre" within the photosynthetic unit. The device has been compared to an antenna, and conduction of the "harvested" energy probably occurs through excitation transfer (see Robinson, 1964; Förster, 1967; Hoch and Knox, 1968; Borisov and Godik, 1973; Seely, 1973). Only the terminal energy acceptor, in the reaction centre, is capable of donating an electron to the primary oxidant (see Rabinowitch, 1959; Kamen, 1963; Clayton, 1966, 1972, 1973; Parson and Codgell, 1975). In this way, maximum use is made of the complex, and therefore biologically expensive, photosynthetic machinery in the reaction centre.

The reaction centre is thought to consist of chlorophyll (chlorophyll a in plants, bacteriochlorophyll a in bacteria) in union with other substances, including a number of redox compounds. The chlorophyll in the centre is, presumably owing to binding to protein, in a modified form. It changes its absorption spectrum when oxidized by light or, nonphysiologically, by treatment with chemicals. The reaction centre is designated by the approximate wavelength of the peak of the action spectrum. For example, in the photoorganotroph *Rhodopseudomonas spheroides* the centre is known as P (for pigment) 870 (nm), but the choice of the wavelength is in any case somewhat arbitrary. In other organisms the action spectrum may be different. Reaction centres have been partly separated from the antennal chlorophyll, and analysed (Clayton, 1966, 1972, 1973; see Fork and Amesz, 1970).

(e) Cyclic Photophosphorylation. Electron Flow

It was long thought that photosynthetic organisms must cover their energy needs by way of the utilization of products of photosynthesis through respiration or fermentation, while it is the direct task of photosynthesis to provide reducing power. A more immediate role for ATP in photosynthesis was first suspected by Lipmann (1941). The formation of energy-rich intermediates was considered a fundamental fact in photosynthesis also by Emerson *et al.* (1944). Ruben (1943) recognized that both ATP and reduced pyridine nucleotide (in present terms: NAD(P)H) are needed for the transformation of the primary product of CO_2 assimilation into final products. The direct production of ATP through the energy of light has been found with chloroplasts, i.e. with plants, by Arnon *et al.* (1954), and with systems from photosynthetic bacteria by Frenkel (1954). The process is called "photosynthetic phosphorylation" or "photophosphorylation".

In the formally simplest case, here described in a highly condensed way, the electron donated by the chlorophyll within the reaction centre returns to it by way of a number of intermediate redox compounds. All these "dark" steps are exergonic, and the synthesis of ATP is coupled to the round trip. Thus no permanent chemical change takes place in any of the members of the redox chain. An external electron donor is dispensable. This is the simplest, and possibly the original, instance of "electron flow" (Arnon, 1959). The generation of ATP due to circular electron flow is called "cyclic photophosphorylation".

Some authors think that ferredoxins, mentioned as electron transfer agents in fermenters [*7d*], are the first electron acceptors in cyclic photophosphorylation by bacteria (Evans and Buchanan, 1965; Buchanan and Evans, 1969; Buchanan *et al.*, 1969 a, b; Arnon, 1969b; Buchanan and Arnon, 1970; Shanmugam and Arnon, 1972) while others implicate coenzyme Q (ubiquinone) or some complex of both compounds (see Vernon, 1968; Fork and

Amesz, 1970; Frenkel, 1970; Gest, 1972; Clayton, 1973; Parson, 1974). Ferredoxin certainly is involved in the cyclic photophosphorylation of plants [*12d*].

It is important that at least the first electron acceptor and the last electron donor are separated spatially. Otherwise the lost electron would have a good chance of returning to chlorophyll rapidly without having been of service for photophosphorylation. The separation is achieved by construction of solid membranes, in which the photosynthetic units are embedded. Electron flow has indeed never been found in solution [*8h*].

Probably 1 molecule of ATP is built up per round trip of 2 electrons, but possibly 2 ATP rather are gained. The site(s) of ATP formation, i.e. the redox step(s) that provide(s) useful energy, is (are) not known with certainty (Baltscheffsky and Arwidsson, 1962; Baltscheffsky and Baltscheffsky, 1963; Vernon, 1964, 1968; H. Baltscheffsky, 1967b; Arnon, 1967 a, b; Avron and Neumann, 1968; Bendall and Hill, 1968; Boardman, 1968; Avron, 1971; Baltscheffsky *et al.*, 1971). However, here—as in other instances of energy-conserving phosphorylation, to be referred to later—the search for definite sites and also for integral values of ATP production per flow of two electrons loses in importance if Mitchell's chemiosmotic hypothesis for energy flow is adopted [*8h, 13b*]. In that case, a "proton-motive force" is set up between the two sides of a membrane first, with an osmotic and an electric term, and "the squiggle" [*6a*] is produced only subsequently. A stoichiometric relationship between electron flow and ATP production cannot be taken for granted (see, for chloroplasts, Trebst and Hauska, 1974).

(f) Non-cyclic Photophosphorylation

For the photosynthetic assimilation of CO_2, NAD(P)H is needed. (For exceptions, where ferredoxin is the direct reductant, see [*10d*].) The redox potentials of organic and of nearly all inorganic compounds, utilized by photosynthetic bacteria as sources of electrons, are not low enough for the "dark" reduction of NAD and even less for that of a ferredoxin. Relevant values of $\Delta G_0'$ will be found in Table 11.1. The electrons must be promoted photochemically to supply NAD(P)H.

In plants, the existence of "non-cyclic photophosphorylation" in addition to that of cyclic photophosphorylation has been securely demonstrated (Arnon, 1959, 1967 a, b; see Nozaki *et al.*, 1963; Vernon, 1964). Here reductant as well as ATP is produced by light. Reduced ferredoxin (Tagawa and Arnon, 1962) transfers its electrons—with hydrogen ions, as hydrogen atoms—via a flavoprotein (Shin *et al.*, 1963) to pyridine nucleotide (NADP) and makes NADPH. The electron must, as in this case it is lost to the flow chain, be replenished from an external source of reductant, in plants in natural conditions from water. Thus permanent reduction and oxidation is achieved: the process is open-ended. Physiologically, the NADPH reduces carbon compounds derived from CO_2, i.e. the "reducing power" serves assimilation. In absence of CO_2, the electrons may be absorbed by artificial oxidants, e.g. ferricyanide or a quinone (Hill, 1939, 1951). The yield in this "Hill reaction" is improved if ADP and P for ATP production are provided (Avron *et al.*, 1958; Arnon *et al.*, 1958; West and Wiskich, 1968); clearly a consequence of the fact that in physiological conditions electron flow and phosphorylation are coupled. This is called "photosynthetic control". A similar phenomenon in oxidative phosphorylation is known as "respiratory control" [*13a*]. An overall term is "electron transport control".

In photosynthetic bacteria reducing power may likewise be produced photochemically with electrons from an external source. There is no other way how organisms can photo-assimilate CO_2. In principle, the external reductant may be organic or an inorganic. For instance, isopropanol (which thereby is oxidized to acetone) may be a source of electrons (Foster, 1940). Another organic reductant is succinate, thereby oxidized to fumarate (see Gest, 1972). Among inorganic electron donors, sulphides, including thiosulphate, are common [*11b*]. It appears that all these reductants supply electrons first, in a dark reaction, to a bacterial cytochrome. Water is never used by bacteria as an ultimate electron source, a point of utmost importance [*12a*].

Some authors, e.g., Arnon (1961), Nozaki *et al.* (1961, 1963), Van Niel (1962) and Amesz (1963), envisage for the production of reducing power in bacteria, as in plants, non-cyclic electron flow. Electrons from the donor are thought to be transported with direct utilization of the energy of light to an acceptor, ultimately for the production of biomass. Again it is not clear whether ferredoxin and/or ubiquinone serves as acceptor [*8e*]. Also it is uncertain whether light-powered non-cyclic electron flow in bacteria, if it exists, is linked to ATP formation.

Alternatively, production of reducing power in photosynthetic bacteria could be attributed to the now well-known phenomenon of "reversed electron flow". [For terminology, see Chance *et al.* (1972).] In connection with the aerobic respiration [*13b*] of mitochondria, i.e. higher organisms, the important discovery, foreseen by Davies and Krebs (1952) and Krebs and Kornberg (1957), has been made that endergonic flow (flow in the direction opposite to that of spontaneous flow) may be forced by an energy-rich compound, typically ATP (Chance and Hollunger, 1960, 1961; Klingenberg and Schollmeyer, 1960; Chance, 1961 a, b; Klingenberg, 1963; Ernster and Lee, 1964; Van Dam and Meyer, 1971). Reversed electron flow is also observed with respiring bacteria (Sweetman and Griffiths, 1971).

While in the respiratory process exergonic electron flow is physiologically coupled with ATP synthesis ("oxidative phosphorylation"), in endergonic flow, on the contrary, ATP is used up ("reductive dephosphorylation"), but reducing power is generated. Thus electrons from an external source are pumped to the level of NADH.

Reversed electron flow would make light-powered non-cyclic electron flow dispensable. Part of the ATP from cyclic photophosphorylation would pump electrons to NAD, and

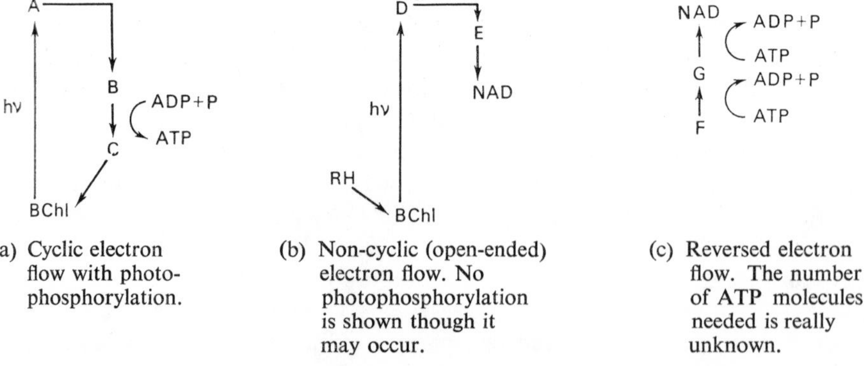

(a) Cyclic electron flow with photophosphorylation.

(b) Non-cyclic (open-ended) electron flow. No photophosphorylation is shown though it may occur.

(c) Reversed electron flow. The number of ATP molecules needed is really unknown.

FIG. 8.3. General schemes of bacterial photosynthesis (Bchl = bacteriochlorophyll a; A, B, etc. = redox compounds; RH = reductant). The arrows give the direction of electron flow. The upward direction is always that of increasing negative standard potential.

the NADH produced would be used for CO_2 assimilation. After Chance and Olson (1960) had suggested the existence of reversed electron flow for the production of reducing power (NADH) in bacteria, this process has indeed been found in photosynthetic bacteria and in systems made from them (Bose and Gest, 1963; Gest, 1966, 1972; Keister and Yike, 1967; Baltscheffsky *et al.*, 1966a, 1971; Jackson and Crofts, 1968; Vernon, 1968; Jones and Vernon, 1969; Baltscheffsky, 1969b; Knobloch *et al.*, 1971). The finding is instructive that the reaction succinate + NAD → fumarate + NADH occurs on addition of ATP to cell-free systems from photosynthetic bacteria in the dark (Löw and Alm, 1964; Keister and Yike, 1967; Vernon and Jones, 1969). The concept of reversed electron flow will also be applied to chemolithotrophic (non-photosynthetic) bacteria [*15*].

The problem is, then, whether photosynthetic electron flow in bacteria should be restricted to the cyclic process. Quite possibly both reversed electron flow and the open-ended non-cyclic process may provide reducing power, the relative contributions depending on conditions (Klemme and Schlegel, 1968; Klemme, 1969; Govindjee and Sybesma, 1970). Multiplicity of reaction pathways may be characteristic of photosynthetic as well as of fermenting bacteria (see Frenkel, 1970). It may have adaptive value for life in different circumstances (Cusanovich *et al.*, 1968).

The formal possibilities of electron flow in bacterial photosynthesis are shown in Fig. 8.3.

One and the same photosystem may serve the cyclic and the non-cyclic process (see, for example, Cusanovich *et al.*, 1968), but some workers (Morita, 1968; Sybesma, 1969; Fowler and Sybesma, 1970; see Hind and Olsen, 1968; Rabinowitch and Govindjee, 1969; Olson, 1970; however, Fork and Amesz, 1969, 1970) have considered the possibility that in bacteria two separate photosystems for the two processes exist. If they exist, their action spectra must be practically identical, as enhancement [*12b*] has not been found with bacteria. The duality would foreshadow the well-established existence of two photosystems in plants. Two photoreactions in series, which clearly are needed in plants for oxygen evolution [*12b*], have, however, not been proposed for bacteria (Frenkel, 1970).

The principal use for the reducing power (with ATP) is the production of biomass through the reductive assimilation of CO_2. The mechanisms for the assimilation will be discussed later [*10*]. The reducing power (again with ATP) may also be used, as in many fermenters [*7*], for the assimilation of nitrogen. Many of the photosynthetic bacteria assimilate nitrogen (Kamen and Gest, 1949; see Stewart, 1973). Finally, if reducing power is generated in excess, photoproduction of hydrogen may occur. The last step is catalysed by hydrogenase, and is subject to regulation (Gest and Kamen, 1949; Gest, 1954, 1972; Arnon *et al.*, 1961a). For instance,

$$\text{succinate} = \text{fumarate} + H_2;\ \Delta G_0' = 20\ \text{kcal}. \qquad (8.1)$$

In Gest's view, the energy is mobilized through reversed energy flow.

In sum, it may now be stated: if promotion of an electron occurs through non-cyclic photophosphorylation, the energy is derived directly from light, absorbed by chlorophyll. If it occurs through reversed electron flow, the contribution of light is more indirect, and consists in the production of ATP as an intermediate.

(g) Origin of Photosynthesis

According to the concept put forward here no light was used directly by eobionts and early organisms, and photosynthesis arose only later. The interval between the time of production of the bulk of the organic solutes by short-wave light before the rise of organisms and the time of exploitation of long-wave light by them might be called a dark age—in a sense different from Gaffron (1965), who referred to our ignorance about the period. In our analysis, the difficulty must be faced that photosynthetic machinery had to be introduced into pre-existing organisms which had no use for light.

The difficulty would be mitigated if a (more modest) role for light were assumed already for the eobionts and first organisms. Thus Granick (1951, 1957, 1965) suggested that energy of light, captured by mineral sensitizers in reducing conditions, produced protoplasm that remained grouped around the active centres. Its first organization would have been an "energy-conversion unit that could perform primitive photosynthesis and respiration". Further evolution is supposed to have consisted in the improvement of this structure. Gaffron (1961, 1962a, 1965) assumed that photochemically produced porphyrin antedated the earliest living cell and that later various biochemical systems, each system serving a different "purpose" of organisms, may have clustered around a complex of pigments. The evolution of these systems was thought by Gaffron to have been complete before the earliest living cell appeared. Calvin (1959), too, suggested that chlorophyll-type photosynthesis began in precellular entities.

However, it is not easy to accept that machinery for the regular transduction of light energy into chemical energy existed in the primitive eobionts; the needed structures, quite sophisticated as far as we know, could surely be built up and maintained only on the basis of a pre-existing smooth and effective metabolism. A third formal possibility, that photosynthetic organisms and fermenters arose independently and developed side by side, must be rejected as implying a polyphyletic origin of life. This is implausible.

The formal simplicity and the unambiguous meaning for the organisms of the cyclic form of electron flow, coupled to photophosphorylation, lead one to consider this cyclic electron flow as more fundamental and more ancient than non-cyclic flow (Arnon *et al.*, 1958, 1961a; Losada *et al.*, 1960; Arnon, 1959, 1961; Olson, 1970; however, Duysens, 1964). In photosynthesis, reducing power is not always essential [*9b*], as has early been emphasized by Gaffron (1935). On the other hand, ATP is always needed. Hence the capacity for making reducing power may have come later than that for making ATP. As far as reducing power was needed, this could be supplied by H_2 as long as the atmosphere was reducing (Losada *et al.*, 1960; Arnon, 1961). Furthermore, reducing power could be provided by reversed electron flow, as explained.

The role, if any, of non-cyclic photophosphorylation in bacteria is still controversial [*8f*]. Moreover, it need not have arisen simultaneously with non-cyclic electron flow, which primarily served reduction. Phosphorylation may have been coupled to non-cyclic electron flow only after the establishment of the latter. Of course, the production of ATP as a by-product in the generation of reducing power is very economical. On the other hand, in view of the priority of photophosphorylation, it might be assumed that the non-cyclic process led to phosphorylation from the beginning.

Baltscheffsky (1974 a, b) has put forward a hypothesis for the origin and the evolution of electron transport chains, with accent on the protein part of the electron carriers rather than coenzymes or prosthetic groups. "The evolution has originated with ferredoxin-like molecules ... and then proceeded ..., resulting in close evolutionary relationships between various neighbor links in the chains thus formed ... I tend to believe that a common ancestry may eventually be traced for all polypeptide structures involved in biological electron transport." Stepwise evolution along the electric potential scale is implied. The hypothesis is based mainly on information on primary and tertiary structural relationships of proteins.

(h) Early Mechanisms

Photosynthesis was required by the bacteria when the primeval amounts of redox compounds useful in energy metabolism ran out [*8a*]. There was a growing need to replenish the stores of useful reductants and oxidants photochemically either from useless substances present from the beginning or from waste left behind by exergonic redox reactions. The photoproducts could then react back to provide energy-rich compounds of the type of ATP. This occurred presumably by processes involving substrate-level phosphorylation; no other way to make ATP was known at that stage.

The required chlorophyll (or a related photocatalyst) was produced by the cells themselves. The absence of porphyrins from clostridia is a strong argument against appreciable amounts of external porphyrins in the primeval soup. May it be recalled that very little porphyrin has been found in carbonaceous chondrites [*4b*], and very poor yields of porphyrins have so far been obtained in experiments on prebiotic chemistry [*4c*]. Neither chlorophyll nor haem, which have the porphyrin moiety in common, appears to have been used by the fermenters although corrinoids (related to vitamin B 12), which also contain four pyrrol rings, are well represented in clostridia and methane formers (see Stadtman, 1967; Ljungdahl and Wood, 1969). The photosynthetic bacteria may have been the first to exploit the porphyrin structure (Lascelles, 1964 a, b).

The back reactions of the photoproducts may at first have taken place in solution, so that they were not spatially segregated. Presumably only processes were retained in evolution where useless back reactions, i.e. mutual annihilation without conservation of energy in energy-rich compounds, were largely prevented by kinetic obstacles (see Gaffron, 1965).

Photochemical reactions leading, to some limited extent, to simultaneous accumulation of an oxidant and a reductant in a common solution have been investigated by many authors (Rabinowitch and Weiss, 1937; Rabinowitch, 1945, 1956, 1961; Krasnovsky, 1948, 1960, 1972; see Kamen, 1963; Daniels, 1972). A relevant example is the light-powered oxidation of ascorbic acid by NAD in pyridine, with chlorophyll as photocatalyst (Krasnovsky and Voinovskaya, 1952). As the experiment showed, the reaction does not go back instantly in the dark in this case.

The obstacle against useless back reactions could be improved many times through spatial separation of the products. To achieve this, the photocatalyst could be inserted into a membrane in such a way that the newly formed reductant had to appear on one side, and the oxidant on the other side. The membrane acted as a barrier against short-circuit annihilation. It was therefore easier to allow electron flow from one side of the barrier to the other side only along particular pathways, and it is imaginable that electron flow could be made dependent on phosphorylation, i.e., be coupled to it. This hypothetical device can be considered as a model for a primitive form of an electron-flow chain. The principle could be elaborated gradually, the original members of the chain improved in their chemical compositions, and additional members fitted into the chain. Parts or extensions of the cell membrane suggest themselves as early photosynthetic membranes. Cell membranes had, of course, preexisted in the fermenters.

(Primitive electron-flow processes were considered for fermenters (Barker, 1956; see Gunsalus and Shuster, 1961; Stadtman, 1966). Thus, it was thought that for a time the growth of *Clostridium kluyveri* cannot be accounted for exclusively on the basis of ATP derived from substrate-level phosphorylation. Therefore it was assumed that in such cases electron transfer reactions from pyridine nucleotide to flavoprotein, presumably in a soluble system, may be coupled to additional phosphorylation. The attraction of the general idea is that it would free bacterial metabolism "from slavish dependence on particular phosphorylatable substrates" (Hall, 1971). Phosphorylation by such a mechanism might be looked at as first step towards photophosphorylation, though no membrane is involved as yet. However, the experimental basis of the idea, in respect to *C. kluyveri*, has since been questioned (Thauer *et al.*, 1968; Schoberth and Gottschalk, 1969; see Anderson and Wood, 1969)).

An evaluation of the role of membranes in bioenergetic processes ought to take into account Mitchell's remarkable chemiosmotic hypothesis [*8e*]. Mitchell (1968; see Mitchell, 1970) has suggested that a

> proton-translocating oxidoreduction system and the reversible proton translocating ATPase may have arisen separately as alternatives for generating the pH difference and membrane potential required for nutrient uptake and ionic regulation . . . in primitive prokaryotic cells, and that the accidental occurrence of both systems in the same cell may have provided the means of starting the free energy of oxidoreduction in ATP synthesized by the reversal of the ATPase, or in some other anhydride, such as pyrophosphate . . ., produced by a similar mechanism.

It may, however, be too early for a further discussion of these hypothetical events.

The simultaneous presence from the beginning of both a photocatalyst and a cytochrome need not be assumed. It is tempting to think that the haem of cytochromes developed from chlorophyll through a gene duplicating mechanism [*5e*].

Inorganic pyrophosphate (PP) can in some respect replace ATP in photosynthetic bacteria. PP is formed, depending on conditions, along with ATP by photophosphorylation, and PP as well as ATP can be used to drive reversed electron flow or to supply energy for CO_2 fixation (Baltscheffsky *et al.*, 1966 a, b; Baltscheffsky, 1969; Pfluger and Bachofen, 1971; Baltscheffsky *et al.*, 1971; Guillory and Fisher, 1972). To some extent, ATP and PP appear to work synergistically (H. Baltscheffsky *et al.*, 1971). Furthermore, PP-driven synthesis of ATP in bacteria has been found along with direct ATP synthesis (Keister and Minton, 1971). PP is, incidentally, also active in chloroplasts (Jensen and Bassham, 1966; Vose and Spencer, 1967) and in yeast (Baltscheffsky, 1968).

It has been pointed out [*6c*] that PP may have been a precursor of the energy-rich nucleotides. But ATP surely existed already in the original fermenters before the invention of photosynthesis, or else the recent fermenters would hardly have it. Thus the participation of PP as found now in the energy metabolism of organisms that arrived much later than fermenters suggests that it may have taken ATP very long in evolution fully to replace PP.

It may be emphasized that all photosynthetic organisms must employ quite a number of quanta for the production of each unit of stable reduced material [(CH_2O)], i.e. that the quantum requirement is always $\gg 1$ [*12e*]. The necessary complicated sequence of events cannot have existed in the beginning of photosynthesis. The photosynthetic bacteria that survive have learned to pile up the energy of several light quanta in an orderly fashion. In this respect they have later been overtaken by the plants [*12a*].

In an ecosystem originally dominated by fermenters, the overall role of light consisted in the reversal of the effects of catabolism. In fermentation energy-yielding substrate was used up, and useless end products accumulated. In light, such trends were reversed, and useful carbon compounds were regenerated from waste. No products had to be discarded. On the basis of the antagonistic dark and light reactions, life could go on indefinitely.

9

PHOTOORGANOTROPHY

(a) Non-sulphur Purple Bacteria

The photoorganotrophs are represented in Nature by the nonsulphur purple bacteria, the athiorhodaceae (see Fromageot and Senez, 1960; Stanier, 1961; Van Niel, 1963; Kondrateva, 1963; Vernon, 1964, 1968; Pfennig, 1967). They were first defined by Molisch in 1907, after Engelmann had found bacterial photosynthesis in 1883. In anaerobic conditions the non-sulphur purple bacteria use organic substrates, indeed a wide range of them, as electron (hydrogen) donors in the light. Fermenters, e.g. clostridia, use the same electron donors in the dark.

The non-sulphur purple bacteria (Fig. 9.1) are, then, closely related to the most ancient bacteria still in existence, and presumably very ancient themselves. However, in some respects they must have changed greatly during the ages. Most of them are now facultative aerobes (respirers). Clearly the machinery for respiration can have been acquired only in a fairly recent, oxygenic, atmosphere. More will be said about bacterial respiration in general and about the aerobic generation of ATP by non-sulphur photosynthetic bacteria in particular later [*14*].

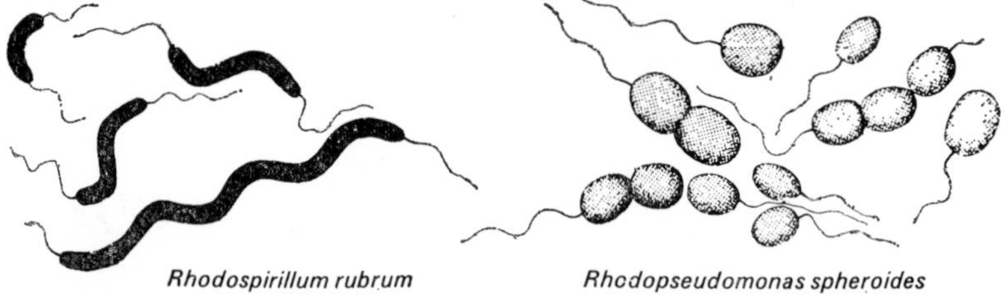

FIG. 9.1. Species of *Rhodospirillum* (left) and *Rhodopseudomonas* (right), both purple non-sulphur bacteria. (In reality much smaller than the purple sulphur bacteria [11*a*].)

Some of the non-sulphur purple bacteria can, in spite of their name, make some limited use of reduced sulphur compounds, along with organic compounds, as electron donors (see Hansen and van Gemerden, 1972). In higher concentrations they may be poisoned by such sulphur compounds. An interesting exception is *Rhodopseudomonas sulfidophila*, which grows well in the light with sulphide as electron donor (Hansen and Veldkamp, 1973). It may be concluded that the usual poor tolerance of the non-sulphur purple bacteria for sulphides is not a matter of great fundamental or evolutionary significance. It has also been reported (Hirsch, 1968) that *Rhodopseudomonas sp.* grows anaerobically under light with CO, an inorganic compound, as a reductant.

The structural basis of photosynthesis—the features of the machinery, and its distribution in the cell—has been much investigated in the non-sulphur and also in the related sulphur purple bacteria (see Gest *et al.*, 1963; Frey-Wyssling and Mühlethaler, 1965; Cohen-Bazire and Sistrom, 1966; Branton, 1968; Lascelles, 1968; Vernon, 1964, 1967, 1968; Echlin, 1970b; Oelze and Drews, 1972; Lampe and Drews, 1972). In the membrane system, the chlorophyll and its reaction partners are contained in "thylakoids" (flattened vesicles with double membranes; Menke, 1962, 1966). The fragmented thylakoids are known as "chromatophores" (Schachman *et al.*, 1952).

(b) Reducing Power and ATP

While purple bacteria in the light always use photosynthetically made ATP, they do not necessarily need photosynthetically made reducing power [*8h*]. A substrate exactly on the oxidation level of the storage material polyhydroxybutyrate (PHB; *7g*), namely, hydroxybutyrate, is photoassimilated without reaction with an oxidant (CO_2) or, for that matter, with a reductant, e.g. H_2 (Stanier, 1961). This beautiful result demonstrates the essential independence of photosynthesis by organotrophs from the generation of reducing power, and consequently also from CO_2 metabolism. To obtain ATP, cyclic photophosphorylation is enough, and a non-cyclic process need not be invoked.

The situation is more complicated when the available substrate is either more reduced or more oxidized than body material. In the former case (for instance, with alcohols or fatty acids) the (anaerobic) bacteria may as a net result in photosynthesis oxidize the substrates with CO_2 to get products of the correct elementary composition. Alternatively, excessive electrons may be removed via ferredoxin. In presence of hydrogenase, the electrons combine with hydrogen ions to free hydrogen [*7d, 8b, f*].

In contrast, when acetate, more oxidized than PHB, is the only substrate (Stanier *et al.*, 1959) electrons for the reduction of part of this substrate may be obtained through disproportionation of another part of the acetate to CO_2 and active (metabolic) hydrogen [H]. In such cases, the electrons for the reductive assimilation of oxidized substrates by photoorganotrophs do not come from excited chlorophyll, but from reductants formed in dark reactions. In other words, no light-powered non-cyclic electron flow is involved. But this need not always apply.

We have now emphasized that in photoorganotrophy utilization of CO_2 is an incidental phenomenon only. Yet we are used to associate mentally the large-scale reduction of CO_2 with photosynthesis, and rightly so. The utilization of CO_2 for the manufacture of biomass has in the end become a main achievement of the phototrophs. The massive gains have been made by the later kinds of phototrophs, namely, by the coloured sulphur bacteria [*11*] and by the plants [*12*]. It is time now to discuss the mechanisms of reductive CO_2 assimilation, i.e. of the processes leading to a net gain of carbon and to an increase of biomass.

10

ASSIMILATION OF CARBON DIOXIDE

(a) Availability of CO_2

Enormous amounts of fully oxidized carbon have been present in the biosphere for very long times [*3c*, *25d*]. Either CO_2 was, as some think, predominant among the volatile carbon compounds almost from the beginning, or CO_2 gradually replaced hydrocarbons when the atmosphere turned from a reducing to a neutral condition. The pressure of the free CO_2 and the related concentration of the bicarbonate ion in the waters were kept fairly low by the Urey reaction with the silicates of the crust [*3c*]. But the carbonate in the rocks has also remained available to organisms provided they had enough ATP and reducing power to attack it. As the biomass increased, and in the waters the reservoir of reduced carbon compounds useful for biosynthesis approached exhaustion, the utilization of the CO_2 as an additional source of organic carbon became a matter of urgency.

(b) Absorption of CO_2 by Fermenters

To what extent could the (heterotrophic) fermenters, the earliest organisms, tap the huge reservoir of carbon in CO_2? Quite probably early fermenters absorbed CO_2, as the contemporary fermenters do. In the present world, absorption of CO_2 is a general property of living matter. Rather remarkably, Lebedeff suggested already in 1908 (see also Lebedeff, 1921) that the utilization of CO_2 might be a common property of all organisms. He started from the consideration that certainly not only some of the photosynthesizers, but also the (non-photosynthetic) strict, i.e. obligate, chemoautotrophs [*15*] obtain all their carbon from CO_2, and generalized this experience.

However, not all uptake of CO_2 is equivalent to net utilization of carbon, leading to a permanent increase in biomass. Such increase is possible only when electrons from some external source are also taken up, and the carbon of CO_2 is reduced to the physiological level. In bacteria that store PHB [*7g*, *9b*] this level lies even below the level of carbohydrate (Van Niel, 1936).

We shall use the neutral terms "absorption" or "uptake", where no statement about the permanence of the gain of carbon from CO_2 in a process is intended. On the other hand, we shall speak of "fixation" or "assimilation", where organic carbon is really gained, i.e. where CO_2 is, in connection with uptake, reduced. We shall not distinguish between the forms CO_2, HCO_3^- and HCO_3^{2-} (see Cooper and Wood, 1971); in all these forms, carbon is fully oxidized.

An absorption of CO_2 by organotrophs (therefore, by heterotrophs) has first been observed, in the nonphotosynthetic propionic acid bacteria, by Wood and Werkman in the thirties (see Wood and Werkman, 1942). Subsequently such CO_2 absorption, by a number of mechanisms, has been established as a widespread property of heterotrophs (see Utter and Wood, 1951; Woods and Lascelles, 1954; Quayle, 1961; Wood and Stjernholm, 1962; Wood and Utter, 1965; Ljungdahl and Wood, 1969). For instance, *Clostridium kluyveri* [*7e*] can obtain one-quarter of its carbon from CO_2 (Tomlinson and Barker, 1954).

An example of an absorption reaction is the carboxylation of pyruvic to oxaloacetic acid (see Wood and Werkman, 1942):

$$CH_3.CO.COOH + CO_2 = HOOC.CH_2.CO.COOH; \; \Delta G_0' = 7 \text{ kcal.} \qquad (10.1)$$

This endergonic reaction is driven by a coupled orthophosphate split of 1 ATP. But there is no concomitant reduction of the carbon, and sooner or later an equal amount of CO_2 is given off again in a decarboxylation reaction. As a consequence, no net gain of carbon comes to the organism.

Other "dark" absorption reactions do involve temporary reduction of CO_2, but still no true assimilation of carbon. Thus the "malic enzyme", which may well date back to early times, catalyses the formation of malate from pyruvate + CO_2:

$$CH_3.CO.COOH + NAD(P)H + CO_2 = CH_3.CH_2.CHOH.COOH + NAD(P); \quad \Delta G_0' = -0.3 \text{ kcal.} \qquad (10.2)$$

This reaction still does not lead to a net gain of organic matter in the biosphere as a whole as long as electrons from organic compounds are needed to provide NAD(P)H. Only if additional electrons were introduced into the system, a net gain of organic matter would follow.

(c) Anaerobic CO_2 Assimilation in the Dark

The only external reductant in the biosphere with a potential low enough for the dark reduction of NAD(P) is free hydrogen:

$$NAD(P)^+ + H_2 = NAD(P)H + H^+; \; \Delta G_0' = -4.3 \text{ kcal.} \qquad (10.3)$$

Moreover, ATP could be made in a dark reaction of H_2 with CO_2, as in methane formers [*7e*]. Thus both reducing power and ATP for the fixation of CO_2 would be available in the dark. But in spite of the thermodynamic possibility no growth of photosynthetic bacteria on hydrogen in the dark has ever been observed (Roelofsen, 1935; Ormerod and Gest, 1962; Gest, 1966).

Can fermenters grow autotrophically on H_2? The best candidates are among the methane formers [*7e*]. In particular, a thermophile, *Methanobacterium thermoautotrophicum*, has been discovered that grows fairly quickly on CO_2 as only source of carbon with H_2 as electron donor (Zeikus and Wolin, 1972). At first glance this interesting organism just seems to reverse hydrogen fermentation. This cannot be the whole truth, however. Another remarkable feature of this organism is that it contains a network of internal membranes (Zeikus and Wolfe, 1973). Such membrane systems are otherwise the privilege of photosynthesizers and the respirers. *M. thermoautotrophicum* is not only an (anaerobic) chemolithotroph, but also an autotroph.

In evolution, anaerobic chemolithotrophs could play a major role only if they had sufficient amounts of H_2 in the atmosphere to exploit. Probably large amounts of H_2 did not persist long, geologically speaking, in the early atmosphere [*3c*]. Moreover, the solubility of H_2 in water is poor, and therefore little H_2 entered the waters and became available to organisms (DeLey, 1968). On the other hand, H_2 was set free all the time in waters in hydrogen fermentations along with CO_2 [*7d*], and this still happens in anaerobic conditions. This living space was and is then available to the anaerobic chemolithotrophs.

(d) Reductive Carboxylic Acid Cycle

Major amounts of CO_2 are fixed by cyclic processes developed for this purpose. On the basis of the evidence from extant organisms, the photoorganotrophs were, with the possible exception of the methane formers just named, the first bacteria to do so. They use organic compounds or hydrogen as source of electrons and light as source of energy for assimilation [*9a*].

In cyclic processes [*1e*] at the end of a series of reactions the acceptor for CO_2 is regenerated. Generally after a time all carbon atoms in the net products are ultimately derived from CO_2. Cyclic processes are efficient. In assimilation by a cyclic process, carbon is fixed on a kind of moving belt. All this would not apply to "linear" processes, e.g. to the carboxylation of pyruvic acid [*10b*]. Here the oxaloacetate formed is largely diverted as a net product, and new pyruvate for its synthesis therefore has to come from a source other than CO_2. Presumably the cyclic processes evolved from linear processes found in fermenters.

Van Niel (1949b) had foreseen the possibility of a fairly simple cyclic process for the reductive assimilation of CO_2, namely, a kind of reversed citric acid cycle. A scheme for the citric acid cycle will be given only later [*13c*], as it has been fully developed and has acquired central importance in aerobic conditions. The overall effects of a reversed full citric acid cycle would be the intake of $2CO_2$ and 8[H] and the production of 1 acetate as AcCoA + $2H_2O$. The acetate (+CO_2) would be further reduced to pyruvate, and this would be the starting material for reversed glycolysis. In fact, a "pyruvate synthase" which catalyses the reduction of acetate + CO_2 by ferredoxin

$$\text{AcCoA} + CO_2 + fd_{\text{red}} \rightarrow \text{pyruvate} + \text{CoA} + fd_{\text{ox}} \tag{10.4}$$

has been observed in clostridia and in photosynthetic bacteria (Bachofen *et al.*, 1964; Andrews and Morris, 1965; Gottschalk and Chowdhury, 1969). This "reductive carboxylation" of acetate is in essence the reversal of the oxidative decarboxylation of pyruvate through which the citric acid cycle is fed.

Crucial importance for a reversed citric acid cycle must further be attributed to an enzyme for the reductive carboxylation of succinyl coenzyme A:

$$\text{succinyl.CoA} + CO_2 + fd_{\text{red}} \rightarrow \text{ketoglutarate} + \text{CoA} + fd_{\text{ox}}. \tag{10.5}$$

A ketoglutarate synthase has in fact been found in some photosynthetic bacteria, namely, *Rhodospirillum rubrum* and *Chlorobium thiosulfatophilum*. On this basis the existence of a "reductive carboxylic acid cycle" has been postulated (Bachofen *et al.*, 1964; Buchanan and Arnon, 1965, 1970; Evans *et al.*, 1966; Buchanan *et al.*, 1967; Arnon, 1969b). The cycle must also include a step where the condensation of oxaloacetate with AcCoA, in the usual representation the start of the citric acid cycle, is reversed.

In the cycle the wheel of the citric acid cycle is indeed turned back (Fig. 10.1). Of course, the role of the series of reactions cannot be any more the supply of metabolic hydrogen and of energy. The series is now anabolic and endergonic. Metabolic hydrogen must be invested as reduced ferredoxin, and energy as ATP. What is gained is AcCoA, i.e. starting material for body substance, notably for carbohydrates and fats; moreover, aminoacids. The cycle is distinguished by the use of ferredoxin rather than NADH as reductant. Ferredoxin is the more direct product of photosynthesis, and its use should be more economical.

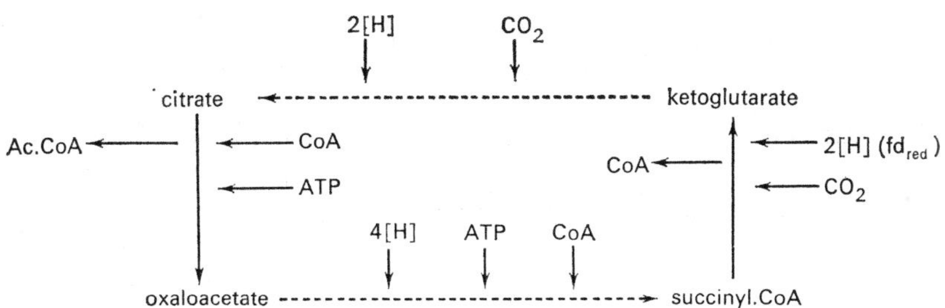

FIG. 10.1. The proposed carboxylic acid cycle (short version).

In addition to this ("short") version of the cycle, a "long" version can also be written down (Arnon, 1969b). Pyruvate synthesis (eqn. (10.4)) may be followed by reaction with ATP and by carboxylation

$$\text{pyruvate} + \text{ATP} = \text{phosphoenolpyruvate (PEP)} + \text{ADP} \qquad (10.6)$$

$$\text{PEP} + CO_2 = \text{oxaloacetate} + \text{P} \qquad (10.7)$$

(PEP has been mentioned in *6b*, and will be mentioned again in *22d*). Conceptually, eqns. (10.6) and (10.7) may be included into the reaction sequence. In the resulting long version 1 oxaloacetate is made from $4CO_2$ and 10 [H] while no AcCo appears as a net product. The necessary supplement to convert the short into the long version is shown in Fig. 10.2.

In some species of green bacteria [*11b*] the cycle may be the only pathway for CO_2 fixation and for biomass production (Evans *et al.*, 1966; Sirevåg and Ormerod, 1970 a, b; Buchanan *et al.*, 1972; Sirevåg, 1974). Others doubt that the sequence that contains the pyruvate and ketoglutarate synthases operates in Nature as a complete cycle on a major scale (Fuller, 1969, 1971; Quayle, 1972; McFadden, 1973). It has been suggested that the reductive synthesis of the ketoacids only serves production by transamination of the corresponding aminoacids, e.g. of glutamate from ketoglutarate (Beuscher and Gottschalk, 1972).

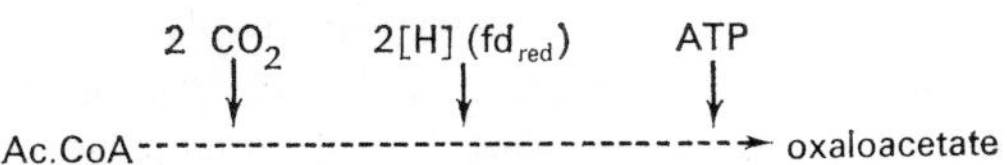

FIG. 10.2. Supplement to the short version, to change it into the long version. The essence is that the oxaloacetate itself is built up, from the product of the cycle, acetyl.CoA.

(e) Reductive Phosphate Pentose Cycle

Long before the reductive carboxylic acid cycle was proposed, the "reductive pentose phosphate cycle" has been found in the plants (see Bassham and Calvin, 1957, 1961; Bassham, 1964; 1971b; Bassham and Jensen, 1967; McFadden, 1973). Here the direct reductant is NADPH rather than ferredoxin. The pyridine nucleotide obtains the electrons for reduction from ferredoxin by way of an enzyme containing flavine [*8f*]. An abbreviated form of the cycle is given as Fig. 10.3. The free energy changes of the reactions of the cycle have been conveniently listed by Bassham (1963) and by Bassham and Krause (1969).

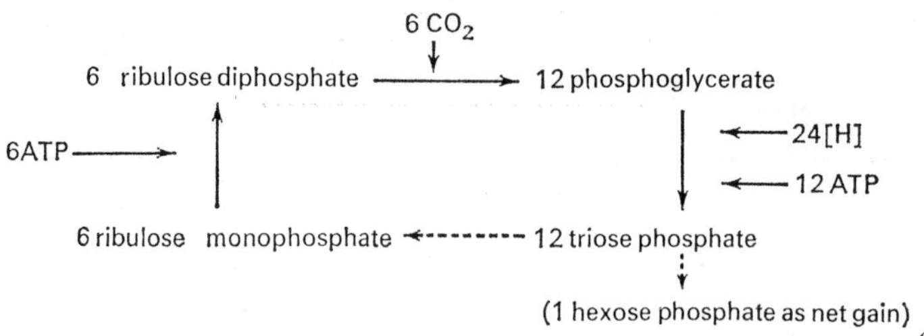

FIG. 10.3. The reductive pentose phosphate (Calvin) cycle, merely from the point of view of bioenergetics.

The (reductive) "Calvin cycle" uses many of the reactions of the (oxidative) pentose phosphate pathway [*7f*], which occurs already in the fermenters, in reverse. It may therefore be assumed to have evolved from the reactions of this pathway. But two new key enzymes appear: phosphoribulokinase (correctly: ribulose 5-phosphate kinase) and ribulose diphosphate (RuDP) carboxylase (carboxydismutase; correctly: ribulose 1,5-diphosphate carboxylase). The latter enzyme accounts for a large part of the protein of chloroplasts. The properties of RuDP carboxylase and also the regulation of assimilation have been discussed by McFadden (1973). Carbohydrate is the main product.

Among bacteria, the Calvin cycle is widely used in photoorganotrophs (Glover *et al.*, 1952; Elsden, 1954; Ormerod, 1956; Stoppani *et al.*, 1965; Fuller, 1969, 1971) and in photolithotrophs and chemolithotrophs (Trudinger, 1956; Aubert *et al.*, 1957; Fuller *et al.*, 1961; Elsden, 1962; Smillie *et al.*, 1962; Vishniac and Trudinger, 1962; Trüper, 1964; Fuller, 1969, 1971; see Klein and Cronquist, 1967; McFadden, 1973). The functioning of the cycle in the chemolithotrophs shows immediately that the cycle contains dark reactions only. As a reductant, the bacteria use NADH. In fermenters, the reductive pentose phosphate cycle has not been found.

The usefulness of the reductive pentose phosphate cycle for the photolithotrophs and chemolithotrophs is clear. They have reducing power, but, when strict, no alternative source of carbon than CO_2. But what is the usefulness for photoorganotrophs, the non-sulphur purple bacteria? As organotrophs, they can reduce CO_2 only at the expense of reduced organic matter, and so net assimilation of carbon does not seem possible. In this respect, the position is similar to that of the CO_2 absorption by fermenters [*10b*].

The usefulness of the Calvin cycle for the photoorganotrophs is more obvious when they apply H_2 as the reductant, i.e. when they live as lithotrophs rather than as organotrophs. Characteristically, the cycle is partly repressed in non-sulphur purple bacteria in presence of an alternative (organic) carbon source, and especially when the bacteria live aerobically by respiration in the dark (Lascelles, 1960; Anderson and Fuller, 1967; Fuller, 1969, 1971; Kelly, 1971; Slater and Morris, 1973).

If it is true that cyclic photophosphorylation, with no external reductant, came first [*8h*], and that CO_2 reduction by the reductive pentose phosphate cycle, with H_2 as external reductant, came later, it follows that the reductive pentose phosphate cycle was acquired by bacteria well after the capacity for photosynthesis. The acquisition involved modification of the preexisting (catabolic) pentose phosphate pathway. Subsequently, the photolithotrophs proper and the plants inherited the cycle.

The facultatively respiring photoorganotrophs, following the anaerobic photoorganotrophs in evolution, maintained the reductive pentose phosphate cycle for life as photolithotrophs. Moreover, as we have said, it is found in the chemolithotrophs, which often cannot grow on any source of carbon but CO_2. But we do not find the cycle in the aerobes that are in no case photosynthetic, e.g. among bacteria in *E. coli*, a facultative respirer, or *Azotobacter vinelandii*, an obligate respirer. Such organisms require in any case organic substrates and have no external source of electrons for net CO_2 reduction. Organic substrates are also needed by desulphuricants [*16a*], although they can use hydrogen as a (non-exclusive) reductant; they have no Calvin cycle. Thus the aerobic and anaerobic non-photosynthetic respirers have lost the cycle.

11

PHOTOLITHOTROPHY

(a) Reducing Power from Inorganic Substances

As we have seen, the further expansion of life required, in addition to ATP, an external supply of electrons for the assimilation of CO_2. But electrons were scarce as long as the only external source was free hydrogen. So the organisms had an incentive to find out additional external inorganic reductants for CO_2. In fact, the photolithotrophs learned to use, through photochemical promotion, inorganic reductants other than H_2. They have been discussed by Fromageot and Senez (1960), Ormerod and Gest (1962), Elsden (1962), Peck (1962), Van Niel (1963), Kondrateva (1963), Drews (1965) and Pfennig (1967). Incidentally, no photosynthetic bacteria are known that are limited to H_2 as a reductant.

In contrast to the photoorganotrophs, the photolithotrophs are two steps removed from the obligate fermenters, e.g. the clostridia: by virtue of the use of light and by virtue of the use of an inorganic reductant. So it is logical to consider them as having arisen later than the photoorganotrophs. First, the bacteria learned to assimilate substrate, also CO_2, photochemically, and subsequently they acquired the ability to use inorganic reductants. The proposed sequence is, then: (İ) fermenters, (İİ) photoorganotrophs, (İİİ) photolithotrophs.

While the transition (İ) → (İİ) must have been fairly sharp, indeed sharp enough to make explanation difficult, the transition (İİ) → (İİİ) probably was rather smooth. The dividing line between photoorganotrophs and photolithotrophs is not absolute among extant organisms. Some photoorganotrophs can to some extent employ inorganic reductants (other than H_2), and some photolithotrophs can use organic reductants (Van Niel, 1944, 1963; Ormerod and Gest, 1962; Anderson and Fuller, 1967; Klemme, 1968; Thiele, 1968). This creates problems for taxonomists (see Pfennig and Trüper, 1971; Kelly, 1971; Van Gemerden, 1972; Hansen and Van Gemerden, 1972). The impression that the transition between non-sulphur and sulphur purple bacteria is gradual has been strengthened by the discovery of *Rhodopseudomonas sulfidophila* [*9a*].

Facultative photoorganotrophy is found rather erratically among the photolithotrophs. Presumably loss of equipment for the intake and/or the utilization of the organic compounds is responsible for inability to use them in energy metabolism. However, as in other cases [*8b*, *15a*], the retention or, in contrast, the loss of this equipment may not be of great evolutionary significance. It is not clear what advantage the bacteria derive from the loss of the capability to use organic substrates, except the very general advantage that comes from throwing overboard redundant machinery.

The standard free energy changes in some reactions of fundamental bioenergetic importance that are performed by photosynthetic bacteria, especially lithotrophs, are shown in Table 11.1. (CH_2O) stands for body substances, with a free energy content one-sixth of that of glucose. Equation (11.1) shows the "thermodynamic distance" between the direct reductant, NAD(P)H, and body substance. As a reminder, eqn. (10.3) has been included, in the reverse direction and multiplied by 2, into the table as eqn. (11.2). Equations (11.3)–(11.5) give the free energy changes with the most common inorganic electron sources. Equation (11.6), reduction of CO_2 by the organic electron donor succinate with oxidation of the latter to fumarate, is given as an example illustrating the position in photoorganotrophs. It must be kept in mind that the table contains standard values. For instance, the pressures of the gases are 1 atmosphere, and the pH values 7.

TABLE 11.1 FREE ENERGY CHANGES IN CO_2 REDUCTION

Reaction	$\Delta G_0'$ (kcal)
$2NAD(P)H + 2H^+ + CO_2 = (CH_2O) + H_2O + 2NAD(P)$	7.7 (11.1)
$2NAD(P)H + 2H^+ = 2NAD(P) + 2H_2$	8.6 (11.2)
$2H_2 + CO_2 = (CH_2O) + H_2O$	−1 (11.3)
$2H_2S + CO_2 = (CH_2O) + H_2O + 2S$	12 (11.4)
$0.5H_2S + CO_2 + H_2O = (CH_2O) + H^+ + 0.5SO_4^{2-}$	28 (11.5)
$2C_4H_4O_4^{2-} + CO_2 = (CH_2O) + H_2O + 2C_4H_2O_4^{2-}$	40 (11.6)

In the assimilation of CO_2 by means of hydrogen, the free energy change (eqn. (11.3)) is not unfavourable, and with high concentrations of reactants and low concentrations of products ΔG assumes negative values. The dark growth of fermenters (not of photosynthesizers!) in presence of H_2 with CO_2 as exclusive carbon source, based on this overall reaction, has been mentioned [*10c*].

$\Delta G_0'$ for the oxidation of sulphide is so strongly positive that practically at no concentrations does any reduction of NAD or CO_2 occur in the dark. The electron flow, as transhydrogenation, must be forced by light. May it be pointed out that in eqn. (11.5) water is decomposed, though no oxygen is set free as such. Liberation of oxygen is the privilege of the plants.

(b) Purple and Green Sulphur Bacteria

The contemporary photolithotrophs include the two sharply divided groups (Fig. 11.1) of the purple sulphur bacteria (thiorhodaceae) and of the green sulphur bacteria (chlorobacteriaceae). Various forms of reduced sulphur are used as reductants by different species, including sulphide, thiosulphate, sulphite and tetrathionate. Oxidation leads generally either to elementary sulphur or further to sulphate. *Chlorobium limicola* disproportionates elementary sulphur photochemically to sulphide and sulphate, and the former serves as a reductant for CO_2 in the light (Paschinger *et al.*, 1974). The mechanisms used for CO_2 assimilation have been mentioned [*10 d, e*].

It is likely that the purple sulphur bacteria (Fig. 11.1) have evolved from purple non-sulphur bacteria. As has been pointed out [*11a*], the taxonomic distinction between the purple sulphur and the purple non-sulphur bacteria, based on the relative ease of the utilization of organic and inorganic reductants, is not clear-cut. From a practical point of view it is important that the sulphur bacteria are generally anaerobes, while many of non-

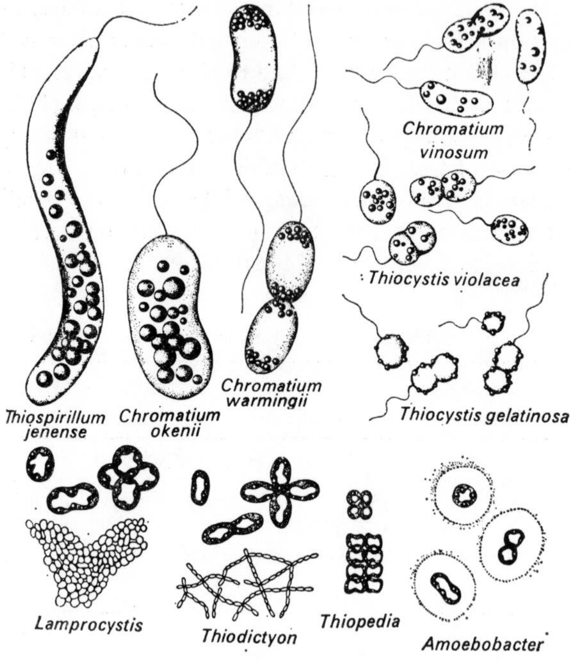

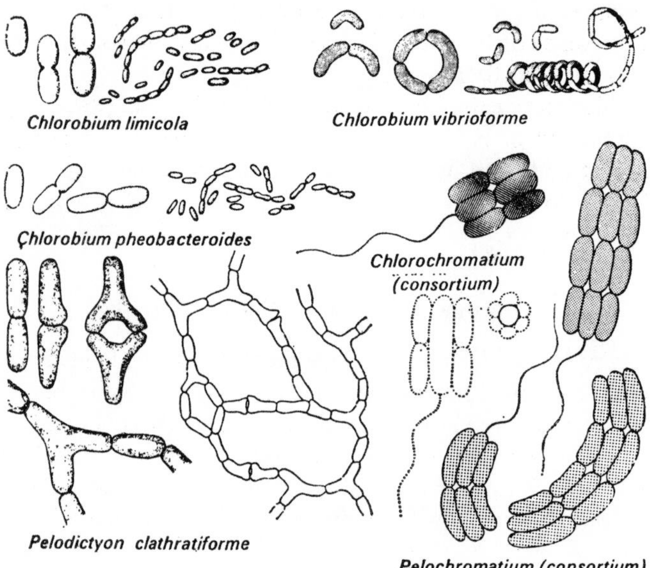

FIG. 11.1. Some purple (top) and green (bottom) sulphur bacteria (Schlegel, 1972). (For absolute size, reference may be made to *T. jenense* (50 × 3.5 μm).

sulphur bacteria are facultative aerobes. However, no reason is known why facultatively aerobic species should not turn up among the purple sulphur bacteria. In fact, microaerophilic behaviour has occasionally been reported [*15f*].

While the green sulphur bacteria (Fig. 11.1) are observed to be more strictly anaerobic and also more strictly autotrophic than the purple sulphur bacteria, they have, as far as is known, a similar energy metabolism. Nevertheless they greatly differ morphologically (Van Niel, 1963; Cohen-Bazire *et al.*, 1964; Cruden and Stanier, 1970; Cruden *et al.*, 1970; Stanier, 1970). While the photosynthetic machinery in the purple bacteria is connected with the cell membrane [*9a*], that of the green sulphur bacteria is contained in separate vesicles not connected with the cell membrane (see Echlin, 1970b). The vesicles are surrounded by single layer membranes, and not, like thylakoids, by double membranes.

The position of the green bacteria on the evolutionary tree remains, for the time being, uncertain. According to Cruden *et al.* (1970) they

> seem to be an isolated and highly specialized group of photosynthetic organisms, remote from the main line of cellular evolution among prokaryotes, which led to the emergence of organisms capable of using water as a photosynthetic electron donor, exemplified by the contemporary blue–green algae,

and according to Stanier (1970)

> these organisms . . . can be plausibly construed as lying at the end of a terminal branch of prokaryotic cellular evolution.

Important evidence about the evolutionary position may be obtained from chemotaxonomy. The distribution of the chlorophylls (Cruden *et al.*, 1970; Stanier, 1970) has been discussed. On the basis of the composition of the ferredoxins, Buchanan *et al.* (1969a) and Hall *et al.* (1973b) suggest, diverging from Stanier, an order fermenters → green bacteria → purple bacteria. This meets with the difficulty that photoorganotrophs are found among the purple, but not among the green bacteria. It is not clear why the capacity for organotrophy should have been lost by the latter, but not by the former. Incidentally, a comparison of the ferredoxins from sulphur and from non-sulphur purple bacteria would be illuminating.

As has been observed by Molisch (1907) and investigated by Van Niel in momentous work (see Van Niel, 1941; Stanier and Van Niel, 1941), photosynthetic bacteria never produce free oxygen, or, what comes to the same, they never extract electrons from water—at least not as a net result. Consequently, after the advent of the photolithotrophic bacteria the biosphere still remained anoxygenic. Nevertheless, an important change did occur: the mild oxidant sulphate became plentiful for the first time.

12

PLANT PHOTOSYNTHESIS

(a) Water as a Source of Electrons

The reserves of useful inorganic external sources of electrons for the reduction of CO_2, among them sulphides, in accessible places of the biosphere were not limitless. The light flux could have supported life on a far larger scale than corresponded to the reserves. This is evident from the subsequent successful evolution of plant photosynthesis, through which much more energy is being tapped. In plants, water is photolysed and oxygen liberated.

Mutations enabling cells to use, in the light, the practically all-present reductant, water, have had tremendous survival value. Operationally we shall call plants all organisms that use water as a reductant, and their colourless, non-photosynthetic, relatives which may be presumed to have arisen from them. Thus the statement that only plants (more precisely: the green plants) are capable of the photolysis of water is nothing but a definition of the plant. According to this definition, prokaryotic and eukaryotic plants exist, as will be seen.

Needing nothing but water, CO_2, minerals and light, the plants are the non-plus-ultra in bioenergetic independence. The plants' way of life is perhaps sufficiently different from photolithotrophy to merit a separate name, photophytotrophy, or, shorter, phytotrophy.

In contrast to earlier ideas (Stanier and Van Niel, 1941), the conclusion has been reached (Van Niel, 1949) and rather generally accepted that there has been only one main line of photosynthetic evolution. This line has led from non-photosynthetic bacteria to photosynthetic bacteria and finally to plants. Of course, this concept does not exclude that extant kinds of photosynthetic bacteria, notably green sulphur bacteria [*11b*], sit at the ends of evolutionary side-branches. Indeed few, if any, of the existing species can be viewed as direct ancestors of the plants.

In spite of formal analogy, it is far more difficult thermodynamically to use H_2O than H_2S as a reductant

$$2H_2S + CO_2 = (CH_2O) + H_2O + 2S; \ \Delta G_0' = 12 \text{ kcal}, \qquad (12.1)$$

$$2H_2O + CO_2 = (CH_2O) + H_2O + O_2; \ \Delta G_0' = 112{\cdot}5 \text{ kcal}. \qquad (12.2)$$

Kinetic and mechanistic studies on oxygen evolution in plants have been undertaken by many (see Kok and Cheniae, 1966; Joliot, 1966; Cheniae, 1970; Witt, 1971; Joliot and Joliot, 1973; Bearden and Malkin, 1974; Fowler and Kok, 1974), but this is still one of the least known areas in photosynthesis research. The path of oxygen from water to O_2 has been considered in more general terms, not restricted to biosystems, by Heidt (1966).

Figure 8.3 has shown that chlorophyll a, in contrast to bacteriochlorophyll a, does not absorb infrared light. Sensitization of chlorophyll a by accessory pigments to long-wave solar radiation does not occur either. Indeed green plants apparently do not utilize infrared rays for photosynthesis. In this way, about one-half of the solar energy flux to ground level is missed by the plants. Nothing is known at what stage in the transition from bacterial to plant photosynthesis this loss occurred, and whether the reason was that long-wave light (smaller quanta) was not good enough for the demands of phytotrophy.

(b) The Z Scheme

For the promotion of electrons from water to NADP, the plants have developed a mechanism to use two quanta in succession for non-cyclic electron flow. Two independent photochemical acts take place, one after the other, and for each of them special photosynthetic apparatus exists at a different site. Of course, the two sites must be suitably coupled in series.

The graphs for the serial process in its possible variants are known as Z schemes. It is thought that the electron is first lifted by light energy in "photosystem 2" and taken up by a specific acceptor. Subsequently the electron undergoes a series of spontaneous (dark) reactions. The electron is thereby handed over stepwise within a chain of redox compounds with less and less negative potentials, i.e. to less and less strong reductants. In the end, the electron fills a hole in "photosystem 1"; this hole must have been made before by removal of an electron from system 1. Only thereafter the second quantum of light energy is applied to the electron, now in system 1, and it is taken up by an acceptor that thereby becomes a far stronger reductant than the acceptor of system 2. (Presumably a less naïve and more realistic picture is that at any given moment a number of available holes exists, distributed statistically, in either of the two photosystems.) The tangible end product is reduced ferredoxin (Tagawa and Arnon, 1962; Whatley *et al.*, 1963; Arnon, 1969b; Evans and Whatley, 1970). This donates electrons by means of a flavin containing enzyme to NADP [*8f*].

The first Z scheme was put forward by Hill and Bendall (1960). The two-step principle has also been thought of or taken up by other authors, and many efforts have been directed towards the detailed elaboration of the scheme (Witt *et al.*, 1961; Duysens *et al.*, 1961; Duysens and Amesz, 1962, 1967; Kok and Jagendorf, 1963; Smith and French, 1963; Duysens, 1964; Witt *et al.*, 1965; Vernon and Avron, 1965; Avron, 1967; Kok, 1967; San Pietro, 1967; Bendall and Hill, 1968; Hind and Olson, 1968; Boardman, 1968, 1970). The members of the electron-flow chain have to be identified and correctly placed. Among the most powerful methods, the investigation of the action spectra (see Fork and Amesz, 1970), of reaction kinetics by means of pulse spectrometry (Witt, 1971, 1972) and of defect mutations (Levine, 1968, 1969) may be mentioned. A simplified recent form of the Z scheme (modified after Boardman, 1970) is shown in Fig. 12.1. As far as is known, the Z scheme applies to all plants (Evans and Whatley, 1970), even though various groups of plants differ greatly in their accessory photosensitizers (see Gibbs, 1967). One of the problems is the order cytochrome → plastocyanine → photosystem 1; the order shown is supported by results of Knaff and Arnon (1970) and Siedow *et al.* (1973b).

The starting-point for the two-quanta hypothesis was the observation of enhanced photosynthetic yields in plants when suitable light quanta of different wave lengths acted together (Emerson *et al.*, 1957; Emerson and Rabinowitch, 1960). The action spectrum for pure lights is given in Fig. 12.2, but in the long-wave range two different quanta synergistically

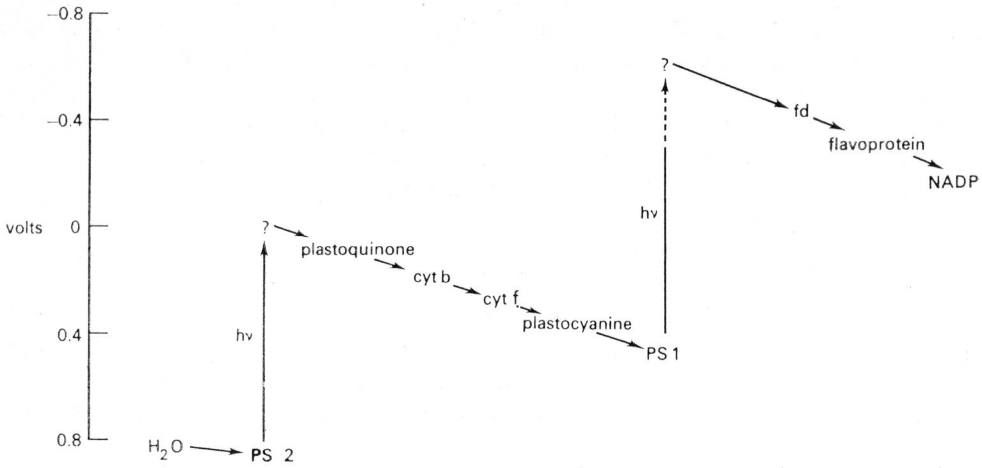

FIG. 12.1. Simplified Z scheme for non-cyclic electron flow. The ordinates indicate the standard potentials of the redox compounds. The numerical values are not to be taken as precise.

give higher yields than corresponds to the sum of the two individual yields (Fig. 12.3). For instance, enhancements of 30% have been observed with a mixture of "red" and "far-red" quanta. Enhancement follows from the difference between the action spectra of the two photosystems. According to Myers (1971):

> A classical description can be made in terms of the following imaginary experiment: a light of wavelength λ_a and another beam of properly chosen wavelength λ_b when presented together give a rate of photosynthesis greater than the sum of the rates when presented separately. More instructively, one can describe enhancement as an increase in quantum yield measured at a wavelength λ_a when a second (unmeasured) beam of properly chosen wavelength λ_b is added.

In the photosystem 1 of plants, the photosensitive substance of the reaction centre [see *8a*] is known as P 700 (Hoch and Kok, 1961). In photosystem 2, the reaction centre absorbs at shorter wavelengths, around 685 nm. Probably both active centres consist of chlorophyll

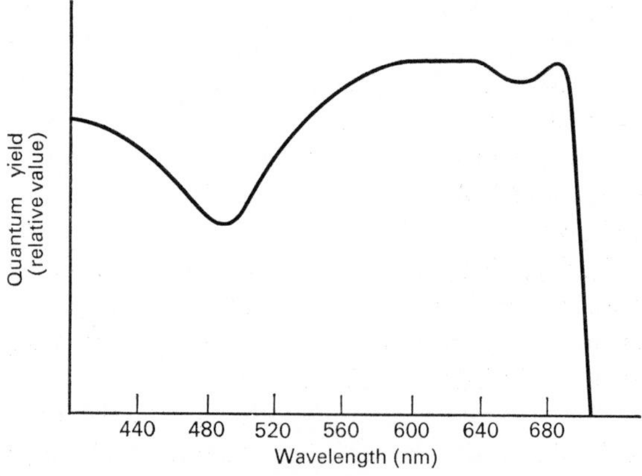

FIG. 12.2. The action spectrum of Chlorella for spectrally pure light (Emerson and Lewis, 1943). The ordinates indicate the amount of oxygen evolved per quantum.

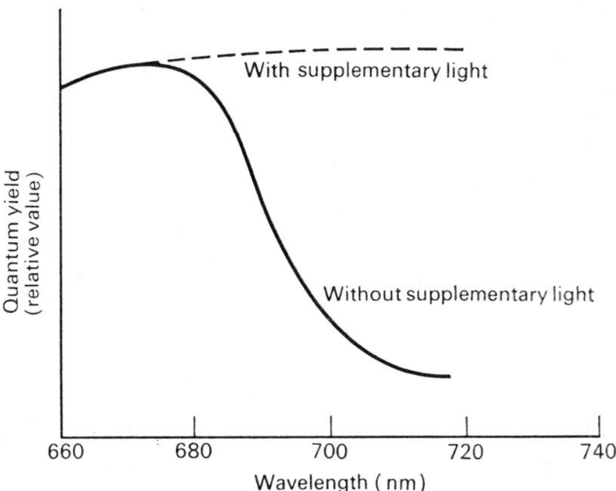

FIG. 12.3. The enhancement effect in Chlorella (Rabinowitch and Govindjee, 1969). Far-red light of the given wavelength supplemented with (near) red light.

a in modified forms. Duysens has termed light active mainly in photosystem 1 or 2 "light 1" or "light 2", respectively. Photosystems 1 and 2 can be partially separated preparatively (Boardman, 1970, 1972; Park and Sane, 1971). Either shearing forces or detergents are used.

No enhancement has been found in photosynthetic bacteria (Blinks and Van Niel, 1963; see Duysens and Amesz, 1967).

(c) The Origin of the Serial Process

Gaffron (1965) expressed the view that a two-quanta process for the utilization of water was developed by bacteria a very long time ago. Only the last step, which leads to the liberation of oxygen, was not taken before the plants arrived. As long as an "enzyme" for this reaction had not been developed, the oxidant produced in the primary photoreaction had to be removed by reaction with an external reductant other than water. The possibility of a transient photolysis of water by bacteria, with immediate removal of the products in chemical reactions, had been considered by Van Niel in his pioneering work on the primary process in photosynthesis (see Van Niel, 1949b).

Speculations on the evolutionary path from photolithotrophy to photophytotrophy were undertaken by Olson (1970). One crucial point is that "the effective utilization of two photochemical reactions in series for the production of a strong reductant (for example, ferredoxin) from electron donors of relatively high potential was achieved by . . . algal (plant. E.B.) prototypes long before they were able to extract electrons from water". Less recalcitrant electron donors than water for protoplants, as suggested by Olson, include hydrazine, hydroxylamine and nitrite. But the presence of the former two compounds in the biosphere at any time is hard to believe. Nor is it easy to see how nitrite could be formed in anaerobic conditions, or, if formed, maintained.

It is possible that the two photosystems of plants arose from separate bacterial systems for non-cyclic and for cyclic electron flow, which, however, never work in series [*8f*].

(d) Non-cyclic and Cyclic Photophosphorylation

An enormous amount of work on photophosphorylation in plants has been carried out (see Simonis and Urbach, 1973; Trebst, 1974). On the relative roles of non-cyclic and cyclic photophosphorylation, it has been stated in Arnon's school (Nozaki *et al.*, 1963) that in contrast to bacteria, where the cyclic process predominates,

> in conventional plant photosynthesis, non-cyclic photophosphorylation would appear to be the dominant photochemical process since, apart from its contribution to ATP, it is the exclusive mechanism for bringing about a hydrogen transfer (via TPN) from water to CO_2. The role of cyclic photophosphorylation in plants would thus appear to be that of supplementing the ATP needs for carbon assimilation which are not fully met by non-cyclic photophosphorylation.

Gest (1966, 1972), a protagonist of the role of the cyclic process in bacteria, admits non-cyclic electron flow and photophosphorylation for the advanced and unique process, photophytotrophy. The first reasonably stable electron acceptor in non-cyclic flow in plants is ferredoxin (Arnon *et al.*, 1964; Arnon, 1965, 1966, 1967a, 1969 a, b), but the search for the primary electron acceptor, of even more negative potential (Kok *et al.*, 1965; Black, 1966) is still going on (Ke *et al.*, 1973; Bearden and Malkin, 1974).

In non-cyclic electron flow, ATP synthesis occurs, coupled to dark reactions, during the electron transfer between photosystems 1 and 2, but possibly also in connection with system 2. It is mostly thought now that several sites of phosphorylation are involved (Horton and Hall, 1968; Izawa and Good, 1968; Böhme and Trebst, 1969; Hall *et al.*, 1971b; Gould and Izawa, 1973, 1974; West and Wiskich, 1973; Reeves and Hall, 1973). 1–2 ATP may be synthesized for each pair of electrons transported to ferredoxin and therefore to NADP (see Hauska *et al.*, 1974; Trebst and Hauska, 1974). Non-integral values of ATP production per electron pair transported are possible on the basis of the chemiosmotic hypothesis [*8e*, *14b*] where coupling is not stoichiometric.

More ATP than NAD(P)H is needed, in plants as in bacteria, alone for carbon assimilation by the Calvin cycle, namely, 3 ATP and 2 NAD(P)H per 1 CO_2. If only 1 ATP were made per NADPH (pair of electrons), the excess needs for ATP would have to be covered otherwise, probably in the main by cyclic photophosphorylation. Ferredoxin may again be an electron acceptor in the cyclic electron flow in plants (Tagawa *et al.*, 1963; Arnon, 1965, 1969b; Curtis *et al.*, 1973). The electrons may in the cyclic process flow from the ferredoxin to one of the members of the "bridge" between PS 1 and PS 2. But the relative contributions of the non-cyclic and the cyclic process to the ATP supply for carbon assimilation are still controversial (see also Tanner *et al.*, 1969; Raven, 1970a; Klob *et al.*, 1973; Trebst and Hauska, 1974).

An argument in favour of the role of the cyclic process is that its selective inhibition in chloroplasts (with antimycin A) severely depresses the yield in photosynthetic CO_2 assimilation, and also disturbs the pattern of the products (Schürmann *et al.*, 1971). Cyclic photophosphorylation is also important to other functions than assimilation (Hoch and Randles, 1971; see Simonis and Urbach, 1973). After inhibition of the non-cyclic process with DCMU (dichlorophenyldimethyl urea), the cyclic process still makes possible manifold cell processes requiring ATP but no reducing power (see Ramirez *et al.*, 1968; Schürmann *et al.*, 1971), like protein synthesis from aminoacids. Assimilation of CO_2 is not possible any more, of course.

In "pseudocyclic" electron flow, formulated by Arnon (see Arnon *et al.*, 1961b; Arnon, 1969b) on the basis of Mehler's (1951) observations, the machinery of non-cyclic electron flow is used, but in final effect nothing but ATP is made. The essence is that photochemically produced reductant is reoxidized at the expense of free oxygen. There is, then, no net electron flow. The experimental distinction between cyclic and pseudocyclic photophosphorylation is made with specific inhibitors. It has been suggested that in presence of air the pseudocyclic rather than the cyclic process provides additional ATP when needed (Heber, 1973).

Photosystem 1 by itself normally allows cyclic photophosphorylation only. It is possible, however, to feed external electrons from a sufficiently strong reductant artificially into photosystem 1 so that net reducing power is obtained from this system. This process in plants has been called "non-cyclic photophosphorylation of the bacterial type" (Losada *et al.*, 1961; Arnon *et al.*, 1965). For example, ascorbic acid in presence of the catalyst (mediator) dichlorophenol indophenol (DPIP) will do (see, however, Böhme and Trebst, 1969).

While most authors think that the electrons flow through photo-system 1 in the non-cyclic as well as the cyclic process, as indicated by the Z scheme as given, Arnon holds that photosystem 1 serves the cyclic process only (Arnon *et al.*, 1965; Arnon, 1967 b, c; McSwain and Arnon, 1968). According to Arnon, photosystem 2 alone looks after the non-cyclic process. Thus the machineries for the cyclic and for the non-cyclic process work in parallel, and are connected at no point. Enhancement is interpreted through synergism of non-cyclic and cyclic electron flow (photosystem 2 and 1), powered to different extents by light 2 and light 1. Non-cyclic flow through system 1 is, according to Arnon, non-physiological.

However, in the present representation (Knaff and Arnon, 1971; Arnon, 1971; Arnon *et al.*, 1971; Malkin, 1971; but see also Malkin and Bearden, 1973) one main feature of the Z scheme is retained: two light quanta in succession are utilized for the non-cyclic process ("photosystems 2 b and 2 a"). Thus altogether three different light reactions are invoked (Fig. 12.4).

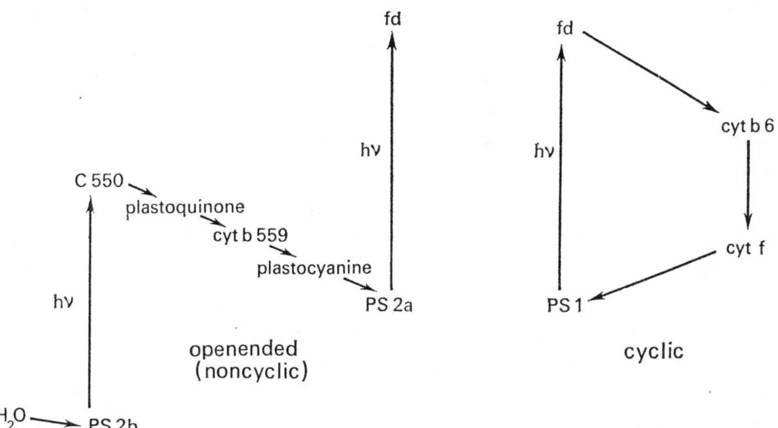

FIG. 12.4. Three light reactions in plants according to Arnon. The non-cyclic and the cyclic pathways are independent. In the figure, no notice is taken of the fact that ferredoxin may be preceded by a primary electron acceptor in the light reaction (see Fig. 12.1). For redox potentials, see also Fig. 12.1. Phosphorylation not shown.

(e) Absolute Quantum Yields

An explanation for the experimental value of the quantum yield (or its reciprocal value, the quantum requirement) in plant photosynthesis is provided by the concept of two light reactions in succession in non-cyclic electron flow. Though the great master Otto Warburg to the end of his life still insisted on lower values (on more efficient use of light), the quantum requirement per atom of carbon is now generally taken to be 8–10 (Emerson and Lewis, 1943; see Gaffron, 1960; Wassink, 1963; Robinson, 1964; Govindjee *et al.*, 1968; Rabinowitch and Govindjee, 1969; Arnon, 1971).

In non-cyclic electron flow, 4 quanta (2 quanta in succession for each electron) are needed to promote fully 2 electrons and so to make 1 NADPH, and 2 NADPH are needed for 1 CO_2. Along with the 2 NADPH, 2 ATP are, in the majority view, obtained in the non-cyclic process. In the reductive pentose phosphate cycle, 3 ATP are required for the reductive assimilation of 1 CO_2 as hexose, namely 2 ATP for the reduction of 2 molecules of phosphoglyceric acid, and 1 ATP for the production of 1 molecule of ribulose diphosphate. Thus, in addition to the 8 quanta used for the non-cyclic process, 1–2 quanta (depending on the yield of ATP per electron) have to be provided to make 1 more ATP by cyclic photophosphorylation.

It used to be emphasized that a similar value of about 9 has consistently been found for the quantum requirement by photosynthetic bacteria, independent of the redox potential of the reductant (Larsen *et al.*, 1952; Larsen, 1954, 1960). This implies that the overall "free energy efficiencies" are lower with the stronger reductants (Baas-Becking and Parks, 1927; see Fromageot and Senez, 1960). The stronger the reductant, the less energy is needed for the promotion of the electron to its position in NADH, and the less well used is the (equal) energy of the light supplied. With hydrogen as a reductant, energy of light should not be needed at all from the standpoint of mere energetics (Table 11.1; see also *10c*). Here the free energy efficiency is obviously worst.

On the basis of the newer knowledge of photosynthetic mechanism, it must rather be asked why in bacterial photosynthesis as many as 9 quanta should be needed. Let us assume first that non-cyclic electron flow exists in bacteria [see *8f*]. Each electron needs only 1 quantum for promotion; this corresponds to 4 quanta per CO_2. If 1 ATP were formed for every 2 electrons promoted, 2 ATP would appear concomitantly. Hence, 1 additional ATP per CO_2 would have to be made through the cyclic process. Altogether then 5–6 quanta (again depending on the yield of ATP per electron) would be needed. It is uncertain, however, whether non-cyclic electron flow in bacteria is coupled to phosphorylation at all [*8f*]. If not, a quantum requirement of 7–10 would be expected (4 quanta for NADH, and 3–6 quanta for ATP).

The situation is even less clear with the alternative assumption that reducing power is made by reversed electron flow. In this case, the expected quantum requirement depends sensitively on the number of quanta needed for 1 ATP in the cyclic process. Moreover, the number of molecules of ATP needed for the production of 1 NADH is unknown. It therefore appears unprofitable to attempt a quantitative estimate now. But whatever the composition of the ATP budget, it is obvious that requirements are, within the frame supplied by thermodynamics, governed by stoichiometry.

(f) Blue–Green and Other Algae

The simplest photophytotrophs are the algae, and among them the most primitive are the blue–green algae (see Geitler, 1960; Fritsch, 1965; Carr and Whitton, 1973; Wolk, 1973; Fogg *et al.*, 1973), which are the only prokaryotes among the algae, and also among all plants (Fig. 12.5). The chlorophyll is distributed along intracellular membranes (Fig. 12.6) derived from the cell membrane (Niklowitz and Drews, 1957; Bergeron, 1963; Cohen-

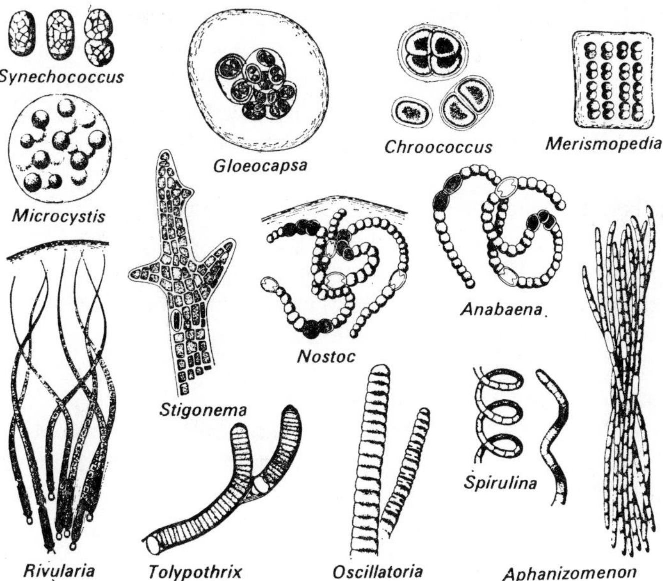

FIG. 12.5. Some blue–green algae (Schlegel, 1972). It is thought that species consisting of single cells are more primitive, species consisting of filaments more advanced. The magnifications differ.

Bazire and Sistrom, 1966; Echlin, 1964, 1966b, 1970b; Lang, 1968), just as it is in purple bacteria [*9a*]. The basic unit is again, as in the bacteria, the thylakoid (Krogmann, 1973; Lang and Whitton, 1973; Fogg *et al.*, 1973; see, however, Rippka *et al.*, 1974).

The blue–greens really have quite diverse colours. Most active light is captured by phycobiliproteins (Stanier, 1974). Together with the photosynthetic bacteria, which are also differently coloured, they can utilize light of all wavelengths, a fact of great ecological significance (Stanier and Cohen-Bazire, 1957). The unequal distribution of light and air is a main factor in the stratification of waters. A curious feature of the blue–greens is their capacity to glide over surfaces rather than to move by swimming [*14g*]. They do not tolerate even mildly acid media (see Brock, 1973 a, b).

The blue–greens differ from photosynthetic bacteria by their power to use water as a photoreductant and therefore to release oxygen. For this reason we have included them

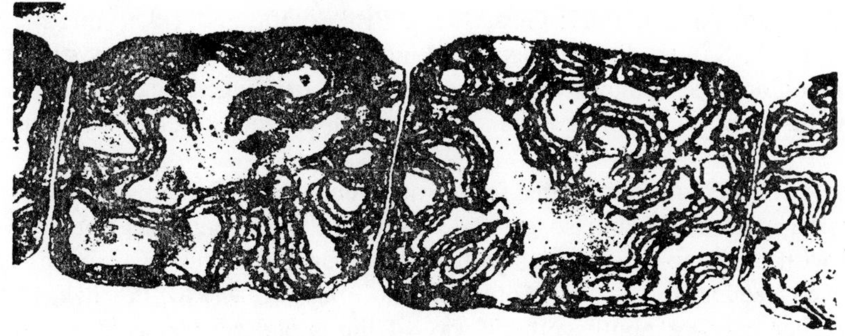

FIG. 12.6. Part of a filament of the blue–green alga *Nostoc muscorum* (Menke, 1961). Diameter 1–2 μm.

among the plants [*12a*]. But otherwise the blue–greens are structurally so close to bacteria (Cohn, 1853, 1874) that a dividing line is hard to draw (see Stanier and Van Niel, 1941, 1962; Van Niel, 1955; Frank *et al.*, 1962; Murray, 1962; Stanier, 1964; Echlin and Morris, 1965; Soriano and Lewin, 1965; Echlin, 1966 a, b; Mägdefrau, 1967; Fuhs, 1969; Leedale, 1970; Brock, 1973a; however, Pringsheim, 1949, 1963b; Punnett and Derrenbacker, 1966).

The difference between blue–greens and bacteria tends to vanish when an alga loses its colour and thereby also the capacity for photophytotrophy in this way. The whole group of the gliding bacteria [*14g*] may have evolved from blue–green algae. Stanier *et al.* (1971) feel that the blue–green algae (German: Blaualgen) might simply be called blue bacteria. Yet we re-emphasize here the striking fact that in spite of the extremely wide gap in cell organization the prokaryotic blue–green algae share the phytotrophic way of life with the green eukaryotes. An explanation is offered by the symbiotic hypothesis [*19 c*, *d*].

The blue–green algae have been enormously successful. They are very widespread in Nature now (see, for example, Shapiro, 1973) and morphologically they have not changed greatly during 3 Gy [*23a*].

(g) Return to Older Processes

Like photosynthetic bacteria, algae can in various ways fall back on more ancient bioenergetic processes. This applies to blue–green and other algae, and in many cases gaps probably point to missing experimentation rather than to differences in capability.

Some green (eukaryotic) algae behave like photoorganotrophic bacteria in depending largely on the photoassimilation of acetic acid (Pringsheim and Wiessner, 1960, 1961; Wiessner, 1963, 1965; Wiessner and Gaffron, 1964; Goulding and Merrett, 1967; see Simonis and Urbach, 1973; Smith 1973a; Fogg *et al.*, 1973). Carbohydrates may similarly be used (for references see Olson, 1970; Simonis and Urbach, 1973; Fogg *et al.*, 1973). A trend back towards the bacterial way of life is also expressed in mutants of the green alga *Scenedesmus* that "forgot" the extraction of electrons from water (Bishop, 1962).

Some kinds of green and of other algae can be adapted to photoreduce CO_2, like photosynthetic bacteria, with hydrogen gas, i.e. an otherwise latent hydrogenase appears (Gaffron, 1939, 1944, 1960; Frenkel and Rieger, 1951; Kessler, 1960; see Spruit, 1962; Bishop, 1966; Stuart and Gaffron, 1971, 1972; Simonis and Urbach, 1973). Such plants photophosphorylate and apply the Calvin cycle (Gingras *et al.*, 1963; Russell and Gibbs, 1968), but they show no enhancement (Gingras, 1966); apparently the hydrogen enters photosystem 1 directly. Without CO_2, some algae may evolve H_2 in the light (Gaffron and Rubin, 1942; Gaffron, 1944; see Arnon, 1965; Gest, 1972). Prokaryotic and eukaryotic plants can to some extent use H_2S in the light as electron donor though apparently they still need photosystem 2 (Knobloch, 1966; Stewart and Pearson, 1970). Prolonged treatment of chloroplasts with red light seems to suppress enhancement in plants, namely, *Elodea* (Punnett, 1966, 1971).

Endogenous fermentation is needed for the maintenance of life in absence of air and light. Exogenous fermentation is not so general among the algae, and has not been much investigated with the blue–greens. Complete reliance on fermentation (growth) has rarely been observed (see Danforth, 1962; Gibbs, 1962; Gibbs *et al.*, 1970; Peschek, 1975).

On the other hand, a photosynthetic way of life is compatible with complete lack of organic nutrients. Inability to use organic compounds is randomly distributed among the algae so that repeated independent loss of the capacity is assumed (see Danforth, 1962; Pringsheim, 1963 a, b, 1964; for blue-greens, Stanier *et al.*, 1971; Smith, 1973a; Stanier, 1973). The strictly autotrophic blue-green algae resemble the strict autotrophs among the photosynthetic bacteria [*11a*], and again strictness probably has no great significance from the point of view of evolution.

Some words will be said later [*15d*] about the utilization of hydrogen by adapted algae through oxidation. The process as such—clearly of no physiological importance—is possible in presence of oxygen only, and could not occur in an anoxygenic biosphere. But—like photoreduction, mentioned above—the process does require action by the ancient enzyme hydrogenase, now normally (if existent) latent in plants.

13

RESPIRATION IN GENERAL

(a) Use of Oxygen

Not too much is known about mechanisms of attack by oxygen that now apply in the case of anaerobes (see Sagan, 1961; Gerschman, 1964; Haugaard, 1968; Gottlieb, 1971; O'Brien and Morris, 1971; Allen *et al.*, 1973). The mechanisms may be manifold. Whatever their details, the high redox potential of O_2 will, of course, be considered as the decisive factor.

Presumably, aerobic organisms have gradually acquired oxygen tolerance. Later, oxygen-tolerant organisms developed enzymes (oxygenases and hydroxylases) through which O_2 was exploited for biosyntheses (see Conn, 1960; Hayashi, 1962, 1974; Mason, 1965; Hayashi and Nozaki, 1969; Hughes and Wimpenny, 1969). In many instances and in the most diverse organisms, preexisting biosynthetic (anabolic) pathways were changed or replaced by others, probably more economical, on transition from anaerobiosis to aerobiosis. For lipids examples have been given by Bloch and Goldfine (1963) and by Lascelles (1964b), for nitrogen compounds by Lascelles (1964b). For example, porphyrin synthesis in eukaryotes requires oxygen (Lascelles, 1964b; Granick, 1967). Organisms dependent on such processes can grow only in aerobic conditions whatever the energy supply. They include many of the aerobic prokaryotes (see Cohen, 1970; Decker *et al.*, 1970), but not, of course, the true facultative aerobes (= facultative anaerobes!), e.g. *E. coli* (Erwin and Bloch, 1964).

Some organisms developed aerobic methods for the facilitation of fermentation and for the improvement of ATP yields (see Dolin, 1961). For instance, *Streptococcus faecalis* can use glycerol as a substrate for fermentation only aerobically (Gunsalus and Umbreit, 1945). It must be dehydrogenated with NAD before entering the glycolytic pathway, and the resulting NADH is reoxidized by O_2 via a flavoprotein. A further case is the aerobic glycolysis of glucose by some lactic acid bacteria, where more ATP is obtained on removal of NADH with O_2 than in ordinary lactic acid fermentation even if no energy is obtained directly from the dehydrogenation of the NADH (Dolin, 1955, 1961). The reason is that aerobically pyruvate need not be used as obligate electron acceptor to regenerate NAD. It is therefore free to be oxidized to AcCoA, yielding one more energy-rich bond per triose.

In an analogous manner, "nitrate fermentation" with nitrate as oxidant is quite common [*16c*]. Elementary sulphur is used as an "incidental oxidant" in some fermentations of marine bacteria (Tuttle and Jannasch, 1973) and ferric iron acts as alternative electron acceptor with nitrate reductase of soil bacteria (Ottow and Glathe, 1971).

(b) The Concept of Respiration

At some stage prokaryotes began to use the oxidation of organic substrates, in the end to CO_2 and H_2O, for the generation of useful energy in the form of ATP. The possible advantage of total oxidation ("combustion") over fermentation of organic compounds is obvious from a comparison of eqn. (7.3) with eqn. (12.2), here written down in reverse for 1 and subsequently for 6 atoms of carbon:

$$(CH_2O) + O_2 = CO_2 + H_2O; \Delta G_0' = -112.5 \text{ kcal}, \quad (13.1)$$

$$C_6H_{12}O_6 + 6O_2 = 6CO_2 + 6H_2O; \Delta G_0' = -675 \text{ kcal}. \quad (13.2)$$

As has been emphasized [*7a*], respiration occurs only in connection with solid membranes. In the prokaryotes, the place of oxidative ATP generation is the cell membrane and its invaginations, in the eukaryotes the inner membrane of the mitochondrion [*18c*]. These mitochondrial membranes show "respiratory control", defined as stimulation by the reactants, ADP + P (see Chance and Williams, 1956; Chance and Baltscheffsky, 1958; Ernster and Luft, 1963; Ernster and Lee, 1964). In some cases, respiratory control has also been observed with prokaryotic systems (e.g. Jones *et al.*, 1971; *14c*). The very similar phenomenon of "photosynthetic control" has been mentioned before [*8f*].

From now on we shall use, somewhat loosely, the short expressive word "respiration" (or, more precisely: "aerobic respiration") for the production of metabolic energy through total oxidation of substrates with oxygen. This is usual in the literature on fundamental bioenergetics and cell metabolism. Essentially, we shall use the term "respiration" as a synonym of "oxidative phosphorylation" [*13d*]. For the history of research in cell respiration the excellent books by Keilin (1966) and Florkin (1972) are available.

However, we must remain aware that respiration when measured—in non-photosynthetic conditions—through O_2 consumption or through CO_2 production, for instance, manometrically, includes the effects of other oxidations, e.g. in peroxisomes [*18f*], where no ATP is generated, and of decarboxylations and carboxylations. It is matter-of-course in traditional plant, animal and human physiology (see, for example, Steen, 1971; Jones, 1972) to call respiration what is directly measured through overall gas metabolism.

A further inconsistency appears in connection with the phenomenon of uncoupling, so important in biochemistry. Some agents, notably dinitrophenol, prevent oxidative phosphorylation and at the same time increase absorption of oxygen and production of CO_2. With the terminology here adopted, such uncouplers are not said to promote respiration. Uncoupling is also known in photosynthesis, but terminological consequences are less obvious there.

In many aerobic respirers, fermentation is suppressed by oxygen. Further reference to this "Pasteur effect" will be made later [*22f*]. The usefulness of such a control mechanism is obvious: waste of substrate for energy metabolism is avoided.

In the phenomenon of "anaerobic respiration", restricted to some kinds of cells, but important from the point of view of evolution, the terminal electron acceptor is an inorganic substance other than O_2. The term "respiration" is justified by the basic similarity of the mechanisms with that of aerobic respiration. This similarity will be pointed out [*16a*].

(c) The Citric Acid Cycle

For the full oxidation of carbohydrates, they are first degraded to pyruvate without participation of free oxygen. Up to this point the whole process, for which characteristically only a soluble system is needed, can still be looked at as a fermentation: glycolysis [*7c*]. The pyruvate is converted to AcCoA through oxidative decarboxylation. The AcCoA condenses with oxaloacetate to citrate [*10d*], and this enters the citric acid cycle (=tricarboxylic acid cycle). This cycle (Fig. 13.1) has been established, as is well known, after the important pioneering work of Szent-Györgyi by Krebs (Krebs and Johnson, 1937; see Krebs and Kornberg, 1957). The operation of the cycle in plants was proposed by Chibnall (1939).

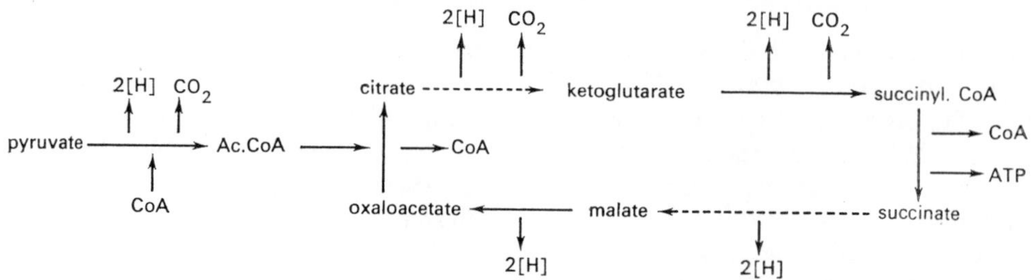

FIG. 13.1. The citric acid cycle, merely from the point of view of bioenergetics.

Some at least of the reactions now included in the citric acid cycle occur already in the fermenters, of course anaerobically. In the clostridia and methane formers the reactions may provide aminoacids (Gottschalk and Barker, 1967; Gottschalk, 1968), notably glutamate through ketoglutarate. Reference to a similar role of reactions of the present cycle in photosynthetic bacteria has been made already [*10d*]. Supply of aminoacids is in the existing oxygenic biosphere still a task of the citric acid cycle (Krebs *et al.*, 1952; Krebs and Kornberg, 1957).

The reactions may gradually have been streamlined to give in the end the full cycle. In each turn of the wheel of the complete citric acid cycle, the two carbon atoms of the acetate that went into the cycle are removed as CO_2 while 8[H] are generated, 6 of them as NADH. Starting from pyruvate, the overall equation is

$$CH_3.CO.COOH + 3H_2O = 3CO_2 + 10\ [H] \tag{13.3}$$

or, starting from glucose,

$$C_6H_{12}O_6 + 6H_2O = 6CO_2 + 24\ [H]. \tag{13.4}$$

The full citric acid cycle may operate anaerobically, e.g. in photosynthetic bacteria, if hydrogen can be set free through a hydrogenase or if a suitable hydrogen acceptor is present (Ormerod and Gest, 1962). Anaerobic variants of the citric acid cycle are also known in animal cells (Krebs and Lowenstein, 1960; Lowenstein, 1967, 1969). In a partial reaction during oxygen shortage, fumarate is reduced to succinate in mammalian organs by NADH, with production of ATP (Wilson and Cascarano, 1970); a reduction of fumarate by a different anaerobic system has been mentioned [*7d*]. Generally, however, in aerobes the metabolic hydrogen, also from succinate, is fed into the respiratory chain [*13d*].

Not only carbohydrates are worked up by the citric acid cycle. Fatty acids are degraded to form AcCoA, which can condense with oxaloacetic acid, and NADH. Furthermore, many of the aminoacids are subjected in the cells to suitable preparatory reactions (transaminations) so that thereafter their carbon atoms can also enter the cycle (see Krebs and Kornberg, 1957).

To replace substances withdrawn for biosynthesis, so-called anaplerotic reactions (see Kornberg, 1966; Krebs and Kornberg, 1957; Kornberg and Quayle, 1970) take place. For instance, oxaloacetate is obtained through the ATP-dependent carboxylation of pyruvate [*10b*]. The "glyoxylate cycle" (Fig. 13.2), which may be understood as a variant of the citric acid cycle, serves bacteria which are forced to live on C_2 substrates and yet need the C_4 compounds—again not only for energy production, but also for biosynthesis (Kornberg and Krebs, 1957; see Lehninger, 1970). In seeds of higher plants and elsewhere, the glyoxylate cycle occurs in the glyoxysomes [*18f*].

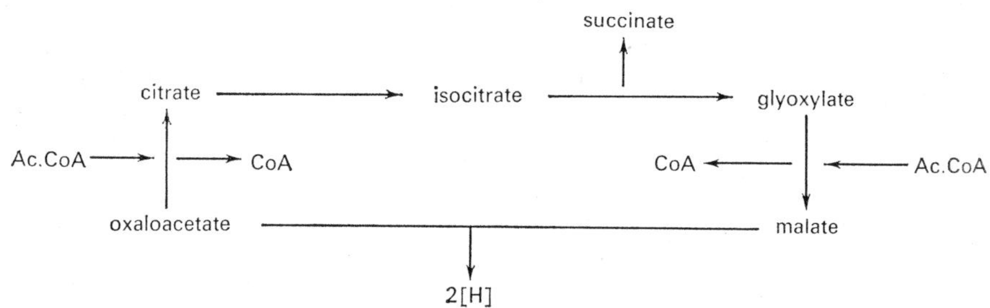

FIG. 13.2. Simplified version of the glyoxylate cycle, merely from the point of view of bioenergetics.

(d) The Respiratory Chain

In aerobic conditions, the NADH hands over its [H] to a chain of redox compounds, the "respiratory chain". The compounds are firmly connected with a membrane and with each other, and they have surely favourable mutual orientations in space (see Green, 1961; Ernster and Lee, 1964; Lehninger, 1964, 1970, 1971; Racker, 1965; Schatz, 1967; Chance *et al.*, 1968; Skulachev, 1971). The respiratory electron-flow chain strikingly recalls the electron flow chain in photosynthesis, where ATP is likewise produced [*8e*]. The principle is much the same.

A self-contained package of the redox compounds is called a respiratory assembly. Step by step, the electrons are transferred, without hydrogen ions (i.e. as such) or together with hydrogen ions (i.e. as hydrogen atoms), to chain members of higher and higher potential, and finally to free oxygen. The flow of the electrons has been clearly demonstrated *in situ* through the changes in light absorption shown by the chain members on reduction or oxidation (see Chance and Williams, 1956; Chance *et al.*, 1968). Important information about the chain is obtained with inhibitors and uncouplers (see Ernster and Lee, 1964).

Although respiratory chains first arose in prokaryotes, it may be best to reproduce here the chain as found in mitochondria, as this has the greatest practical importance, and therefore is best known. A simplified version (after Lehninger, 1970) is:

$$\begin{array}{l} \text{Fp 2, etc.} \\ \qquad\searrow \\ \qquad\quad \text{CoQ} \rightarrow \text{Cyt b} \rightarrow \text{Cyt c}_1 \rightarrow \text{Cyt c} \rightarrow \text{Cyt (a + a}_3) \rightarrow \text{O}_2 \\ \qquad\nearrow \\ \text{NAD} \rightarrow \text{Fp1} \end{array}$$

So the chain has five different cytochromes. Coenzyme Q is also known as ubiquinone [*8e*], cytochrome ($a + a_3$) as cytochrome oxidase (see Lemberg, 1969; Malmström, 1973). The flavoproteins other than Fp 1 collaborate with dehydrogenases specific for other electron donors than NADH, e.g. succinic dehydrogenase. Some words will be said about the somewhat aberrant respiratory chains in the mitochondria of higher plants later [*22e*]. For clarity, no account is taken here of the participation in the respiratory chain of non-haem Fe–S proteins [*7d*], whose role, however, is appreciated increasingly.

The electron flow through the chain from NADH to O_2 is coupled, at definite steps, to the synthesis of 3 ATP from ADP and P per 2 electrons, i.e. per atom of free oxygen. An equivalent statement is: the "P/O ratio" is 3. This is the result of "oxidative phosphorylation" [*13b*]. After phosphorylation as a consequence of fermentation and respiration had been investigated by many workers in Meyerhof's (see Meyerhof, 1930, 1937) laboratory and elsewhere, Engelhardt (1930, 1932) found respiration to give ATP in nucleated red blood cells, Kalckar (1937) found it to give ATP in cell-free systems from animal tissue, and Belitser and Tsybakova (1939) established that respiration provides more than 1 atom of ATP per atom of oxygen (see Kalckar, 1941, 1966b, 1969, 1973; Engelhardt, 1973). With $P/O = 3$, each molecule of glucose in its respiration provides for 38 ATP, against only 2 ATP in its fermentation. The bulk of this ATP is made by oxidative phosphorylation rather than by the substrate-level phosphorylation in the preparatory steps.

It is widely known that three groups of hypotheses try to account for oxidative phosphorylation. They have largely been developed on the basis of work on mitochondria. Of course, the same hypotheses also compete for the explanation of photophosphorylation.

1. the chemical hypothesis (Slater 1953, 1968, 1971; Chance and Williams, 1956);
2. the chemiosmotic hypothesis (Mitchell, 1961, 1966, 1967, 1968, 1969, 1970, 1971, 1973; Skulachev, 1972) and
3. the conformational hypothesis (Boyer, 1965; Green *et al.*, 1968; Green and Ji, 1972a, b; Boyer *et al.*, 1973; Green, 1974).

Mitchell's hypothesis has been mentioned before in connection with the origin of photosynthesis [*8h*].

It is impossible, and in view of the rich literature unnecessary, to explain or discuss these competing hypotheses here (see Griffiths, 1965; Pullman and Schatz, 1967; Greville, 1969; Deamer, 1969; Lardy and Ferguson, 1969; Racker, 1965, 1970; Storey, 1970; Harold, 1972). It might be emphasized, however, that a successful hypothesis must account for the role of membranes. Neither oxidative nor photophosphorylation has ever been observed in a liquid medium, i.e. in absence of membranes, as has been pointed out [*7a*]. Membranes can separate compartments of differing ion concentrations and electric potentials, and phosphorylation may depend precisely on the gradients of these magnitudes.

14

AEROBIC RESPIRATION OF PROKARYOTES

(a) Respiration of Blue–green Algae

Blue–green algae [*12f*] were the first organisms to liberate oxygen into the atmosphere, up to then essentially anoxygenic. Obviously these prokaryotic algae were also the first organisms to develop defences against the aggressive element, oxygen, and to become oxygen-tolerant. If the early oceans contained a lot of ferrous ion [*24a*], this captured the oxygen initially set free, and thereby provided a long time span for the algae to get used to the new poison.

One would expect the early "protoalgae" (DeLey, 1968) not yet to have known respiration and to have been just oxygen-tolerant. Apparently such organisms have died out. At present, no blue–green algae (or other plants) are known that cannot gain energy from respiration, though the citric acid cycle of the blue–greens, as far as is known, is deficient (A. J. Smith, 1973a). Even more, true growth of plants without oxygen has rarely been observed. Even green algae (Nührenberg *et al.*, 1968) that are hydrogen-adapted and do not set oxygen free [*12g*] cannot grow without oxygen. It may be, however, that the oxygen is needed for biosynthesis rather than for respiration. Some plants prefer subnormal oxygen pressures (Stewart and Pearson, 1970). However that may be, prokaryotic and eukaryotic plants can, after abandoning photosynthesis and chlorophyll, entirely rely on respiration [*12f*, *14f*; see Pringsheim, 1963 a, b, 1964; Alexopoulos and Bold, 1967).

The treatment of respiration might be focused at first on the blue–green algae. Unfortunately, not so many data are available (see Holm-Hansen, 1968; Carr and Whitton, 1973; Fogg *et al.*, 1973; Peschek, 1975). A striking result, admittedly obtained so far with a few organisms only, is that the metabolic hydrogen to enter the respiratory chain appears to be derived from the pentose phosphate pathway rather than the (full) citric acid cycle [*7f*], and that the hydrogen comes in as NADPH.

Be it recalled that at least many of the present photosynthetic bacteria operate, in various conditions, the citric acid cycle as a full cycle [*13c*]. It is to be assumed that the common ancestors of these bacteria and of the blue–greens, which must have been anaerobes, lacked this capability and that the bacteria came to use it only after their transition to aerobiosis. But it is not obvious why the blue–greens did not acquire the full citric cycle, or, if they did, why they lost it again.

Because much more is known about the oxygen relations of bacteria than about those of the blue–greens, and because the eukaryotes developed these relations from those of the bacteria, if the symbiotic hypothesis [*19 b, d*] is true, we shall refer more often to the respiration of bacteria than of blue–greens.

(b) Protection against Oxygen

After the arrival of free oxygen in the atmosphere as a by-product of plant photosynthesis, the bacteria had, like the blue–green algae, to deal with it. To protect themselves, some bacteria just withdrew into an anoxygenic environment, e.g. into the bottom mud of lakes. They remained anaerobes. Other bacteria developed special mechanisms to cope with the oxygen problem.

In this connection, bioluminescence may be mentioned. Bioluminescence of bacteria probably arose many times independently, and the taxonomy of these creatures is complicated (Hendrie *et al.*, 1970). McElroy and Seliger (1962, 1963; see also Seliger and McElroy, 1965) have proposed that the luminescence of bacteria was evolved as an early mechanism of defence against oxygen. The luminescent bacteria are thought to make substances that readily react with free oxygen at the ambient temperature in such a way that the energy of the reaction is largely or fully removed as light. Thus the oxygen is diverted from other targets, in the reaction of which the organisms would be damaged. Characteristically, low pressure of oxygen is sufficient for luminescence.

(In many advanced organisms, including insects and even fishes, bioluminescence serves special purposes. For example, the "glowworms" or "fireflies" recognize each other for mating. Also, the luminescence of bacteria may be useful to higher animals whose symbionts they are. But these are different stories.)

Objections have been raised against this "vestigial" hypothesis (Hastings, 1968). Non-luminescent mutants are frequent, and so it is not clear why luminescence has nevertheless been maintained for such an enormous time. Furthermore, the needed bacterial enzyme, luciferase (a general term), is capable of induction and repression, which would again be hard to understand if luminescence were not useful now (Nealson *et al.*, 1972). Possibly light is a by-product in the formation of an active form of oxygen (see Fridovich, 1975), which could attack compounds not readily metabolized otherwise. No specific induction of an enzyme for the utilization of each substrate would be needed, and the concentration of the substrate could be low.

(c) Oxygen in Bacterial Energy Metabolism

Facultative and obligate (strict) respirers occur among nearly all morphological types of bacteria, Gram positives and Gram negatives (Kluyver and Van Niel, 1936). The citric acid cycle is commonly found (see Kornberg, 1959; Krampitz, 1961), and the concentration of the enzymes of the cycle, e.g. in *E. coli*, is far higher in aerobic than in anaerobic conditions (see below).

Just as in the mitochondria of the cells of the various eukaryotes, metabolic hydrogen ([H]), mostly as NADH from the citric acid cycle, enters membrane-bound redox chains where the electrons flow to flavoproteins and cytochromes and finally to free oxygen (*13d*). The electron flow is again coupled to synthesis of ATP by oxidative phosphorylation (see Conn, 1960; Dolin, 1961; Smith, 1961, 1968; Gelman *et al.*, 1967; Aleem, 1970; Harold, 1972). According to Schatz (1967) "all essential conclusions (about oxidative phosphorylation. E.B.) . . . apply also to bacteria, as the coupling between respiration and ATP synthesis appears to obey a unitary principle in all cases so far investigated".

However, the respiratory chains in bacteria differ. A frequent feature are terminal cytochrome oxidases other than that of the cytochrome a + a_3 type (see Bartsch, 1968; Meyer and Jones, 1973). Horio and Kamen (1970) emphasize the differences compared to higher organisms: "no one ever presented . . . a respiratory complex from bacterial sources completely identical with that of mitochondria". Not surprising. The striking fact is rather that respiration should be so similar fundamentally in all bacterial groups and in the eukaryotes. This fact requires explanation.

The primitivity of bacteria is probably reflected in the generally low values of P/O ratios, i.e. in the lesser efficiency of energy conservation. While in a few cases values up to 3 have been reported, results are contradictory even with one and the same organism (Hempfling, 1970; Baak and Postma, 1971; Van der Beek and Stouthamer, 1973; Meyer and Jones, 1973). For the detailed investigation of the chains and the phosphorylating sites, specific inhibitors are, as in mitochondria, of great value (see Rieske, 1967; Harold, 1970).

Relatively good data about the composition of the respiratory chain exist, e.g. for *Mycobacterium phlei*, where the general similarity of the chain to that in mitochondria has been noted, and three sites of phosphorylation have been detected (Asano and Brodie, 1965; Pullman and Schatz, 1967; Orme *et al.*, 1969; Phillips *et al.*, 1970; Asano *et al.*, 1973). Such similarity has also been pointed out for *Mycobacterium flavum* (Erickson, 1971), for *Micrococcus denitrificans* grown with O_2 or nitrate as terminal electron acceptor (Scholes and Smith, 1968; Sapshead and Wimpenny, 1972) and for *Micrococcus lutea* (Erickson and Parker, 1969).

Azotobacter vinelandii has a complex, branched cytochrome system (Bruemmer *et al.*, 1957; Daniel and Erickson, 1969; Ackrell and Jones, 1971; Jones *et al.*, 1971). Three phosphorylating sites have been found, and respiring membranes with a P/O ratio about 1 have been prepared. These membranes are subject to respiratory control [*13a*], and their capacity for phosphorylation is influenced by the whole environment within the cell.

Complex respiratory chains containing quinones, flavoproteins and cytochromes have also been found in mycoplasmas (Van Demark and Smith, 1964). These organisms are the smallest bacteria known (0.2–0.3 μm) and without cell walls (see Smith, 1971; Razin, 1973). They normally live as parasites. In *M. arthritidis*, oxidative phosphorylation is present and is sensitive to the usual uncouplers, but the P/O value is low (see Razin, 1973). In respect to the rickettsiae, obligate parasites, it has been suggested that they meet their endogenous ATP requirements by respiration, but depend on the host for the supply of key metabolites (see Weiss, 1973).

Some studies refer to facultative aerobes. The membrane-bound electron transport chain of *E. coli* has been fractionated and analysed (Baillie *et al.*, 1971). The concentration of the cytochromes as well as that of the citric acid cycle enzymes (see above) depends on the "redox environment" during the transition from anerobiosis to aerobiosis (Gray *et al.*, 1966; Wimpenny and Necklen, 1971; Sapshead and Wimpenny, 1972). In the paper by Wimpenny and Necklen many references to literature about the influence of oxygen on facultative aerobes are given. The study of mutants of *E. coli* has proved particularly valuable (Cox and Gibson, 1974).

Branched electron transport systems, as mentioned for *A. vinelandii*, are quite common in bacteria. According to White and Sinclair (1971)

> Some bacteria can form a membrane-bound electron transport system that is considerably more complex than the system found in the mitochondria of eukaryotic cells. The systems are more complex as they contain many more membrane-bound dehydrogenases, which are the donors of electrons to the respiratory system, and the many reductases by which electrons reduce oxygen and alternative terminal electron acceptors. The multiple primary dehydrogenases and oxidases are linked by pathways which overlap and interconnect by various degrees. These interactions result in "branching" in the electron

transport systems. Electrons can enter the system through numerous dehydrogenases and leave by way of various oxidases . . . At the risk of being teleological, it appears that bacteria gain evolutionary efficiency in having the capacity to synthesize branched electron transport systems. For example, *Haemophilus parainfluenzae* produces a more extensive and complex electron transport system as the oxygen tension in the growth medium is lowered. By means of this process, a respiratory system with a progressively greater affinity for oxygen and the ability to utilize alternative electron acceptors enables the organism to maintain a rapid growth rate. This capacity for modification of the composition of the electron transport system is unnecessary in the carefully controlled intracellular environment in which the mitochondria of eukaryotic cells must function.

Several aerobic bacteria limit themselves to partial (incomplete) oxidation of organic substrates. For instance, acetic acid bacteria obtain energy from the oxidation of ethanol, ultimately derived from carbohydrates, to acetic acid. At first, metabolic hydrogen is obtained in the dehydrogenation of the substrate. The metabolic hydrogen is used for oxidative phosphorylation, and in the end converted to H_2O. Thus O_2 is consumed, but no CO_2 is set free. In technology, the process is, in contradiction to Pasteur's definition [*7a*], called an "oxidative fermentation", and in the particular case "acetic acid fermentation". In science, it is preferable to term the process an incomplete form of respiration. The general principle is the same as in complete oxidation, and ATP is made by oxidative phosphorylation.

(d) The Origin of Bacterial Respiration

How could respiration in all the aerobic bacteria develop independently from that in blue–green algae? How could it develop in parallel in so many groups of bacteria? Be it also noted the effects of uncouplers, inhibitors and pH are generally similar. How can convergence on such a large scale be made plausible?

True, within a respiratory chain the place of each redox compound, once it is present at all, is more or less determined by its standard potential. Thus pyridine nucleotide must appear at the beginning, flavoprotein in the middle, and cytochromes at the end of the chain. Moreover, the presence of some of the redox compounds is plausible. Pyridine nucleotide appears in any case in all bacteria, including the strict anaerobes, and flavoproteins are found in many anaerobes as well as in all aerobes.

One major problem is the presence of cytochromes. Many kinds of cytochromes have been found in the aerobic bacteria (first by Keilin, see Keilin, 1966; Fujita and Kodama, 1934; Smith, 1954; Kamen and Vernon, 1955; Newton and Kamen, 1961; Bartsch, 1968; Horio and Kamen, 1970; Kamen and Horio, 1970), and indeed all respiratory chains contain cytochromes. Yet cytochromes are never present in (anaerobic) fermenters. If the aerobes were thought to have evolved directly from fermenters, either a monophyletic descent of all respirers from fermenters or an independent origin of the cytochromes in every one of the bacterial groups that made the transition from anaerobic to aerobic life would have to be assumed. Both assumptions are implausible.

The problem is not even limited to the parallel emergence, in many kinds of bacteria, of a new class of compounds. An even deeper problem is that in every case the whole principle of the respiratory chain seems to have appeared anew. As has been emphasized, in oxidative phosphorylation the redox compounds are not just mixed, but they form an orderly assembly, the electron-flow chain. The spatial order, the particulate and periodic structure of the respiratory assemblies is clearly indispensable. So the additional question must be asked: Did this membrane structure that involves, among other compounds, pyridine nucleotide, flavoproteins and cytochromes, in each case arise *de novo*? Hardly likely.

(e) The Conversion Hypothesis

A solution may be sought through derivation of all respiring bacteria from photosynthetic bacteria. The general similarity of the electron-flow chains in photosynthesis and respiration has often been noted. Like the machinery for respiration, that for photosynthesis is always contained in ordered assemblies in connection with membranes. Moreover, the photosynthetic electron flow chain involves, similar to the respiratory flow chain, flavoproteins, cytochromes, quinones and non-haem Fe-S proteins. Not surprisingly, many authors (see, for examples, Duysens, 1964; Gaffron, 1965; Olson, 1970) have proposed that the photosynthetic chain, once it existed, could be taken over, modified and adapted to respiration according to needs.

It is a further step to claim (Broda, 1970, 1971) that all and every kind of respiring bacterium has descended from one or the other of the photosynthetic bacteria. This might be called the "(obligate) conversion hypothesis". It would not be too difficult to imagine that such a transition may have occurred many times in parallel, involving many photosynthetic and respiring species, if the postulated achievement were limited to a conversion of existing machinery from photosynthesis to respiration. This hypothesis accounts for the remarkable diversity in detail of the respiratory chains in the different bacteria. It also accounts for the flexibility of respiration. In one and the same organism can, as probably is true for bacterial photosynthesis [*8f*], different pathways be used in different conditions for bacterial respiration. The diversity presumably again has adaptive value [*14c*].

There is no reason why in blue–green algae, as in bacteria, respiration should not have arisen repeatedly, and therefore mechanisms should not show a similar diversity. Unfortunately, as has been pointed out, not much is known about respiratory chains in blue–greens.

It has been reported that oxygen can be used by flavin systems [*13a*]. However, no ATP is generated, and so Dolin (1961) sees the cytochromeless "respiration" through flavins, which often proceeds at considerable rates, as an early and essentially unsuccessful attempt to deal with O_2. Now it is conceivable that cytochromes, as soon as they had arisen in connection with photosynthesis [*8c*], supplemented short flavin-based "respiratory" chains (Lascelles, 1964b; Singer, 1971) in the oxygenic conditions that were a late consequence of photosynthesis. Incidentally, in view of the absence of oxidative phosphorylation it might be better here to speak of redox chains rather than of respiratory chains.

Certainly haemoproteins other than cytochromes are found outside the photosynthetic membranes. Thus in aerobic conditions haemoproteins became useful in the disposal of poisonous H_2O_2, a product of oxygenases: catalase and peroxidase. At some stage(s), haemoglobins appeared in higher organisms [*20 a, b, 22e*]. In the microsomes from the liver of vertebrates, i.e., likewise in eukaryotes, cytochromes are found [*19b*].

However, the fundamental difficulty with any hypotheses that seek the origin of complete respiration in flavin-based processes is that the principles of the membrane-bound respiratory chain and of oxidative phosphorylation would have to arise *de novo* many times in the different bacteria. So it appears better to think that the membrane-bound photosynthetic electron flow chains were converted to respiratory chains. All the haemoproteins outside the photosynthetic and respiratory chains may well have been derived from the cytochromes within these chains.

(f) The Origin of the Aerobes

On the basis of the conversion hypothesis, bacteria intermediate between the photosynthesizers and the respirers may be thought generally to have resembled the extant facultative aerobes among the photoorganotrophs, the non-sulphur purple bacteria [*9a*], e.g. *Rhodopseudomonas*. These organisms therefore may serve as models to understand the transition from photosynthesis to respiration. Indeed much thought has been given to the structural and functional relationship of the photosynthetic and of the respiratory system in photoorganotrophs.

To deal with the relationship between photosynthesis and respiration, structural terms independent of function have been coined by Drews. A distinction is made between the cell membrane (CM) and the intraplasmatic membrane (ICM) system. This may take up a large part of the cell volume, and has probably arisen by invaginations of the cell membrane. The photosynthetic system of the purple bacteria is found preferentially, but not exclusively, in the ICM, the respiratory system in or near the CM. Thus the topographical separation of the two ATP generating systems is incomplete.

Some authors tend to think of the two systems as more or less separate (Smith and Baltscheffsky, 1959; Smith and Ramirez, 1959; Ramirez and Smith, 1968). Others envisage complex cycles serving both photosynthesis and respiration (Horio and Kamen, 1962; Nishimura and Chance, 1963; Vernon, 1964). However that may be, the interaction between the two systems is certainly strong (Van Niel, 1941; Cohen-Bazire and Kunisawa, 1960; Gaffron, 1960; Smith and Ramirez, 1968; Thore *et al.*, 1969; Drews *et al.*, 1969; Oelze and Drews, 1970; Melandri *et al.*, 1971; Lien and Gest, 1973 a, b; Garcia *et al.*, 1974). Phenomena of induction and repression, largely reversible, are pronounced.

Interestingly, mutants of a non-sulphur purple bacterium (*Rhodopseudomonas spheroides*) have been obtained that grow photosynthetic pigments aerobically in the dark (Lascelles and Wertlieb, 1971). Some of these mutants are nevertheless unimpaired in their aerobic growth. On the other hand, a respiration-deficient mutant (of another *Rhodopseudomonas* species) is a valuable test organism for the mechanism of suppression of chlorophyll formation by O_2 (Marrs *et al.*, 1972).

(It has been pointed out that non-sulphur purple bacteria, e.g. *Rhodopseudomonas capsulata*, capable both of photosynthetic and oxidative phosphorylation, offer unique advantages for the study of the energetics of bacteriophage growth [Schmidt *et al.*, 1974].)

Mutual influence of photosynthesis and respiration has also been observed in blue–green algae (Brown and Webster, 1953; Pelroy *et al.*, 1972; Peschek, 1975; see Fogg *et al.*, 1973). The looser coupling of these processes in the eukaryotic plants is clearly a consequence of the separation of mitochondria and chloroplasts, both of which have sharply defined functions. Eukaryotic plants can simultaneously photosynthesize and respire at full speed and make ATP by both processes.

Later, many of the photosynthetic bacteria that had become aerobic lost the capacity for photosynthesis and turned into pure respirers, the strict aerobes among the bacteria. The loss of the capacity for photosynthesis is well known also among the blue–green algae, from which colourless gliding "bacteria" are derived, dependent on respiration [*14g*]. In analogy, the (respiring) colourless sulphur bacteria [*15*] could be derived from (photosynthetic) coloured sulphur bacteria [*11*].

The question of the temporal order of the appearance of strict and of facultative respirers after the conversion of photosynthesizers into respirers remains. Photosynthetic bacteria with a capacity for alternative growth through mere fermentation are now quite uncommon [*8b*]. It is nevertheless possible that facultative respirers evolved from such organisms. Or else obligate respirers descended from strict photosynthesizers, and they subsequently accentuated and redeveloped the ever-present, but often hidden, capacity for fermentation.

The lactic acid bacteria [*7c*] are unique among the anaerobes (it is rather a question of terminology) in being quite tolerant to oxygen. Moreover, hints of a respiratory system and of haemoproteins are found in some of them. They are microaerophilic. On these and other grounds it has been suggested that the lactic acid bacteria have retrogressed from aerobic life (Whittenbury, 1963, 1964; Bryan-Jones and Whittenbury, 1969). The propionic acid bacteria, which may also have regressed, have more haemoproteins than lactic acid bacteria, but they are only microaerotolerant (see Decker *et al.*, 1970; Stanier *et al.*, 1971; Schwartz and Sporkenbach, 1975). It is likely that at one time "primary" oxygen-tolerant bacteria existed that had not regressed from bacteria capable of using oxygen.

(g) The Gliding "Bacteria"

Like the blue–green algae [*12f*], many Gram negative organisms traditionally considered as forming a subclass of the bacteria show characteristic gliding, rather than swimming, movement. This kind of locomotion presupposes a special structure of the outer cell layers (see Castenholz, 1973). An example is the filamentous organism *Leucothrix mucor*, a colourless, strictly aerobic, organotroph. Its respiratory chain contains a number of cytochromes and has been compared with that in blue–greens (Biggins and Dietrich, 1968).

Most authors now derive the gliding "bacteria" from the blue–green algae. This possibility had been considered, with a varying degree of enthusiasm, many times (Pringsheim, 1949, 1963b; Harold and Stanier, 1955; Edelman *et al.*, 1967; Stanier *et al.*, 1970; Schlegel, 1972; Brock, 1973b; Soriano, 1973). As is true so often, the taxonomy of the gliding organisms does not even pretend to correspond to evolutionary relationship, and a pragmatic attitude prevails (Soriano and Lewin, 1965). It has been said that their classification is in a state of disorganization (Soriano, 1973; see also Lewin, 1969; Lewin and Lounsbery, 1969).

Rather remarkably, *L. mucor* is closely related to the filamentous organisms *Thiotrix* and *Beggiatoa*, both of which oxidize H_2S to S and even to sulphate. At least some strains of *Beggiatoa* require organic nutrients so that they must be classed as heterotrophs, although Winogradsky (see Stanier *et al.*, 1970; Schlegel, 1972) and others (Keil, 1912) had described *Beggiatoa* as an autotroph. However that may be, the strains are stimulated by H_2S, and presumably they cover much or all of their energy needs as lithotrophs (Scotten and Stokes, 1962; see Postgate, 1969; Roy and Trudinger, 1970).

If the descent of the gliding "bacteria" from blue–greens is accepted, they must be distant from the true bacteria. The filamentous sulphur "bacteria" must then be nearer to the prokaryotic algae than to the bacteria. Has the capacity for utilization of reduced sulphur compounds for energy metabolism developed here independently from the true sulphur bacteria, or has this capacity emerged anew after being largely (but not entirely; *12g*) latent in the blue–greens? A comparative study of the detailed mechanisms will produce an answer.

Bewilderment must be caused by the report (Pierson and Castenholz, 1971, 1974; Castenholz, 1973) that some gliding organisms, previously classed as flexibacteria, contain bacteriochlorophyll rather than chlorophyll a. They are now considered by their discoverers as photosynthetic bacteria. This *Chloroflexus* would then be a glider that has not descended from blue–green algae. *Chloroflexus* may contribute to the formation of stromatolites [*23a*].

15

CHEMOLITHOTROPHY

(a) General Features

Like the respiring organotrophs (chemoorganotrophs), the nonphotosynthetic lithotrophs (chemolithotrophs), discovered by Winogradsky (see Winogradsky, 1888, 1949), use oxidation in air to make ATP, but the reductants are inorganic (see Fromageot and Senez, 1960; Gibbs and Schiff, 1960; Lees, 1960, 1962; Elsden, 1962; Gelman *et al.*, 1967; Peck, 1968; Aleem, 1970). Internal membrane systems recalling those of photolithotrophs are generally found in the chemolithotrophs (Wallace and Nicholas, 1969; see Kelly, 1971). The chemolithotrophs are, like the photolithotrophs, Gram negative. Normally, they are aerobic and apply free oxygen as terminal electron acceptor. Sometimes oxygen is replaced by nitrate ion.

Important groups of chemolithotrophs are the thiobacilli, the nitrificants and the hydrogen (Knallgas) bacteria. They oxidize, respectively, compounds of S or of N or free H_2. In absence of an external substrate, organic storage material is respired (Bömeke, 1939; see Kiesow, 1967). Clearly the chemolithotrophs have kept at least some of the equipment for organotrophy.

In many cases, external organic substrates can also, to a greater or lesser extent, be exploited for energy production. In other cases, external organic substances are used for biosynthesis, but not for energy production, as can be shown with isotopes. The strictness of the lithotrophy or even autotrophy in some of the organisms has been much discussed (see Rittenberg, 1969; Roy and Trudinger, 1970; Kelly, 1971). Clearly the lithotrophs that cannot use external organic substrates have lost the capacity to do so. Therefore, in the present context this is not a problem of greater importance for the chemotrophs than for the phototrophs [*8b, 11a*]. Rittenberg (1972) has stated:

> Examples remain of bacteria that have not been cultured in the absence of an inorganic energy source or light. Such forms are appropriately described as obligate chemolithotrophs or obligate phototrophs. The available evidence . . . suggests that none of the bacteria is, at the same time, an obligate autotroph. From ecological and evolutionary considerations, absolute dependence on CO_2 for all carbon makes little sense, and bacteria with such a requirement would be an anachronism on the earth as it now exists.

It seems sound to class the chemolithotrophs along with the aerobic chemoorganotrophs, within the context of evolutionist considerations, as respirers. This is true even though in chemolithotrophy no CO_2 is produced. On the contrary, CO_2 is, as a source of carbon, absorbed. The essential feature is provision of energy by oxidation with O_2 or nitrate.

The consideration of the chemolithotrophs as respirers is justified by the fact that they make ATP by oxidative phosphorylation. This has also been found with cell-free extracts, e.g. from nitrificants (Aleem and Nason, 1960) and from thiobacilli (Hempfling and Vishniac, 1965). As always, the machinery for electron flow and oxidative phosphorylation is membrane-bound and not soluble. (Aberrant results were obtained by Cole and Aleem (1970) with *Thiobacillus novellus*.) The experimentally determined P/O ratios and the sites of phosphorylation vary greatly between genera and species (Aleem, 1972).

The reductants transfer electrons to a member of the electron-flow chain (Aleem, 1965, 1970; Trudinger, 1961; Kiesow, 1967; Peck, 1968; Roy and Trudinger, 1970; Kelly, 1971). Because of their insufficiently negative redox potentials, the reductants (except H_2) cannot directly supply electrons to NAD (Table 11.1). For example, in the oxidation of thiosulphate by *Thiobacillus neapolitanus*, cytochrome c is reduced first (Saxena and Aleem, 1972). The NADH needed for CO_2 assimilation is thereafter generally provided (Fig. 15.1) by ATP-powered reversed electron flow [*8f*] to NAD (Aleem *et al.*, 1963; Aleem, 1966a, b, c, 1968; Kiesow, 1967; Peck, 1968; Sewell and Aleem, 1969; see Forrest and Walker, 1971). But many questions in respect to the composition of the electron flow chains and the points of entry of the electrons remain open (see Kiesow, 1967; Wallace and Nicholas, 1969; Aleem, 1970).

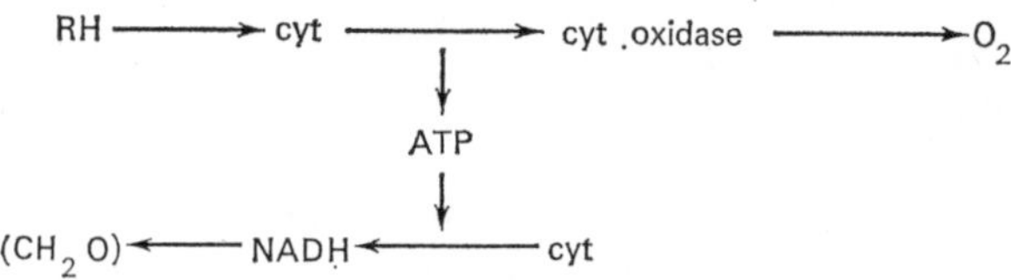

FIG. 15.1. One of the possible schemes for energy-linked electron transport in chemolithotrophs. The primary reductant (RH), pyridine nucleotide (NADH) and final electron acceptor (organic matter, shown as CH_2O) are written down in the reduced states.

The chemolithotrophs obtain carbon for biosynthesis by assimilation of CO_2. In the main, the reductive pentose phosphate cycle is used (see Peck, 1968), as in photolithotrophs [*10e*]. In heterotrophic growth, the cycle enzymes tend to be repressed, again as in photolithotrophs (see Rittenberg, 1969; Kelly, 1971). Storage materials [*7g*] include PHB, carbohydrates, and, in some hydrogen bacteria (Dreysel *et al.*, 1970), fats. At least some of the enzymes of the citric acid cycle are found (see Kelly, 1968; Peeters *et al.*, 1970). They are partly used for biosynthesis [see *13c*], i.e. anabolically (see Gottschalk, 1968; Kelly, 1971).

(b) Colourless Sulphur Bacteria

The best known chemolithotrophs are the colourless sulphur bacteria or thiobacilli. They oxidize inorganic sulphur compounds, including sulphide, sulphur, sulphite and thiosulphate (see Baalsrud, 1954; Bisset and Grace, 1954; Vishniac and Santer, 1957; Trudinger, 1967, 1969; Peck, 1968; Roy and Trudinger, 1970; Suzuki, 1974). Oxidation may, but need not, go as far as sulphate. H_2S oxidation to sulphate is found with membrane preparations (Adair, 1966). In *T. denitrificans*, the electrons are transported to nitrate (Baalsrud and Baalsrud, 1954; Milhaud *et al.*, 1958; Peeters and Aleem, 1970). So this organism is both a chemolithotroph and a nitrate respirer [*16a*]. Consequently, it is a respirer that in this way of life neither absorbs O_2 nor produces CO_2!

Curiously, some chemolithotrophic bacteria can use either sulphur compounds or ferrous ion as reductants (see Peck, 1968; Postgate, 1968b; Roy and Trudinger, 1970). *Thiobacillus ferrooxidans* = *Ferrobacillus ferrooxidans* oxidizes both components of pyrite (Postgate, 1968b, 1969). *T. ferrooxidans* has been reported to change, when kept in an organic medium, into an obligate chemoorganotroph (Shafia *et al.*, 1972). If this result were confirmed, the ease of the transition would again strikingly demonstrate the essential similarity of lithotrophs with organotrophs.

In a particular case, namely, the aerobic oxidation of sulphite, a specific phosphorylation reaction at substrate level has been observed in some thiobacilli, the only such reaction known which is based on the dehydrogenation of an inorganic reductant (Peck, 1962, 1968; Trüper and Rogers, 1971). The reaction has been written as

$$SO_3^{2-} + AMP = APS + 2e^-, \quad (15.1)$$

$$APS + P = ADP + SO_4^{2-}, \quad (15.2)$$

$$2ADP = AMP + ATP, \quad (15.3)$$

where APS stands for (energy-rich) adenosine phosphosulphate or adenyl sulphate. (APS was detected by Robbins and Lipmann (1958) and by Wilson and Bandurski (1958) in a different context.) The three enzymes are APS reductase, ADP sulphurylase and adenylate kinase. In the fermentation-type reaction (15.1) the electrons are taken up into the respiratory chain, in natural condition probably by cytochrome c, and only thereafter serve oxidative phosphorylation to make additional ATP. In parallel, AMP-independent sulphite oxidation also occurs (see Peck, 1962; Roy and Trudinger, 1970; Aminuddin and Nicholas, 1974). In *T. concretivorus* two different, but similar, flow chains are used for electrons from sulphide, on the one hand, and from sulphite, on the other hand (Moriarty and Nicholas, 1970).

(c) Nitrificants

The nitrifying bacteria oxidize (*Nitrosomonas*) ammonia to nitrite or (*Nitrobacter*) nitrite to nitrate aerobically (Kluyver and Donker, 1926; see Lees, 1954; Delwiche, 1956; Nason and Takahashi, 1958; Nason, 1962; Takahashi *et al.*, 1963; Kiesow, 1967; Peck, 1968; Wallace and Nicholas, 1969; Aleem, 1968, 1970; Suzuki, 1974). ATP is made by oxidative phosphorylation. *Nitrobacter* has also been grown on acetate, without oxidized nitrogen or CO_2, as a heterotroph and organotroph (Smith and Hoare, 1968).

(d) Hydrogen Oxidizing Bacteria and Algae

Some chemolithotrophs use hydrogen as reductant of oxygen, or, sometimes, nitrate (see Gest, 1954, Gray and Gest, 1965; Schlegel, 1966, 1969; Peck, 1968; Schlegel and Eberhardt, 1972; Suzuki, 1974). Like the aerobic chemoorganotrophs, the "Knallgasbakterien" (Knallgas $= 2H_2 + O_2$) are polyphyletic. They are Gram positive or negative (Baumgarten *et al.*, 1974; Bernard *et al.*, 1974). H_2 is never the exclusive reductant. So they are facultative organotrophs, but some mutants need CO_2 along with organic reductants (Ahrens and Schlegel, 1972). Many of the Knallgas bacteria are pseudomonas-like, and have been given the generic name *Hydrogenomonas*. The pathways of electron flow are greatly influenced by pretreatment. Thus in *Pseudomonas sacharophila* they depend on whether the organism was grown with H_2 or succinate (Donawa *et al.*, 1971). In H_2-grown *Hydrogenomonas eutropha*, P/O was found to be 2 with H_2 (Ishaque and Aleem, 1970).

The hydrogen oxidizers are the only chemolithotrophs that can reduce NAD directly (and reversibly), i.e. without reversed electron flow, just as in photolithotrophs NAD can be directly reduced only by H_2 [*11a*], and H_2 is the only electron donor to make possible anaerobic chemolithotrophy [*10c*]. The soluble enzyme for reduction by H_2 (in *Pseudomonas sacharophila* and *Hydrogenomonas ruhlandii*) has been called hydrogen dehydrogenase (Bone *et al.*, 1963). Other hydrogenases transfer hydrogen to other members of the respiratory chain (see Eberhardt, 1969; Bernard and Schlegel, 1974).

At least in some of the hydrogen bacteria, as in other chemolithotrophs [*15a*], the reductive pentose phosphate cycle operates (see Schlegel, 1966, 1972), though they also accept organic reductants. The hydrogen hydrogenase presumably provides NADH for the cycle (Schlegel and Eberhardt, 1972).

Some facultative aerobes (=facultative anaerobes, of course), notably non-sulphur purple bacteria [*9a*] and *E. coli*, can also obtain their energy by burning hydrogen. Likewise, algae adapted to photoreduction [*12g*] can alternatively use H_2 as fuel (Gaffron, 1942, 1944; Bishop, 1966; Gibbs *et al.*, 1970). Here again O_2 (at low partial pressure), but no light, is needed to make ATP and to fix CO_2. The adapted algae still employ the reductive pentose phosphate cycle (Gingras *et al.*, 1963; Russell and Gibbs, 1968).

Knallgas bacteria, grown on industrial hydrogen, are being considered as feedingstuff or food (Schlegel, 1969; Schlegel and Lafferty, 1971). Through proper choice of species and breeding, care must be taken that as much as possible of the biomass is digestible (preferably carbohydrate and protein), and that the content of nucleic acid, unwholesome in excess, is kept down.

(e) Oxidizers of C_1 Compounds

Some aerobic bacteria oxidize CO (Kistner, 1954; Hirsch, 1968; see Quayle, 1972; Hubley *et al.*, 1974; Schlegel, 1974). Others use C_1 compounds with hydrogen (see Ribbons *et al.*, 1970; Whittenbury *et al.*, 1970 a, b; Wilkinson, 1971; Quayle, 1972). Methane and its oxidation products, all compounds with one carbon atom only, are counted as organic substances by the chemist. The aerobic bacteria that use them, often as the only source of carbon and energy, are not so sure. In some ways, they resemble chemolithotrophs (Whittenbury, 1971).

The methane oxidizers require methane or methanol as substrate. The oxidation is believed to proceed by the steps (see Quayle, 1972)

$$CH_4 \rightarrow CH_3OH \rightarrow HCHO \rightarrow HCOOH \rightarrow CO_2$$

While not all steps have been elucidated, it is clear that NAD linked dehydrogenases take part (see Kornberg and Quayle, 1971). The first step, however, may involve oxygenases. ATP is presumably obtained by oxidative phosphorylation, beginning with NADH. Cytochromes are present (Anthony, 1975).

Biosynthetic utilization of carbon occurs either through a "serine pathway" or through a "ribulose phosphate cycle" (Lawrence and Quayle, 1970; Quayle, 1972; McFadden, 1973; Strøm *et al.*, 1974). The two groups are also distinguished by differences in their internal membrane systems, which are always well developed (Davies and Whittenbury, 1970; Patt *et al.*, 1974). The dichotomy is so sharp that separate origins of the groups are suspected.

In the serine pathway, glycine, of still controversial origin, is hydroxymethylated to serine, and this C_3 compound is anabolized to hexoses, etc. In the ribulose phosphate cycle, which is better known, formaldehyde is condensed with ribulose-5-phosphate to hexulose-6-phosphate. This is subsequently epimerized to fructose-6-phosphate, and after a few further steps ribulose phosphate is regenerated. The similarity of the cycle with the reductive pentose phosphate cycle, where ribulose diphosphate is the acceptor [*10e*], is unmistakable. But because of the lesser level of oxidation of CH_2O compared to CO_2, only ATP, and no NAD(P)H, is needed for the assimilation of carbon and for the production of carbohydrate.

Pseudomonas oxalaticus (Quayle and Keech, 1959; see Quayle, 1961, 1972; Blackmore *et al.*, 1968; Kelly, 1971) appears as an extreme case. The CO_2 produced by oxidation of formate is reduced after fixation by carboxydismutase. It may be recalled that this key enzyme of the reductive pentose phosphate cycle [*10e*] is not found in ordinary aerobic chemoorganotrophs, i.e. in dark respirers of organic compounds. Thus organic matter is made by reduction of CO_2 as if an inorganic substrate (say, H_2S) had been used. However, doubt has recently been thrown on the ability of *P. oxalaticus* to grow on formate without a cosubstrate (Höpner and Trautwein, 1971). Certainly *P. oxalaticus* can, in contrast to the methane oxidizers, also be grown on other organic substrates. In this case, it behaves as an ordinary heterotroph and chemoorganotroph.

Perhaps *P. oxalaticus* is related to organisms like the non-sulphur purple bacterium *Rhodopseudomonas palustris* that also assimilate carbon from formate via CO_2 (Stokes and Hoare, 1969). A similar case to that of *P. oxalaticus* is *Bacterium formoxidans* (Sorokin, 1961, 1966b).

(f) The Origin of the Chemolithotrophs

The lithotrophic like the organotrophic bacteria had to learn oxidative phosphorylation independently of the blue–green algae. Quite likely here again electron-flow chains preexisting in photosynthetic ancestors were converted to respiratory chains, i.e. the conversion hypothesis [*14e*] applies. So (colourless) thiobacilli evolved from (coloured) thiorhodaceae. Knallgas bacteria, too, might be directly derived from photosynthetic bacteria, which likewise utilize free hydrogen, though derivation of other genera from aerobic organotrophs could be considered.

A few photolithotrophs, species of the purple sulphur bacteria *Chromatium* and *Thiocapsa*, are partly aerobic, and capable of oxygen-dependent ATP formation (Gibson, 1967; Pfennig, 1970). These species might be transitional forms, to a limited extent analogous to the facultatively aerobic non-sulphur purple bacteria [*9a*].

Peck has proposed for early times a mixed fermentation with oxidation of reduced sulphur by organic substances, coupled to phosphorylation at substrate level, probably similar to reaction (15.1). It is doubtful whether organic compounds with a sufficiently high redox potential for such (dark!) fermentation reactions existed in the primeval soup (Broda, 1974). But whether they existed or not, there is no obvious way from substrate level phosphorylation to electron flow and to oxidative phosphorylation. It appears preferable to derive oxidative phosphorylation from photophosphorylation in every case.

How did those chemolithotrophs arise that do not correspond to existing photolithotrophs? Oxidizers of ferrous iron might, in view of their use of sulphur compounds as alternative reductants, be considered as a variant of the thiobacilli. On the other hand, it is not impossible that photosynthetic (anaerobic) iron bacteria existed at the time when the oceans contained ferrous ion [*24a*], but are extinct now. Similarly, photosynthetic bacteria obtaining electrons from the reductant NH_3 may have died out after giving rise to the nitrificants, which oxidize NH_3 aerobically. The possibility has further been mentioned (Quayle, 1972) that a CH_4 oxidizer could have developed from an aerobic NH_3 oxidizer.

It has been suggested (Peck, 1968) that growth on formate was an intermediate step in the evolution of autotrophy. Evidence for this is that some chemolithotrophs have formate dehydrogenase, and *Nitrobacter* can grow on formate. But assumptions of this kind are not essential if the chemolithotrophs can be derived from the photolithotrophs.

Generally, then, respiration (oxidative phosphorylation) is a common phenomenon among bacteria. The preference accorded to mitochondrial respiration by most workers is merely due to the importance to us of the mitochondria as the "power houses" (Siekevitz, 1957; Lehninger, 1964) of the higher organisms. If the symbiotic hypothesis is accepted, and moreover a monophyletic origin of the mitochondria is assumed, they must be derived from one particular, but unidentified, kind of respiring bacterium [*19b*].

16

BACTERIAL ANAEROBIC RESPIRATION

(a) Nitrate and Sulphate Respirers

Some kinds of bacteria use nitrate or sulphate instead of oxygen as terminal electron acceptors for the oxidation of substrates, allowing oxidative phosphorylation to proceed anaerobically. Oxidative phosphorylation of ADP takes place, as in aerobic respiration, through an electron-flow mechanism. All these organisms have, like aerobic respirers, membrane-bound electron-flow chains. This is why they have been aptly called respirers by Sato and Egami (see Sato, 1956; Taniguchi *et al.*, 1956). Anaerobic respiration has not been found among eukaryotes.

In some instances, both the electron donor and the electron acceptor are inorganic, as in chemolithotrophy. Examples are *Micrococcus denitrificans* and *Thiobacillus denitrificans* [*15b*], when they use H_2 or H_2S as electron donor and nitrate as electron acceptor. They are then to be considered not only as anaerobic respirers, but also as anaerobic chemolithotrophs.

Table 16.1 shows relevant free energy changes. H_2 has been chosen for the table as standard electron donor, and in most cases complete reduction of the electron acceptor has been assumed. The values of *G* for the gases refer, of course, also to the standard states.

TABLE 16.1. FREE ENERGY CHANGES IN SOME TYPES OF RESPIRATION

Reaction	*Go'* (kcal)	
$H_2 + 0.25SO_4^{2-} + 0.5\,H^+ = 0.25H_2S + H_2O$	−14	(16.1)
$H_2 + 0.25NO_3^- + 0.5H^+ = 0.25NH_4^+ + 0.75H_2O$	−41	(16.2)
$H_2 + 0.4NO_3^- + 0.4H^+ = 0.2N_2 + 1.2H_2O$	−58	(16.3)
$H_2 + 0.5O_2 = H_2O$	−57	(16.4)

The enzymes involved in these "dissimilatory" nitrate and sulphate reductions, which serve energy metabolism, differ from those mediating the "assimilatory" nitrate and sulphate reductions, whereby amino groups and sulfhydryl groups for body substances are made. The terms are Kluyver's (1953) and Postgate's (1959); for nitrate assimilation see, for example, Takahashi *et al.* (1963), Nicholas (1963), and Payne (1973), for sulphate assimilation, Schiff and Hodson (1973) and Payne (1973). The enzyme systems for assimilation are, in contrast to those for dissimilation, soluble.

Nitrate respiration is fairly common (Kluyver, 1953; Baalsrud and Baalsrud, 1954; Delwiche, 1956; Nason and Takahashi, 1958; Nason, 1962, 1963; Takahashi *et al.*, 1963;

Peck, 1968; Payne, 1973). Yet no strict nitrate respirers ("denitrificants") are known. They all accept oxygen as an alternative terminal electron acceptor, i.e. they are capable of aerobic respiration. In fact, they prefer it, and nitrate respiration is easily repressed by oxygen. Substrates are oxidized to CO_2 in nitrate respiration, as in oxygen respiration. Work with inhibitors has been undertaken (e.g. Radcliffe and Nicholas, 1970) to find the forking point between electron transport to O_2 and to nitrate (see Horio and Kamen, 1970).

Nitrate respirers include, in addition to the thiobacilli mentioned, species of non-sulphur purple bacteria and many strains of *E. coli*. In some organisms, the dissimilatory reduction of nitrate stops at the level of nitrite, while others produce free nitrogen or nitrous oxide ("denitrificants" in a more restricted sense). Nitrite itself may act as terminal electron acceptor. All the nitrate respirers are Gram negative.

All sulphate respirers ("desulphuricants") are, in contrast, strict anaerobes. Best known is the genus *Desulfovibrio* (Baars, 1930; Postgate, 1959, 1965; Peck, 1962, 1968; Wilson, 1962; Trudinger, 1969; Roy and Trudinger, 1970; Le Gall and Postgate, 1973). The organisms make ATP by reduction of sulphate or other oxidized sulphur compounds by means of H_2 and organic substances, e.g. lactate. The organic compounds are not fully oxidized to CO_2. A typical reaction (after Postgate, 1969) is

$$2CH_3.CHOH.COOH + SO_4^{2-} = 2CH_3.COOH + 2CO_2 + H_2O + HS^- + OH^-. \quad (16.6)$$

The desulphuricants cannot be grown as autotrophs (Mechalas and Rittenberg, 1960; Postgate, 1969). The reductive pentose phosphate cycle has not been found in the sulphate respirers, and they are not known to use the reductive carboxylic acid cycle (Gottschalk, 1968). Possibly they take up CO_2 rather like other organotrophs (see Roy and Trudinger, 1970). But it is hard to explain in this way that up to one-third of the biomass may be derived from CO_2 (Sorokin, 1966 a, b).

The substrate, sulphate, is activated by means of ATP, i.e. energy-rich adenine phosphosulphate (APS) is formed first (Fig. 16.1). The electron transport chains of the desulphuricants contain several cytochromes (see Horio and Kamen, 1970; Le Gall and Postgate, 1973). Yet they are, compared with that in mitochondria, "somewhat truncated" (Maroc *et al.*, 1970). It is not certain whether in each organism only one chain exists. The electrons from the reductants may enter the respiratory chains via non-haem Fe–S proteins (*7d*; see Roy and Trudinger, 1970; Le Gall and Postgate, 1973).

In absence of sufficient sulphate, *Desulfovibrio* can live as a fermenter. For instance, *D. desulfuricans* ferments pyruvate (Postgate, 1952, 1963, 1965; Akagi, 1964, 1967; Rittenberg

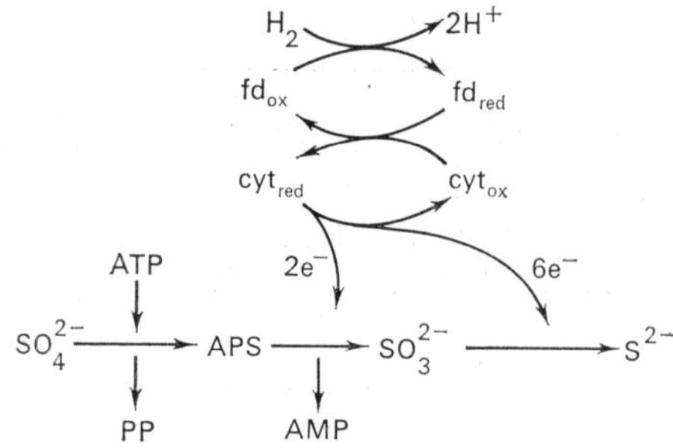

FIG. 16.1. Pathway of dissimilatory sulphate reduction in *Desulfovibrio*, after Roy and Trudinger (1970). fd = ferredoxin [*7d*], and APS = adenosine phosphosulphate [*15b*, *16a*].

1969). The citric acid cycle is incomplete, as in fermenters, and at any rate often in lithotrophs and in blue–green algae, but it serves anabolic reactions, notably production of glutamate (Gottschalk, 1968). In contradistinction to the clostridia, with which some sulphate respirers used to be confused, they are Gram negative.

(b) Sulphureta

In a natural contemporary anaerobic environment, far more sulphate is reduced in dissimilation than in assimilation (see Van Gemerden, 1967; Postgate, 1968b). While the desulphuricants reduce sulphate, and some marine bacteria are capable of the dissimilatory reduction of elementary sulphur [*16a*], the coloured sulphur bacteria (photolithotrophs), also anaerobes, do the opposite: they oxidize reduced sulphur compounds. In presence of oxygen (not tolerated by desulphuricants, i.e. at the fringes of anaerobic ecosystems), the colourless sulphur bacteria (thiobacilli) and gliding sulphur oxidizers (*Thiotrix*, *Beggiatoa*) likewise act as oxidizers.

Together, all these bacteria operate most of the sulphur cycle of the biosphere (Baas-Becking, 1925; Butlin, 1953; Postgate, 1968b, 1969; Alexander, 1971). Concurrently with the oxidation and reduction of sulphur, carbon is reduced and oxidized; in this way, a carbon cycle is opposed to the sulphur cycle. The biotic sulphur cycle has, of course, always been influenced by volcanic activity, and is increasingly modified now by man (see Junge, 1972; Kellogg *et al.*, 1972).

Anaerobically, the photosynthetic sulphur bacteria and the sulphate respirers form, if carbon compounds are available, an ecosystem that is in principle complete, and is kept going by the energy of light, a "sulphuretum" (Baas-Becking, 1925; Peck, 1966, 1974; Postgate, 1969). Sulphureta of limited size now occur in natural waters (Fig. 16.2). It may be assumed that such ecosystems were common in Nature before the atmosphere turned aerobic.

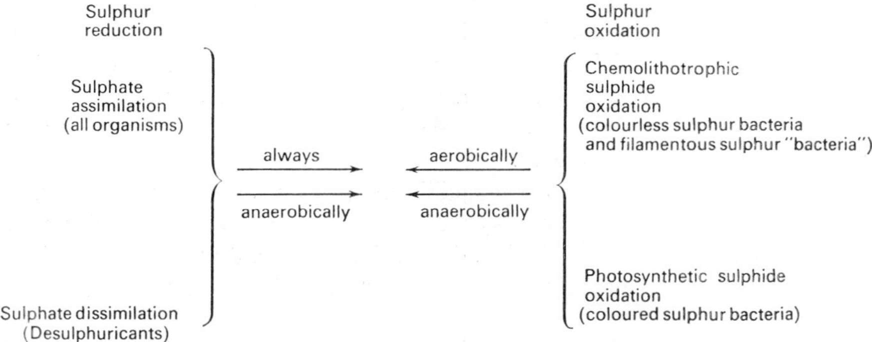

FIG. 16.2. The natural sulphur cycle of prokaryotes. For simplicity, only totally reduced and totally oxidized compounds are considered, with sulphur on the oxidation level either of sulphide or of sulphate.

(c) The Origin of the Nitrate Respirers

What is the evolutionary position of anaerobic respiration? In accordance with the conversion hypothesis [*14e*] it is assumed that these respirers arose from other organisms capable of electron flow along membranes, unless the principle of electron flow appeared in themselves first.

Let us first consider the nitrate respirers. They are obviously very closely connected with the pure oxygen respirers. Some think that nitrate respiration originally was a step on the way to O_2 respiration (Egami, 1957, 1973, 1974; Ishimoto and Egami, 1957; Nason, 1962; Takahashi *et al.*, 1963; Yamanaka, 1964, 1967; Yamanaka and Okunuki, 1964; George, 1964, Hall, 1971, 1973 a, b). It has further been suggested by the Japanese authors that the nitrate respirers descended from a particular group of bacteria called "nitrate fermenters".

The term "nitrate fermentation" (Sato, 1956; Takahashi *et al.*, 1963) is used for the insufficiently known reaction of some anaerobic fermenters, especially *Clostridium welchii* (=*perfringens*), with nitrate (Woods, 1938). Like all clostridia, this organism lacks cytochromes, and yet it reduces nitrate to nitrite, without assimilation of nitrogen. Further "soluble" nitrate reductases (see Nason, 1963) for energy metabolism have also been described, e.g. in aerobically grown *E. coli* (Taniguchi and Itagaki, 1960) and in *Spirillum itersonii* (Gauthier *et al.*, 1970). Of course, it always ought to be ensured that the nitrogen is really not assimilated in the system (see Wallace and Nicholas, 1969).

The question of the solubility of the nitrate reductases has not been investigated enough. Some of the reducing enzyme systems are said to be quite complex and partly bound by the cytoplasmic membrane (Pichinoty *et al.*, 1969; Pichinoty and Chippaux, 1969). "Reductases" may also consist of fragments of respiratory chains, which would account for the reported presence, e.g. in *S. itersonii*, of cytochromes (Bandurski, 1965).

In nitrate fermentation, the nitrate could have a role similar to that of free oxygen used to dispose of excess reductants [*13a*], i.e. as a non-essential hydrogen acceptor, and the reduction need not serve ATP production directly. Kluyver (1953), Verhoeven (1956) and Nicholas (1963) have considered such processes as "incidental dissimilatory nitrate reduction", presumably analogous to the incidental reduction of O_2, S and Fe_2O_3 [*13a*].

According to Egami, nitrate fermentation antedates nitrate respiration, and an evolutionary sequence ordinary fermentation → nitrate fermentation → nitrate respiration → oxygen respiration has been proposed (Ishimoto and Egami, 1957; Egami, 1973). One difficulty with this concept is that the intimate relationship between photosynthesis and respiration, which is indubitable and has led to the conversion hypothesis, has no place in it. Furthermore, there is a lot of doubt whether the early, anaerobic, biosphere could contain any nitrate. This point has already been made by Nason (1962), though that author was generally sympathetic towards the idea of a descent of aerobic from nitrate respirers, and also by Peck (1966) and DeLey (1968).

Hall (1973b) has invoked various non-equilibrium processes for the early production of nitrate, e.g. synthesis by lightning from N_2 and photolytic O. But even if appreciable amounts of nitrate were made in this way, which is doubtful, there would still be the problem of its protection from reductants, notably ferrous iron and sulphides. No satisfactory suggestion was offered.

But if the anoxygenic biosphere really did not contain nitrate, organisms could obviously learn to use nitrate for fermentation or respiration only after the arrival of free oxygen. In the aerobic biosphere, nitrate respiration became possible, logically, before organisms learned to use oxygen for respiration. But it appears more natural to reverse the order, and to assume that after oxygen had appeared oxygen respiration came first. The only known process for nitrate production in the biosphere is chemolithotrophic oxidation of NH_3 with oxygen [*15c*], i.e. a kind of aerobic respiration was a precondition for the appearance of nitrate.

The famous physical chemists Lewis and Randall (1923) have pointed out that combination of nitrogen with oxygen and water would be thermodynamically possible in the biosphere. In standard conditions,

$$0.5N_2 + 1.25O_2 + 0.5H_2O = HNO_3 \text{ aq.}; \Delta G_0' = 3.6 \text{ kcal.} \tag{16.7}$$

Consequently, with existing pressures of the gases the reaction is exergonic as long as the concentration of the nitric acid is less than 0.1 molar. As Lewis and Randall remark, happily no catalyst for the reaction appears to be present in Nature.

In reality, the situation is more complicated, as atmospheric nitrogen is assimilated all the time, and the products are, in the present atmosphere, oxidized to nitrate. There is no doubt that the speed of this process is ample to establish equilibrium in relatively short time. The lack of equilibrium is to be attributed to the disappearance of nitrate through assimilation and respiration. The problem has been pointed out, in chemical rather than biochemical terms, by Sillén (1965, 1966, 1967). Certainly we have here on a terrestrial scale an example of a non-equilibrium that is maintained through constant input of free energy [*1e*]. The terrestrial nitrogen cycle has recently been discussed by Delwiche (1970).

(d) The Origin of the Sulphate Respirers

The evolutionary connection between all kinds of sulphur bacteria (photolithotrophs, chemolithotrophs and desulphuricants) must be close. An interesting detail is that the enzyme system involved in sulphate activation for entry into dissimilation is similar to that responsible for the terminal stages of sulphur oxidation by lithotrophs, notably *Chromatium* (see Peck, 1962, 1974; Trüper and Peck, 1970; Roy and Trudinger, 1970; Schiff and Hodson, 1973; Le Gall and Postgate, 1973; Kirchhoff and Trüper, 1974).

It has been suggested that the sulphate respirers were the first organisms with cytochromes, and that the photosynthetic bacteria descended from them and added chlorophyll to the cytochromes (Peck, 1966; Klein and Cronquist, 1967; Gelman *et al.*, 1967; Postgate, 1968b; Hall, 1971; Le Gall and Postgate, 1973). Arguments in favour of this scheme include: (1) the presumed similarity with the clostridia. However, the clostridia are Gram positive, and the sulphate respirers Gram negative. (2) The presumed simplicity of the complement of cytochromes in the sulphate respirers. However, more and more different kinds of cytochromes are being found in them.

And was there much sulphate on the early Earth? This is doubtful. Sulphate from space, as found now in carbonaceous chondrites [*4b*], some of which, to judge from water content, etc., were never heated, would hardly have survived accretion in reducing conditions, probably with considerable heating. In the secondary terrestrial atmosphere, which remained reducing for a period of the order of 0.5 Gy (Holland, personal communication), sulphide, and not sulphate, was the stable form of sulphur. Consequently, sulphate could not be produced in fermentations either (Broda, 1974). On the contrary, sulphate, as far as it existed, could perhaps be used as an incidental electron acceptor or as a source of sulphur for assimilation.

The only remaining possibility for formation and regeneration of sulphate in the early biosphere was photochemical oxidation of H_2S, maybe

$$HS^- + 4H_2O = 4H_2 + H^+ + SO_4^{2-}; \Delta G_0' = 47 \text{ kcal.} \tag{16.7}$$

It is doubtful whether significant stationary concentrations of sulphate could be obtained in this way, with investment of the energy of light, in the existing reducing conditions.

As the atmosphere turned from reducing to neutral [*3c*], the free energy change of the reaction (16.7) improved, i.e. sulphate increased in stability (Holland, 1962, and private communication; Sillén, 1965). Even so, the velocity of this reaction, exergonic at pH 9 only below hydrogen pressures of $10^{-6.2}$ atmospheres, may always have been small. It must not be forgotten that the sulphide was largely locked up in compounds of minute solubility. Unfortunately, geological evidence is poor. In the oldest sediments some (not much) sulphate is found [*24a*]. but they may not have been laid down before the arrival of the photosynthetic sulphur bacteria [*11b*].

In balance, it appears likely that sulphate in bulk was first formed by these bacteria. But there is no doubt that sulphate preceded free oxygen. Therefore, the sulphate respirers may well have antedated the oxygen respirers, and have been components of the natural sulphureta [*16b*]. We are then back to the sequence, here preferred: photosynthesis → sulphate respiration. Thus chlorophylls, not cytochromes, were the first porphyrin derivatives to be used by organisms [*8h*]. It is probable that at one stage photosynthetic bacteria existed that could reduce sulphate for ATP generation in the dark, just as the facultative aerobes among the extant photoorganotrophs reduce oxygen. Those sulphate reducers seem to have disappeared, but may have been the ancestors of the desulphuricants.

We have suggested a polyphyletic origin of aerobic respiration in bacteria [*14d*]. Possibly some of the aerobic respirers have descended from sulphate respirers. There is, however, no need for such an assumption, and there are no obvious candidates for descent from sulphate respirers among the oxygen respirers. It is perhaps better to look for the ancestors of all aerobic respiring bacteria directly among the photosynthetic bacteria.

Incidentally, the concept, here doubted, of precedence of the sulphate respirers before the photosynthesizers would imply a tremendous time interval of the order of gigayears between the start of sulphate respiration and that of oxygen respiration. This would be true whether the latter were derived from the former or not.

The reductive pentose phosphate cycle, present in the photosynthetic bacteria, may have been lost by the sulphate respirers, which in any case require an abundance of organic carbon; in the photoorganotrophs it is repressed when organic substrates are supplied [*10e*].

Addendum in proof. Peck (1974) has now stated that "sulfate reducing bacteria were not antecedents of photosynthetic bacteria, but rather evolved from ancestral types which were photosynthetic bacteria. Although initially surprising, this evolutionary relationship is consistent with the idea that the accumulation of sulfate, the obligatory terminal electron acceptor for the sulfate reducing bacteria, was the result of bacterial photosynthesis."

(e) Carbonate Respirers?

Some authors (see Kluyver and Van Niel, 1956; Stadtman, 1967; Peck, 1966, 1968; Ljungdahl and Wood, 1969; Stanier *et al.*, 1970) emphasized the analogy between the anaerobic respirers, on the one hand, and bacteria that reduce carbonate anaerobically without light on the other hand. Such bacteria are the methane formers [*7e*, *10c*]. The term "carbonate respiration" was coined. However, in contrast to the real respirers these bacteria have no machinery for electron flow and no cytochromes (Postgate, 1968b). In a few cases, the energy metabolism has been elucidated at least partly. It differs widely from that in the respirers. The organisms therefore may be considered as fermenters rather than respirers. This is also Schlegel's (1972) view. It is possible that the methane formers are more ancient than the photosynthesizers and, *a fortiori*, than the respirers.

17

A SCHEME FOR PROKARYOTIC EVOLUTION

(a) General Features

A tentative scheme for the evolution of prokaryotes, represented by characteristic groups, may be drawn up (Fig. 17.1). The organisms named serve as examples only. The scheme has been derived by application of the criteria mentioned before [*2a*]. The transitions must be biologically useful, thermodynamically possible, and mechanistically plausible. The conversion hypothesis has been heeded.

Clearly, the scheme will need revision on the basis of further biochemical, physiological and morphological data. It will be particularly interesting to check the proposed evolutionary sequences, or improved versions, against the data for compositions of DNA and of selected proteins [*2d*]. Other authors have proposed quite different phylogenetic trees for the prokaryotes, among them Hall (1971), Horvath (1974) and Uzzell and Spolsky (1974). These trees are incompatible with the results of the considerations given in the preceding chapters of the present book.

The arrows indicate the direction of evolution, mostly by acquisition and rarely by loss of bioenergetic potential. Roman numerals indicate organisms that are not known, but whose present or past existence is postulated. Organisms can sometimes—but because of the possibility of permanent loss not always—in special conditions return, against the arrow, to the bioenergetic processes of their ancestors. For instance, *E. coli* can in absence of oxygen grow by fermentation of external substrates, but most blue–green algae cannot do so.

Actual descent is not meant to be implied. The particular organisms or genera selected may well have branched off the main lines. As stressed by Pringsheim (1964):

> . . . müssen wir uns . . . durchaus von der Vorstellung freimachen, als könne man die uns heute umgebenden Lebensformen stammbaumartig miteinander verknüpfen. Die heute lebenden Organismen sind nur die verschwindend wenigen übrig gebliebenen Abkömmlinge einer unvorstellbar großen Zahl verschiedenartiger Lebensformen, deren weitaus größter Teil spurlos verschwunden ist . . . es ist unwahrscheinlich, daß eine Art sich über lange Zeiten unverändert fortgepflanzt hat, nachdem eine andere durch Mutation aus ihr hervorgegangen ist.

To keep the scheme transparent, branching has been indicated explicitly only where intelligibility is thereby improved.

Mostly, bioenergetic evidence only has been used, but the Gram reaction (+ and − in circles) has been taken into account for placing aerobic chemoorganotrophs. While a few Gram variable bacteria are known (see Bartholomew and Wittwer, 1952), no particular difficulty follows in connection with our scheme. The transition Gram (+) → Gram (−) occurs once only in the scheme, the reverse transition not at all; we shall return to the Gram problem below. The Gram reaction of important bacteria has been listed by Aaronson and Hutner (1966).

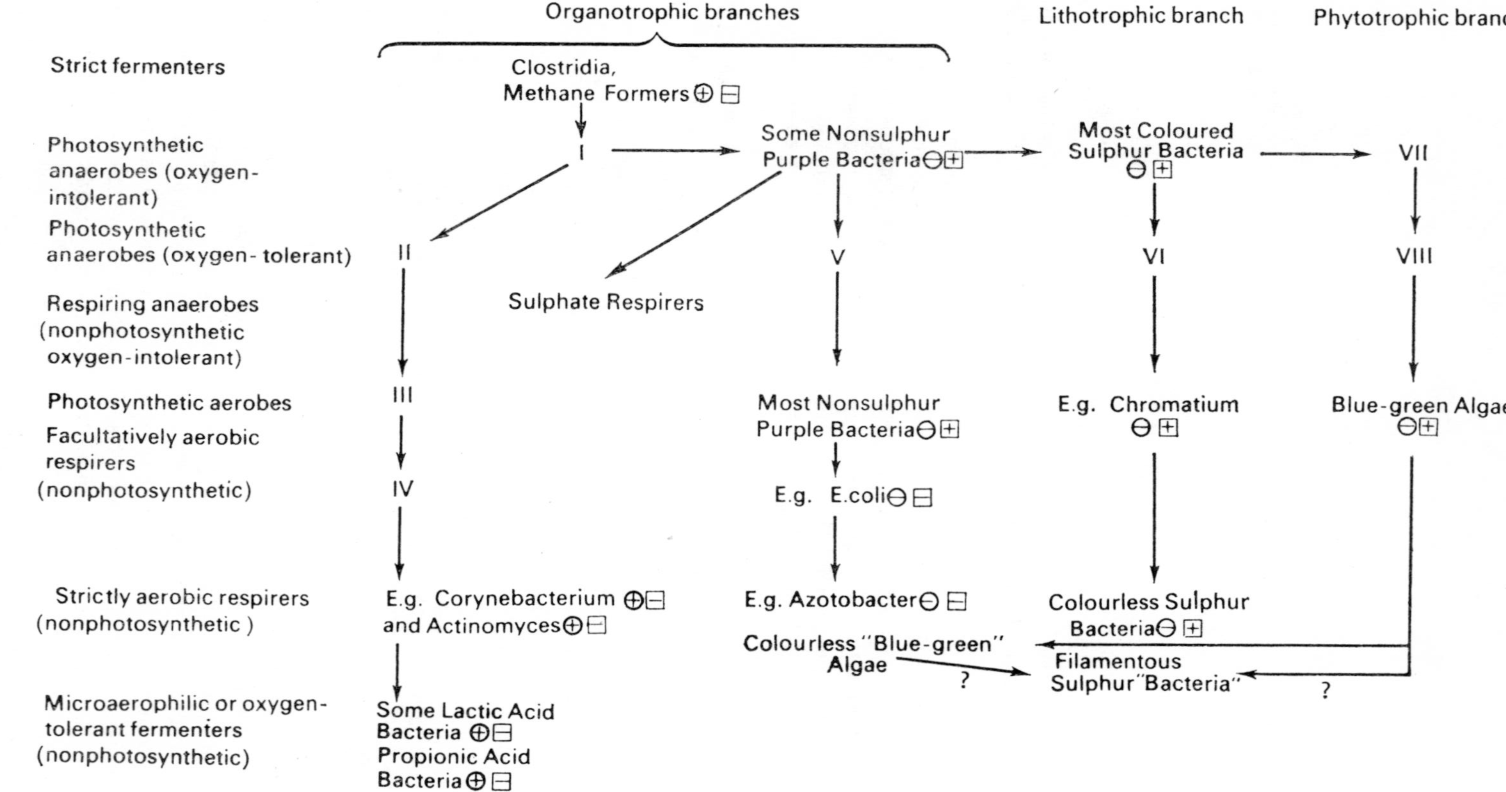

FIG. 17.1. General scheme for the bioenergetic evolution of prokaryotes.
(⊕ Gram positive, ⊖ Gram negative; ⊞ Calvin cycle present, ⊟ absent)

(b) Special Features

For clarity, the facultative replacement of free oxygen by nitrate as terminal electron acceptor [*16a*]—e.g. in *Escherichia coli* or *Thiobacillus denitrificans*—has not been explicitly mentioned in the scheme. Further, for reasons explained [*8b, 11a, 15a*] no notice has been taken of the strictness or otherwise of autotrophy among the lithotrophs.

It is remarkable that Gram positive and negative bacteria are cleanly separated provided Gram positive respirers, e.g. *Actinomyces* and *Corynebacterium*, are placed as they are. On the basis of the conversion hypothesis, the former existence of Gram positive phototrophs is implied. The hypothetical extinct organisms II, III and IV have been inserted to explain the presence of cytochromes and microaerotolerant or microaerophilic behaviour [*14f*] in lactobacilli and propionic acid bacteria, both Gram positive.

A further interesting feature is the clean separation between organisms with and without the reductive pentose phosphate cycle (+ or –, in squares). It has been assumed that lithotrophs have the cycle, and that organotrophs, including organotrophic respirers, generally have not. A partial exception are the photoorganotrophs where the Calvin cycle operates when inorganic reductants are used [*10e*].

The hydrogen oxidizing bacteria, formally chemolithotrophs, are surely a very mixed lot, and have therefore not been considered at the present stage. Nor has the curious organism *Pseudomonas oxalaticus* [*15e*] been included. May it at the end be recalled that *Chromatium*, which has been included, may be taken as only partly aerobic [*15f*].

Addition in proof. In contrast to some previous authors, Drews (1973) has called the blue-greens Gram positive. However, the similarities of the envelope with that of Gram negative bacteria are strong, as noted by Drews and his colleagues themselves (Drews, 1973; Katz *et al.*, 1974). So a sophisticated analysis may still produce justification for grouping the blue-greens with the Gram negative rather than the Gram positive bacteria also in respect to envelope structure.

18

EUKARYOTES AND THEIR ORGANELLES

(a) Prokaryotes and Eukaryotes

The closeness of blue–green algae and bacteria [*12f*] is expressed in the fact that they both are prokaryotes, and not eukaryotes. This fundamental distinction has been introduced in 1928 by Chatton (see Chatton, 1937). According to Stanier and Van Niel (1962) and to Stanier (1964, 1970), the difference between prokaryotes and eukaryotes is the greatest evolutionary discontinuity found in the present-day living world. The division between prokaryotes and eukaryotes is absolute, and no intermediate forms exist. In respect to the algae, Stanier *et al.* (1966) said: "The existing gap between the blue–green algae and the various specialized groups of eukaryotic algae is an extremely wide one, and consequently we have no clues how the eukaryotic type of cell may have arisen. The later emergence of the higher plant kingdom from eukaryotic algae, in contrast, can be much more clearly reconstructed." The eukaryotes (Fig. 18.1) include all algae except the blue–greens, the higher plants, which we shall call metaphyta, the fungi, the protozoa and the metazoa.

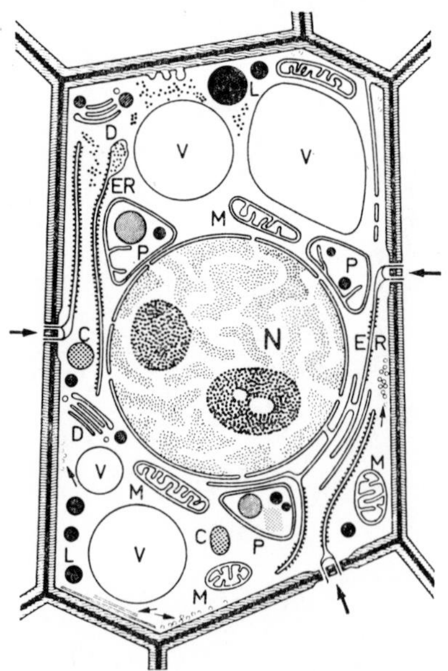

FIG. 18.1. Scheme of an eukaryotic (plant) cell (Sitte, 1965). N = nucleus,, ER = endoplasmatic reticulum, in places with ribosomes, D = dictyosomes, V = vacuoles, M = mitochondria, P = plastids (here as proplastids), C = peroxisomes, L = lipid droplets.

In these circumstances, it may not be unnecessary to insist that prokaryotes and eukaryotes have common roots, i.e. that in a wide sense they are monophyletic (Stanier, 1970). The fundamentals of the biochemical processes are the same. Moreover, it is generally thought that the prokaryotes are more ancient. This is also the view of Stanier (1970) though he thinks that the contrary view is arguable. Lone voices to the contrary have been those of Bisset (1973; see below), of Reanney (1974) and of Brooks and Shaw (1973). The last-named authors [*24b*] propose that the Earth was colonized by eukaryotic algae from space, which later degenerated into prokaryotes and viruses!

The difference between prokaryotes and eukaryotes has been described by Stanier and Van Niel (1962):

> Within the enclosing membrane of the eukaryotic cell, certain smaller structures, which house sub-units of cellular function, are themselves surrounded by individual membranes, interposing a barrier between them and other internal region of the cell. In the prokaryotic cell, there is no equivalent structural separation of major sub-units of cellular function; the cytoplasmic membrane itself is the only major bounding element which can be structurally defined.
>
> The difference is most universally expressed in the organization of the nuclear region and of the enzymatic machinery responsible for respiration and photosynthesis. Whereas the nucleus of the eukaryotic cell in the interdivisional state is characteristically separated from the surrounding cytoplasm by a nuclear membrane, no such boundary appears to exist in the prokaryotic cell. In the eukaryotic cell, the enzymatic machinery of respiration and of photosynthesis is housed in specific organelles enclosed by membranes, the mitochondria and chloroplast, respectively. Homologous, membrane bounded organelles responsible for the performance of these two metabolic functions have not been found in the prokaryotic cell.

Mitosis is the most important feature of the contemporary eukaryotic cell. Many eukaryotes live without chloroplasts, and some eukaryotes appear to have lost the mitochondria, but mitosis is never missing in cells capable of reproduction. In contrast, the prokaryotic cell shows neither mitosis nor meiosis. To emphasize the primitivity of the organization of the genetic material in the prokaryotes, it has been suggested to replace for them the term "chromosome" by the term "genophore" (Ris, 1961; Stanier, 1970).

The concept of the prokaryote and the eukaryote have revolutionized the broad division of the living world into kingdoms. Haeckel (1866, 1894; see Lwoff, 1944; Stanier, 1966; Stanier *et al.*, 1970) had called simple organisms that do not form differentiated tissues, protists. Naturally, the protists included prokaryotes and eukaryotes. While taking into account Haeckel's pioneering work, it appears reasonable now, extending a scheme of Copeland (1956), to recognize five kingdoms (Whittaker, 1959, 1969; see Margulis, 1971; Leedale, 1974; Swain, 1974). Viruses [*5a*] are excluded.

(A) Monerans (=Prokaryotes)
(B) Protists (all undifferentiated animals, fungi and eukaryotic plants, i.e. essentially unicells).
(C) Higher Animals
(D) Higher Fungi
(E) Higher Plants

An assignment of all organisms either to the Plant Kingdom or to the Animal Kingdom, which corresponds to the traditional, and in many places still maintained, dichotomy Botany–Zoology in the universities, is hopelessly out of date.

Of course, from some points of view, including that of bioenergetics, we must cut across the frontiers of the kingdoms. Notably, plant photosynthesis (phytotrophy) is an enormously important feature so that in this respect the joint treatment of some monerans (blue–greens), some protists (algae) and the higher plants is justified.

(b) Intracellular Organelles

What is a useful definition of an (intracellular) organelle, as mentioned, for example, by Stanier and Van Niel (1962)? It has been seen that prokaryotes use the cell membrane and its extensions, organized into thylakoids, for photosynthesis and respiration [*9a*]. The membranes and the structures which they compose might be considered as organelles. They set apart a group of interlocking functions within the cell (Fig. 18.2).

In eukaryotes, far more elaborate structures are found within the cells. Most complicated are, if we leave out nuclei [*18 a, e*], the energy-transducing organelles (mitochondria, chloroplasts). Indeed, suborganelles (thylakoids, etc.) are contained within them. The mitochondria and chloroplasts are genetically semiautonomous and are surrounded by double membranes, as will be seen. But it must be admitted that a natural and really satisfactory terminology for organelles, which will clearly distinguish between structures within prokaryotes and eukaryotes, is not yet available.

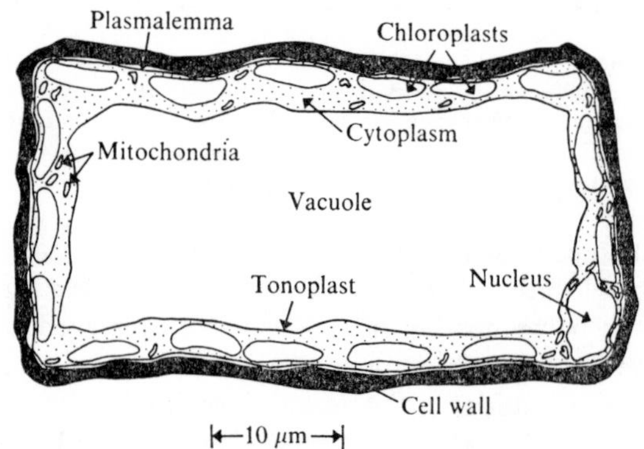

FIG. 18.2. Scheme of mature leaf cell from a higher plant, showing the presence of many membrane-enclosed subcellular organelles. (From *Plant Cell Physiology: a physicochemical approach* by Park S. Nobel. W. H. Freeman and Company. Copyright © 1970.)

The cytoplasm of eukaryotic cells outside the organelles is always criss-crossed by a system of membranes, mostly consisting of flattened vesicles, the endoplasmic reticulum (see Stanier, 1970). After the membranes are broken in the process of isolation, the resulting particles are called microsomes; they contain the protein-making ribosomes. In some regions, membranes are more densely and regularly packed, forming "Golgi bodies" of fairly well-defined shape. The endosecretory "lysosomes" are probably derived from the Golgi bodies. There are also peroxisomes [*18f*] and flagella [*19e*].

The complete separation of photosynthetic and respiratory machinery (and for that matter, of other organelles, e.g. peroxisomes) in eukaryotic plants makes it possible that these different functions proceed, and can be regulated, quite independently. In prokaryotes they are, when they coexist, inevitably much mixed up.

(c) Mitochondria

The mitochondria were first observed by Altmann (see Altmann, 1894), but could, because of their smallness, not really be investigated by light microscopy. The dimensions of mitochondria are of the order of a few micrometres only; in rat liver, about $1 \times 1 \times 3\ \mu$m.

While in a few organisms each cell holds only one mitochondrion, there may be thousands of mitochondria in a vertebrate cell. In the most diverse eukaryotes (fungi, animals, plants) mitochondria are built according to the same general plan (Fig. 18.3), though mitochondria from different organisms differ in details (see Bonner, 1965; Lieberman and Baker, 1965; Anderson, 1967 (protozoa); Hanson and Hodges, 1967; Chance *et al.*, 1968; Lance and Bonner, 1968; Öpik, 1968; Korman *et al.*, 1970; Racker, 1970; Smoly *et al.*, 1970). For the mitochondria or microorganisms, see Lloyd (1974).

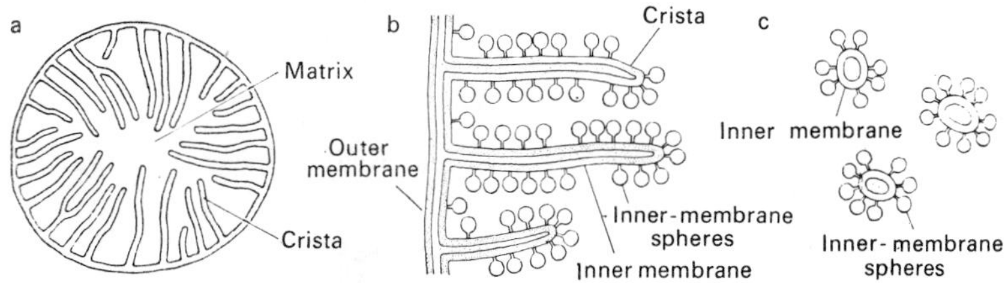

FIG. 18.3. Scheme of a mitochondrion (Racker, 1968). Either of the two membranes seen on the left is 6 nm thick. The inner membrane is deeply folded into "cristae" covered with spheres. The inner membrane, shown more enlarged in the centre, is the site of oxidative phosphorylation. This is still possible after fragmentation by ultrasound into submitochondrial particles (right).

The mitochondria are responsible for the oxidative phosphorylation in eukaryotes (Kennedy and Lehninger, 1948). In respect to energy metabolism, the mitochondrion is an integrated and streamlined whole in the sense that substrate is fully oxidized at great speed. The rates of respiration are far larger than could be expected if the redox compounds were present in solution (Ogston and Smithies, 1948). Like the respiratory chains in bacteria, the chains at least in the mitochondria of microorganisms are adaptable. For example, in *Euglena gracilis* at least three respiratory pathways, depending on growth conditions, exist (Linnane *et al.*, 1972).

Mitochondria have many other tasks in addition to phosphorylation; for instance, the synthesis of fatty acids. Thus they are multifunctional organelles (see Smoly *et al.*, 1970; Flavell, 1971), but other functions tend to remain in the background wherever efficient energy production is crucial. Thus muscle mitochondria differ from liver mitochondria.

The literature on the mechanisms of oxidative phosphorylation in mitochondria is enormous. Important books are those by Racker (1965, 1970) and by Lehninger (1964, 1970, 1971). Newer articles are by Van Dam and Meyer (1971), by Palmer and Hall (1972) and by Wilson *et al.* (1973). General features of respiratory chains have been outlined before, and a recent version of the mitochondrial electron flow chain has been given [*13d*].

The mitochondria are enclosed by two distinct membranes, with an intermembrane space between. The outer membrane is permeable for all smallish molecules, while the permeability of the inner membrane is much less, and selective. The substrates must, of course, enter, and the ATP must in effect get out. The inner membrane with its extensive invaginations, the cristae, is the place of the respiratory chain. More precisely, metabolic hydrogen (e.g. from the citric acid cycle) reacts with the inner membrane only at the inner surface, i.e. from the "matrix" side. Mitochondrial synthesis of nucleic acid and protein [*19a*] on ribosomes is also localized in the inner space. While the intermembrane space does not directly take part in the reactions culminating in oxidative phosphorylation, it may very well have indirect and regulatory functions; it contains many enzymes.

Most fundamental work on mitochondria has been done with preparations from animal (rat liver) and fungal (yeast) cells. In eukaryotic plants, the mitochondria operate along very similar lines, but at any rate in many higher plants their biochemistry has suffered some modifications [*22c*]. In practice, photophosphorylation is for plants a more plentiful source of ATP than is respiration. But ATP is needed in darkness, too. This may be the reason why all eukaryotic plant cells have, as mentioned, retained the mitochondria in evolution (Guilliermond *et al.*, 1933; see Clowes and Juniper, 1968).

Mammalian red blood cells have no mitochondria, and the same may be true of the cells of some anaerobic parasites [*20a*, *22f*]. In such cases the mitochondria have clearly been lost.

(d) Aerobic and Anaerobic Yeast

In other eukaryotic cells the mitochondria seem to disappear temporarily. The most famous organisms with this property are, of course, yeasts (fungi). Anaerobically grown yeast lacks a normal respiration and the classical cytochrome complement; these are adaptively regained in presence of oxygen (Ephrussi and Slonimski, 1950; Slonimski, 1953; see Lindenmayer and Smith, 1964; Stanier, 1970; Margulis, 1970; Linnane *et al.*, 1972).

Anaerobically grown yeast lives by alcoholic fermentation. Yet the genetic continuity of the mitochondria is maintained by semilatent, dedifferentiated, organelles, "promitochondria" (Morpurgo *et al.*, 1964; Schatz, 1965, 1967). The promitochondria are precursors of mitochondria, they contain DNA, they are visible with the electron microscope, and they can be concentrated by centrifugation (Criddle and Schatz, 1969; Plattner and Schatz, 1969; Plattner *et al.*, 1970).

The sensitivity of the mitochondria to the presence or absence of oxygen shows that they are determined not only by the genotype, but also by the environment. Just as in some prokaryotes, namely, the facultatively aerobic bacteria, the machinery for respiration is profoundly affected by anoxia.

In aerobically grown "petite" mutants, found to be incapable of making ATP by respiration, non-functional mitochondria occur (Ephrussi and Slonimski, 1955; Yotsuyanagi, 1962; see Wilkie, 1968; Linnane *et al.*, 1972). Concerning the lesion in this mutant (biosynthesis of a haem), Tuppy and Berkmayer (1969) can be consulted. The hereditary change is now known to be related to the deletion of variable, but large, segments of the mitochondrial genome (Mahler and Bastos, 1974; Slonimski, 1974). With chloramphenicol, the phenotypic expression of the mitochondrial genome may be reversibly inhibited (Clark-Walker and Linnane, 1967). Incidentally, it has been pointed out that in the last analysis even anaerobic yeast depends on free oxygen. The yeast needs sterols, and these are made with participation of oxygen [*13a*]. But the sterols can be supplied with the medium.

(e) Nuclei

In respect to nuclei, attention can here be drawn only to phosphorylation. While in the eukaryotes the mitochondria are the principal organelle for oxidative phosphorylation they are apparently not the only place for oxygen-dependent ATP synthesis. At least some kinds of nuclei, notably of thymus cells, have mechanisms not only for glycolysis, but also for aerobic ATP production (Osawa *et al.*, 1957; McEwen *et al.*, 1963, 1964; Penniall *et al.*, 1964; Bereznev *et al.*, 1970; see Conover, 1967, 1970; Siebert, 1968). The nuclei contain cytochromes, but the mechanism of ATP production may deviate from that in mitochondria.

Fatty acid is reported to be the principal substrate (Konings, 1970). While the process may be of the utmost importance for biosynthetic processes in nuclei, it cannot be considered a major contributor to the energy pool of the cell (Racker, 1965).

(f) Peroxisomes

Among the "microbodies" obtained from cells by centrifugation most kinds are distinguished by their content of oxidases and of catalase (DeDuve and Baudhuin 1966; DeDuve, 1969 a, b; Hruban and Rechcigl, 1969; Tolbert, 1971 a; Gerhardt, 1974; for terminology see especially DeDuve, 1969b). These "peroxisomes" are generally a little smaller than mitochondria. They have no cristae, and are surrounded by a single membrane. Especially the liver peroxisomes have been well investigated (Baudhin, 1969; Sies, 1974). Peroxisomes have also been found in protozoa, fungi and plants (see Richardson, 1974).

The peroxisomes always contain fairly many additional enzymes, but the pattern varies rather strongly. The peroxisomes contain flavoproteins, but no cytochromes. They are non-phosphorylating, i.e. they do not conserve energy as ATP, but at least some of them contribute to CO_2 production. In the eukaryotes, catalase is found only in the peroxisomes. The evolutionary position of peroxisomes will be considered later [*19f*].

Peroxisomes are present in all leaves (Tolbert *et al.*, 1969; Yamazaki and Tolbert, 1970) where they contain enzymes for "photorespiration" [*22d*]. The "glyoxysomes", which most authors include among the peroxisomes, occur, for example, in germinating fatty seedlings and in the protozoon *Tetrahymena pyriformis* (Hogg, 1969; Müller, 1969; *20a*). They serve the synthesis of C_4 from C_2 compounds [*13c*], and therefore ultimately the production of carbohydrate from fatty acid (Breidenbach and Beevers, 1967; Beevers, 1969; DeDuve, 1969a; Gerhardt, 1974). The degradation of the fatty acids to C_2 compounds occurs within the glyoxysomes, but for the process as a whole cooperation with the mitochondria and the cytosol is required. Moreover, the NADH obtained in the degradation of the fatty acids is, in net result, utilized within the mitochondria for ATP production (see Gerhardt, 1974).

(g) Chloroplasts

In all eukaryotic plants the chlorophyll is contained together with the other members of the electron-flow chain and with photosensitizers, e.g. carotenoids, within the highly organized chloroplasts (Granick, 1961; Kamen, 1963; Park, 1966; Weier *et al.*, 1966; Goodwin, 1966; Bogorad, 1967; Menke, 1967; Leech, 1968; Mühlethaler, 1967, 1971; Echlin, 1970b; Whittingham, 1970; Park and Sane, 1971). The chloroplasts by themselves are capable of the whole process of photosynthesis (see Arnon, 1966; Jensen and Bassham, 1966; Gibbs, 1971).

The chloroplasts have no cytochrome oxidase, and are therefore incapable of respiration (Arnon, 1956; Hill, 1956; James and Das, 1957; Warburg *et al.*, 1959; Lundegardh, 1961). But chloroplasts contain many enzymes involved in other processes than photosynthesis (see Smillie and Scott, 1969; Goodwin, 1971a).

The dimensions of the chloroplasts (Fig. 18.4) in different plants are very different. Already in the algae, the sizes and shapes vary greatly, and there is no obvious connection with phylogenetic position. In higher plants, chloroplasts are, typically, $2 \times 2 \times 5$ μm. The number of chloroplasts within a cell may be in some algae one only, in other cells, including some algal cells, it may be up to a few hundred.

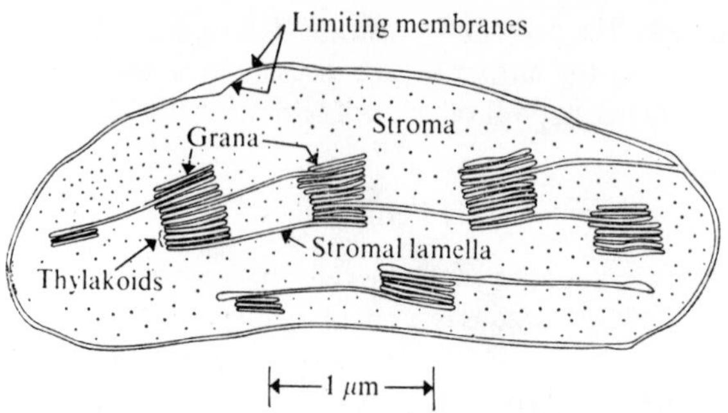

FIG. 18.4. Chloroplast from a leaf mesophyll cell (From *Plant Cell Physiology: a physicochemical approach* by Park S. Nobel. W. H. Freeman and Company. Copyright © 1970.)

The ultrastructure of chloroplasts in different organisms is often characteristic (see Gibbs, 1970; Dodge, 1974), but even in one and the same organism the morphological features of the chloroplasts are strongly influenced by conditions (see Clowes and Juniper, 1968; Echlin, 1970b). The control and regulation of photosynthetic carbon metabolism in eukaryotic plants is being actively studied (Bassham and Jensen, 1967; Bassham and Krause, 1969; Kanazawa *et al.*, 1970; Bassham, 1971 a, b).

The photosynthetic machinery (thylakoids) within the chloroplasts is contained in the long and slim "lamellae". In the plants higher than the algae, the lamellae are largely organized to stacks, so-called grana. The matrix, in which the lamellae are embedded, is called stroma. The stroma contains the enzymes of the dark reactions for CO_2 assimilation. The double membrane around the chloroplast is easily fractured in preparation. It holds the stroma back, but is not essential for the primary (light) reaction of photosynthesis. The double membrane is only poorly equipped with enzymes, and the enzymes associated with the thylakoids are missing (Douce *et al.*, 1973). It has, however, transport functions (Heldt and Sauer, 1971).

Chloroplasts may lose their pigments, and therefore their photosynthetic power; they are then referred to as leucoplasts. (Sometimes this term is confusingly used in a wider sense.) The leucoplasts still retain some of their functions. For instance, they may still produce and accumulate starch.

The loss of pigment is mostly reversible. Thus the well-known *Euglena*, to which we shall return later [*21a*], bleaches on cultivation in the dark, but greens in the light (see Schiff and Epstein, 1966; Bogorad, 1967). Plants that are kept in the dark from the beginning develop poorly differentiated "etioplasts", which transform into chloroplasts on illumination (see Kirk and Tilney-Bassett, 1967).

In other cases, the loss of photosynthetic power is irreversible, and is presumably due to loss of functional genes. For instance, *Euglena* may be "cured" of its chloroplasts by treatment with streptomycin (Provasoli *et al.*, 1948; Bovarnick *et al.*, 1974 a, b) or other agents (see Schiff and Epstein, 1967; Walles, 1971). After loss of the organelle, *Euglena* is indistinguishable from a protozoon, *Astasia* [*21a*].

19

ORIGIN OF MITOCHONDRIA AND CHLOROPLASTS

(a) Mitochondria as Semi-autonomous Genetic Systems

The mitochondria have their own inherited and functional equipment for protein production: DNA, RNA and ribosomes. They may therefore be considered as semi-independent genetic systems (see Gibor and Granick, 1964; Jinks, 1964; Gibor, 1967 a, b; Granick and Gibor, 1967; Linnane *et al.*, 1972). The DNA in mitochondria (and in chloroplasts) is a carrier of "non-Mendelian" inheritance. This had been studied long before the discovery of organellar DNA (see Sager, 1972).

The semiindependence rather than independence is expressed in the fact that only some of the mitochondrial proteins are coded for by mitochondrial DNA (see, for example, Cohen, 1973). The majority of the mitochondrial proteins are synthesized on extramitochondrial ("cytoplasmic") ribosomes on the basis of information laid down in nuclear DNA and transported to the ribosomes by messenger RNA from the nucleus. The proteins built up on the ribosomes must then somehow get into the mitochondria. In spite of their differences [*18c*], all mitochondria of one and the same organism contain, as far as is known, only one kind of DNA.

Some authors proposed that mitochondria can arise *de novo* from cytoplasm. References to such papers have been given by Lehninger (1964) and Baxter (1971). Evidence against an origin *de novo* has been produced, for example, by Luck (1963) who showed that labelled choline contained in preformed mitochondria is specifically transferred to newly formed ("young" mitochondria). The idea of *de novo* synthesis has probably been generally abandoned after the discovery of mitochondrial DNA (Nass and Nass, 1963; Luck and Reich, 1964; see Nass *et al.*, 1965; Borst and Kroon, 1969; Linnane *et al.*, 1972). The continuity of mitochondria is now widely accepted.

The mitochondria may be assumed to multiply by division, although direct information on this point is limited (see Baxter, 1971; Linnane *et al.*, 1972). The DNA presumably replicates in the same way as in the nucleus, but the division of the whole membrane system of the mitochondria perhaps ought not to be pictured as a dramatic short-time event (Ashwell and Work, 1970). In support of this view, the authors emphasized that the turnover times of the different components of the mitochondria, as measured through the rate of the incorporation of isotopes, vary greatly. Baxter (1971) has pointed out that "quite complex structures . . . can be reconstituted from their substructures *in vitro* . . . The spontaneous formation of mitochondria by the coordinated integration of substructures is not out of the question, and such a process begins to reapproach, in a sense, the concept of *de novo* synthesis."

(b) Mitochondria as Endosymbionts

As long as cytology was restricted to the light microscope, and biochemical investigation of separated mitochondria also was not possible, detailed ideas about them could not be developed. Even so, Altmann (1894), Wallin (1927) and other early authors put forward the concept that the mitochondria could be endosymbionts of bacterial origin; for literature about early speculations see Lederberg (1952), Novikoff (1961), Lehninger (1964), Sagan (1967), Nass (1969c) and Margulis (1970). Occasionally, it was even thought that the symbionts could be resuscitated to independent life; for references, see Buchner (1966) and Margulis (1970). This is, of course, not considered possible now, if only because of the dependence of the mitochondria on nuclear genes.

After the discovery of DNA in mitochondria, the hypothesis of their bacterial origin has been revived, and it now commands wide support (see Nass *et al.*, 1965; Nass, 1967; Edelman *et al.*, 1967; Sagan, 1967; Küntzel and Noll, 1967; Margulis, 1968, 1970; Küntzel, 1969; Noll, 1970; Raven, 1970b). Monographs and survey articles on the origin of the mitochondria, considering these new data, but not necessarily accepting the endosymbiotic hypothesis, were written by Roodyn and Wilkie (1968), Nass (1969c), Wagner (1969), Ashwell and Work (1970) and Taylor (1970, 1974).

The endosymbiotic hypothesis is meant to account for the wide gulf between prokaryotes and eukaryotes. It is remarkable that (aerobic) bacteria are about the same size as mitochondria, that they contain phosphorylating respiratory chains with cytochromes in their surrounding membrane, as do the mitochondria, that they respond in a like way to inhibitors and uncouplers, and that they show remarkably similar swelling-contraction cycles and ion transport activities (Lehninger, 1964). There are reasons for thinking that the inner membrane of mitochondria, the locus of respiration, stems from the bacterial cell membrane, while the outer membrane has been contributed by the host cell.

(A few words ought to be said about cytochromes outside the mitochondria. In microsomes, i.e. in the broken endoplasmatic reticulum, cytochromes are found (Klingenberg, 1958; Garfinkel, 1968; Archakov *et al.*, 1975). They are part of hydroxylases to biolipids, but also to drugs (see Yu *et al.*, 1974). Similar haemoproteins are found in bacteria, for instance, in pseudomonads (Yu *et al.*, 1974). But it is perhaps significant that according to the sequence analysis of the protein component the cytochrome of the endoplasmatic reticulum of mammalian cells, i.e. of the microsomes, is only very distantly related to mitochondrial cytochrome c (Mathews *et al.*, 1971). If these extra-mitochondrial cytochromes were derived in evolution from the cell membrane, this would point to a photosynthesizer as the original host [*14e*, *19f*].)

Some more strong evidence for the endosymbiotic hypothesis comes from the recent investigation of the machinery for protein production in mitochondria (see Wilkie, 1970b; Baxter, 1971; Schnepf and Brown, 1971; Tewari, 1971; Kroon and Saccone, 1974; Schatz and Mason, 1974). The mechanisms of transcription of information from DNA to RNA and of translation from RNA to protein much resemble those in bacteria. At least in animal cells, mitochondrial DNA consists, like bacterial DNA, of double-stranded rings. The ribosomes have the same weight as bacterial ribosomes, and inhibition of protein synthesis is observed with substances active against protein production in bacteria, and not against nucleus-based protein production in eukaryotic cells.

The amount of DNA in some lower eukaryotes (yeast, *Neurospora*) is far larger than in the mitochondria of higher animals. This may be explained through loss of genes and transfer of their duties to the nuclear genome in the evolution from lower to higher eukaryotes. Evolution may even depend on this reduction (Noll, 1970). Perhaps the remaining rest is, for some unknown reason, an irreducible minimum. Or else the mitochondrial genes are entirely "on their way out" (Borst, 1970).

While the total amounts of DNA in the mitochondria of higher organisms are similar, compositions differ. This is still consistent with a monophyletic origin of the mitochondria, if it were assumed that different parts of the mitochondrial DNA were lost by different organisms. In view of the basic similarity of all mitochondria, a polyphyletic origin should be viewed with reluctance [*19e*].

The original endosymbiont may have evolved from an endoparasite that resisted digestion. Similar recent cases are known. For instance, endosymbiotic bacteria are found in amoebae (Pangborn *et al.*, 1962). Also, bacteria of the genus *Rhizobium* enter the root cells of leguminoses and establish there the well-known cooperation for the assimilation of free nitrogen. The protozoon *Paramecium aurelia* contains "organelles" of bacterial origin and of uncertain significance that are distinguished by Greek letters (Beale *et al.*, 1969). Be it noted that the known hosts are all eukaryotes.

Lehninger (1964) has pointed out the possible advantages of endosymbiosis leading to the formation of mitochondria:

> It seems not impossible that such intracellular parasites may have become nonpathogenic in the course of evolution of the host cell and achieved a truly symbiotic relationship with it . . .
>
> The basis of metabolic symbiosis may lie in the possibility that a very large cell, such as in mammalian tissues, which might have a several thousandfold larger volume than a bacterium, might not be able to survive if its phosphorylation processes and respiratory chains were limited to the outer plasma membrane, as is the case in simple bacteria. The energy requirement of the large cytoplasmic mass could not be adequately supplied with ATP from respiratory assemblies in the plasma membrane alone, either because of the relatively long diffusion paths or because the plasma membrane could not possibly contain enough respiratory assemblies to manufacture the amounts of ATP required.
>
> . . . . Normally, bacteria make ATP but retain it within the cell; it is not secreted. The bacterial membrane might, however, have acquired the property, which mitochondria do have, of making possible transfer of high-energy phosphate from internal ATP to external ADP via adenylate kinase or ATP–ADP exchange enzymes in the membrane.
>
> It is of some interest that the rickettsiae [see *14c*. E.B.] which parasitize cells of some insects and higher animals, catalyse oxidation of the Krebs cycle and also oxidative phosphorylation. These organisms have never been cultivated successfully in a synthetic medium, suggesting they are quite dependent on factors that may be provided by the cytoplasm of the host cell. In fact, there is some evidence that the rickettsiae are unable to catalyse glycolysis and possibly require a cytoplasmic medium in which the products of glycolysis are provided for them.

It has been pointed out that the respiratory chain in bacteria is polyphyletic [*14d*]. It will be interesting to establish in which group of bacteria the respiratory chain is most similar to that in mitochondria. In this way information as to the kind of original endosymbiont, if any, might be obtained. *Micrococcus denitrificans* is a candidate (John and Whatley, 1975).

(c) Chloroplasts as Semi-autonomous Genetic Systems

The origin of the chloroplast is seen by many authors in a similar light as that of the mitochondrion, though the chloroplasts are more specialized. Like mitochondria, chloroplasts are semiindependent genetic systems (Renner, 1929; Gibor and Granick, 1964; Kirk, 1966; Kirk and Tilney-Bassett, 1967; Bogorad, 1967a; Carr and Craig, 1970; Stubbe, 1971; Walles, 1971; Tewari, 1971).

The amount of DNA is far larger in the chloroplast than in the mitochondrion (see

Smillie and Scott, 1969). So it may have more genetic autonomy, though the DNA is still by far insufficient to code for all proteins in the chloroplast (see Cohen, 1973). The machinery for protein synthesis in chloroplast again resembles that in prokaryotes, and consequently also that in mitochondria (see Goodwin, 1971; Tewari, 1971; Schnepf and Brown, 1971; and, with a more sceptical outlook, Woodcock and Bogorad, 1971).

In the past, many authors believed in a *de novo* formation of chloroplasts from cytoplasm, in a transformation of mitochondria into chloroplasts (e.g. Calvin, 1957, 1959) or in a formation of plastids from nuclei. (For reference to these authors, see Diers, 1970.) Now the continuity of the chloroplasts as such and their independence from mitochondria cannot be doubted any more (see Weier, 1963).

Chloroplasts divide by fission (see Granick, 1961; Frey-Wyssling and Mühlethaler, 1965; Kirk and Tilney-Bassett, 1967; Clowes and Juniper, 1968; Cavalier-Smith, 1970; Ridley and Leech, 1970; Stubbe, 1971). In algae, chloroplast fission is obvious. In higher plants, where a sexual stage intervenes between generations, continuity is through undifferentiated "proplastids", visible under the electron microscope only. "Fission of mature plastids occurs, if at all, in vegetative tissue. The best information about plastid replication comes from plastid inheritance" (Clowes and Juniper, 1968).

(d) Chloroplasts as Endosymbionts

Chloroplasts may be derived from endosymbiotic blue–green algae. This idea was first put forward by Schimper (1885), was supported by Mereschkowsky (1905) and Wallin (1927) and considered with some favour by Geitler (1923). Following the discovery of DNA in chloroplasts, which preceded its discovery in mitochondria, the endosymbiotic idea has been taken up again with great vigour (Ris and Plaut, 1962). It is being discussed by many authors (see Edelman *et al.*, 1967; Stutz and Noll, 1967; Holm-Hansen, 1968; Echlin, 1970b; Cohen, 1970, 1973; Schnepf and Brown, 1971).

The similarity of chloroplasts with blue–green algae is indeed striking. Moreover, recent symbioses of (complete?) blue–green algae ("cyanelles") with eukaryotic cells have been reported for a long time (Lauterborn, 1895; Pascher, 1929; see Pringsheim, 1958; Geitler 1960; Buchner, 1966; Edelman *et al.*, 1967; Echlin, 1970b; Alexander, 1971; Whitton, 1973).

For instance, blue–green endosymbionts have been described within protozoa (e.g. Hall and Claus, 1963; see Ball, 1969), within fungi (Echlin, 1966a) and also within eukaryotic algae. Uptake of algae by colourless eukaryotic plant cells (e.g. Hall and Claus, 1967; Echlin, 1970b) would essentially mean reacquisition of chloroplasts. The division of host (again always eukaryotic) and endosymbiont is broadly synchronized.

It is hard to distinguish reliably between cyanelles and chloroplasts (see Echlin, 1967; Stanier, 1970; Schnepf and Brown, 1971; Taylor, 1973, 1974). Possible criteria include: the cyanelle must contain a complete genome (DNA). It may contain the full complement of enzymes for respiration and nitrogen fixation. It may have a cell wall. It ought to grow outside the host. But all these properties might be lost.

The mutual benefit expected from symbiosis is obvious. The blue–green alga could assimilate CO_2 and also N_2, while the host contributes not only its capacity to obtain energy by fermentation, but above all its superior organization as an eukaryote. In the development of a chloroplast from the endosymbiont, assimilatory power for CO_2 at least has remained, while that for N_2 [*7d*] seems to have been lost. The capacity for respiration, universally shown by contemporary blue–greens [*14a*], has also disappeared [*18g*].

Acquisition of cyanelles has never been reported for cells of higher plants. They may be excessively differentiated. Alternatively, it may not have been possible to distinguish cyanelles from chloroplasts in this case. It appears that chloroplasts can live for weeks as symbiotic organelles in cells of the digestive tract of slugs (Kawaguti and Yamasu, 1965; Taylor, 1970; Trench, 1971).These animals have special mechanisms for the placement of the chloroplasts. It is also possible to introduce chloroplasts (from spinach or violets) artificially into mammalian cells in culture, where they remain viable during some cell divisions at least (Nass, 1969b). In hens' eggs even a division of chloroplasts has been observed (Giles and Sarafis, 1974).

Not only prokaryotic but also eukaryotic algae, e.g. species of *Chlorella*, occur as endosymbionts of many invertebrates (for protozoa, see, for example, Ball, 1969; Karakashian, 1970; Taylor, 1973). Under conditions of normal food supply they also divide more or less in step with the host cell (Karakashian, 1970; see Smith, 1973a). (Their chloroplasts would then have to be considered as (degenerated) symbionts of a second order!) In fact, the symbiosis of (green) algae with protozoa appears so welcome to the latter that the absorption of the algae may be genetically regulated by the host. Thus Chlorella can multiply inside *Paramecium bursaria* (Margulis, 1970, on the basis of Karakashian [1963] and Karakashian and Siegel [1965]):

> . . . . only until the normal, genetically regulated number of algae per Paramecium is attained. The multiplication then stops. Should the protozoen encounter free-living Chlorellae, they are promptly digested. Its own algal partners, however, are totally immune. Somehow the Paramecium recognizes its symbiont, although with the electron microscope it is not easy to see any morphological difference between the free-living Chlorella and the symbiotic one.

The usefulness of symbiosis between algae (also blue–green algae) and fungi has been discussed with lichens (Ahmadjian, 1963, 1967, 1970; Ahmadjian and Hale, 1973; Smith, 1973b). It appears that here again the non-photosynthetic organism, the fungus, is the senior partner who may pick up algae according to need. These fungi have never been found free of symbionts.

(e) Events Leading to Symbiosis

The mitochondria of lower eukaryotes (yeast, *Neurospora*) differ from those of the metazoa and metaphyta [*19b*]. Among the higher eukaryotes, notably among the metaphyta, again differences in mitochondrial structure and function are observed. Nevertheless, the basic features of all mitochondria are so similar that the working hypothesis has gradually been adopted [*19b*] that all mitochondria have a monophyletic origin (see also Raven, 1970b).

The most detailed picture of the ways in which endosymbiosis may have been established and may subsequently have led to the formation of mitochondria and chloroplasts has been given by Margulis (L. Sagan (=Margulis), 1967; Margulis, 1970). She has covered other aspects of the development of the eukaryotic cell fundamentally from the same point of view, namely, the acquisition of motility through the so-called (9 + 2) flagella or cilia (akin to the flagella), and on this basis the evolution of the machinery for mitosis. We shall limit ourselves mostly to the discussion of the organelles directly concerned with the bioenergetic processes, the mitochondria and chloroplasts, but shall also assess the peroxisomes.

According to Margulis (Fig. 19.1), respiring bacteria entered larger bacteria that were still relying on fermentation. The host must have been in the habit of ingesting particulate matter and must have had features that helped to make the union successful. No such prokaryotes are known now. The endosymbionts ("protomitochondria") contributed capacity for efficient energy production. The endosymbiosis became obligate and resulted in the evolution of an aerobic "amoeboid" organism, still amitotic.

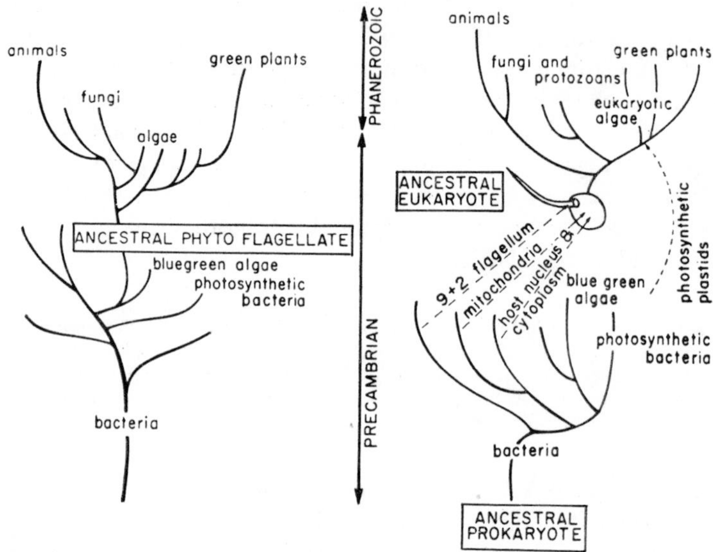

FIG. 19.1. Comparison between the classical (left) and the symbiotic (right) views of the evolution of cells (Margulis, 1970).

Gradually, the respiring endosymbionts developed into what we now know as mitochondria. As the remaining genome of the mitochondria is small indeed, most genetic functions, namely the production of most proteins, must sooner or later have been transferred to the host. The reason was probably that in this way centralized control of the work of the mitochondria has been insured. According to Stanier (1970): "An evolutionary change in the structure of the eukaryotic genetic system assured the permanence of the captivity of the respiratory endosymbionts. This was the abstraction of some determinants of endosymbiotic function from the genophores of the endosymbionts and their incorporation into the nuclear genophore of the host."

In the view of Margulis, the amoeboid host with its endosymbiont acquired motility through further union, this time with a spirochaeta-like (motile) prokaryote: "The partners were aided in their quest for food: a second group of symbionts, flagellum-like bacteria comparable to modern spirochetes, attached themselves to the host's surface and greatly increased its motility." This new symbiont was gradually converted into a flagellum (or cilium) and intimately associated with its host. Most protozoa are covered with flagella or cilia or have flagellated or ciliated stages. All eukaryotes have, always according to Margulis, some common flagellated ancestor. Later, the machinery for motility was used in the "amoeboflagellate" to build up equipment for mitosis, i.e. an eukaryotic nucleus. This evolved by many steps.

A different version of the origin of mitosis was put forward by Goksöyr (1967). Several anaerobic prokaryotic cells established a coenocytic relationship and the DNA of the several cells concentrated in the centre; the result was an anaerobic, but mitotic, eukaryote. No detailed hypothesis about the mechanism of concentration was given. After the appearance of oxygen in the atmosphere, the eukaryote formed a symbiosis with respiring prokaryotes (protomitochondria). Nass (1969c) proposed that in a colony of prokaryotes fusion and extreme specialization took place. In a benevolent discussion, Stanier (1970) nevertheless called this hypothesis unattractive, and found that it smells of science fiction.

Though she agrees "that chloroplasts and mitochondria may well have arisen from free-living prokaryotic cells which participated in the origin of eukaryotes", Sager's (1972) views on the origin of the eukaryotic nucleus also diverge greatly from those of Margulis:

> The central issue in the evolution of the eukaryotic cell is the origin of the nucleus, a question which has hardly yet been touched upon. The nucleus, totally lacking the machinery for protein synthesis, may well be the most recently evolved organelle of all. The tight integration we have seen between the nucleus and the organelles may then be the consequence of evolutionary shifts of some DNA from the cytoplasm to a new organelle, the nucleus, principally concerned with centralized genetic regulation.

At one stage, photosynthetic endosymbionts were taken up. They were blue–green algae, and not photosynthetic bacteria. It stands to reason that those endosymbionts were most useful that could use water as source of electrons for the assimilation of CO_2. The photosynthetic endosymbionts evolved into chloroplasts.

(f) The Role of the Peroxisome

It has been assumed by Margulis that the host of the protomitochondrion was an anaerobe. According to DeDuve and his colleagues (DeDuve and Baudhuin, 1966; Müller *et al.*, 1969; DeDuve, 1969 a, b, 1973) the host contained some ancestral form of a peroxisome [*18f*], and consequently was an aerobe, but peroxisomes were, and are, incapable of oxidative phosphorylation. So the acquisition of a protomitochondrion was nevertheless a great advantage to the host. The original machinery of the host for oxidation may subsequently have been diverted to other uses.

In his detailed considerations, partly based on Dolin (1961), DeDuve has argued that all peroxisomes in all eukaryotes are related, and therefore must go back to a common ancestor. The ancestral peroxisome is viewed as an oxidizing particle of general metabolic importance, comparable in some respects to a mitochondrion. It was, as has been pointed out, not capable of energy conservation, but it harnessed the production and destruction of H_2O_2 to the disposal of excess electrons and "would release the severe constraints imposed upon [fermentative. E. B.] energy metabolism by the necessity of using metabolites as final electron acceptors" [*13a*]. The ways in which excess electrons (in NADH or other reductants) could enter the peroxisome have been lucidly discussed by Gerhardt (1971).

After the acquisition of the mitochondrion, the peroxisome gradually declined. In divergent evolution, different functions were deleted in different branches. Some functions were lost irretrievably, but others were repressed only. The fact that peroxisomes still appear numerous and active in some animal tissues at least, notably in liver and kidney, points to their continued albeit more restricted importance, for instance for detoxification. In some plant tissues, they are responsible for gluconeogenesis from C_2 compounds (derived from fatty acids) through the glyoxylate cycle [*13c*]. One might speak of a degree of division of labour between mitochondria and peroxisomes. The hypothesis of DeDuve appears to answer the question how a peroxisome, bioenergetically inefficient, could arise in competition with a mitochondrion at all. It was there before!

DeDuve further has discussed the origin of the peroxisome. Intracellular particles may have formed from infoldings of the cell membrane, from which they separated at some stage. During separation they may have maintained digestive power previously exerted through exoenzymes. Through the capacity for internal digestion, the cell may have emancipated itself from a stagnant environment, increased in size and acquired amoeboid motility. In this way, endocytosis and the lysosome system as well as the secretory endoplasmatic reticulum, with which the peroxisomes are still connected, may have originated (DeDuve and Wattiaux, 1966). (In endocytosis [phagocytosis and pinocytosis] major objects are taken up.) The peroxisome became the main oxidative organelle of a primitive phagocytic organism of large size.

Another line of evolution among bacteria, also in aerobic conditions, may according to DeDuve have led to increased specialization of the cell membrane for oxidative phosphorylation. This organism, still a prokaryote, later became the symbiont of the phagocytic host. Some sort of competition, but also collaboration, between mitochondria and peroxisomes ensued. For the study of the interaction of peroxisomes with mitochondria, yeast is being used as an object (Szabo and Avers, 1969), as it can be obtained with or without functional mitochondria [*18d*]. Of course, in plants the peroxisomes must adapt to coexistence with the chloroplasts.

The rather bold suggestion has been made by DeDuve (1969b) that maybe eukaryotes devoid of mitochondria (e.g. certain trypanosomes) but equipped with peroxisomes might allow an improved study of the supposed ancestral functions of these particles. However, at least with one organism (*Tritrichomonas foetus*) the isolation of such microbodies has shown that they are not typical peroxisomes, as they do not contain catalase, but hydrogenase (Lindmark and Müller, 1973).

(g) Chloroplasts in Evolution

Why did in the Margulis scheme the acquisition of respiration precede that of motility through flagella or cilia? Presumably the improved energy supply from ıespiration was a condition for this kind of motility and therefore also for the development of mitosis. As respiring bacteria could exist only in the oxygenic atmosphere, i.e. after the blue–green algae had produced free oxygen in bulk, the symbiosis with these bacteria must have been formed after plant photosynthesis (phytotrophy) arose. But no photosynthetic organisms were symbiotic at first. Non-photosynthetic eukaryotes evolved without ever having had any photosynthetic eukaryotes as ancestors. In this way, following Margulis, the temporal order was

$$\text{Mitochondrion} \rightarrow \text{Mitosis} \rightarrow \text{Chloroplast}$$

That the remaining part of the genome of chloroplasts is much larger than that of mitochondria is consistent with the view that the acquisition of protochloroplasts occurred much more recently than that of protomitochondria.

Stanier (1970) rather thinks that mitosis arose before symbiosis as a consequence of the increase in size and functions of a prokaryote. Moreover, for him the presumed length of the time interval between the birth of the blue–greens and that of the respirers (which needed at least a minimum content of oxygen in the atmosphere) is a reason to doubt that non-photosynthetic eukaryotes were formed first. Therefore, the order would be

$$\text{Mitosis} \rightarrow \text{Chloroplast} \rightarrow \text{Mitochondrion.}$$

The concept of Margulis of the temporal precedence of non-photosynthetic before photosynthetic eukaryotes is, of course, opposed to the traditional view that all protozoa descended from eukaryotic algae by loss of the power for photosynthesis and for autotrophy. The concept of Stanier corresponds more closely to traditional thought. The question will come up again in the discussion of the relationship of plants and animals [*21a*].

While we see no reason to assume a polyphyletic origin of the mitochondria, there is a case for doing so in respect to chloroplasts. A polyphyletic origin of chloroplasts in algae, to account for their diversity (see below), has been proposed by Sagan (1967), Gibor (1967a, b) and Margulis (1970). Later, the chloroplasts of different origin may have evolved in parallel, in similar ways lost the cell walls, the machinery for respiration and nitrogen fixation, etc.

It follows from a polyphyletic origin of the chloroplasts that the classification of plants on the basis of the pigments in their chloroplasts is basically wrong (see also Dodge, 1974). Preferable features for taxonomy include the nature of the mitotic process, the position of the flagella, etc. (Lederberg, 1952; Margulis, 1970).

Raven (1970b) has emphasized the diversity of the pigments in algal chloroplasts. All extant blue–green algae merely have phycobilins as accessory pigments. Among eukaryotic algae, only the red algae contain phycobilins as sensitizers, while the green algae in the botanists' sense, the ancestors of the higher plants [*21a*], have chlorophyll b. The prokaryotic ancestors of the chloroplasts in algae with other accessory pigments, notably chlorophyll b, are assumed to have died out. This is already implicit in Sagan's (1967) hypothesis, as pointed out by Stanier (1970). Stanier (1974) further emphasized that "the unique organization of the photosynthetic apparatus, which bears a special light-harvesting organelle, the phycobilisome, a common mechanism of energy capture and transfer in this organelle. . ." make a cyanobacterial origin (origin from blue–greens) of the chloroplast of red algae virtually certain. This does not apply to the chloroplasts in other groups of eukaryotic algae. Of course, a polyphyletic origin of the chloroplasts implies a gene transfer from chloroplast to nucleus that occurred independently in different kinds of eukaryotic plants.

Lee (1972), however, questions the past existence of prokaryotic algae with other pigments, and offers a phylogenetic tree in which all eukaryotic algae descended from hosts of cyanelles by changes in pigments and other features. Lee argues that the steps needed for such changes are not too difficult to envisage. Thus the symbiotic origin, but in contrast to Sagan not its multiplicity, is maintained. Then one could return to a classification of algae based on their pigments. For a further discussion of Raven's and Lee's views see Taylor (1974).

If the chloroplasts are polyphyletic, Stanier's concept that the acquisition of mitochondria follows rather than precedes that of chloroplasts meets with the difficulty that all mitochondria in all eukaryotes are so much alike. They are presumably monophyletic, as has been explained. This is impossible if they had to evolve in a multitude of photosynthetic hosts.

(h) Some Objections Against the Endosymbiotic Hypothesis

Opponents of the endosymbiotic view generally assume that the membrane system of the mitochondria arose through the infolding of pre-existing membranes, notably the plasma membrane, in prokaryotes (Robertson, 1960, 1964; Uzzell and Spolsky, 1974). Thus the internal membrane system of aerobic prokaryotes would be the forerunner of the mitochondrial membrane system. At some point, the mitochondria would separate from the plasma membrane. Through some mechanism, the mitochondria would take up some of the DNA of the cell. A series of mutations would lead, stepwise, from prokaryotes to eukaryotes.

Haldar *et al.* (1966) proposed that the mitochondria may have evolved from vesicles derived from areas of the cell membrane similar to the cytochrome-rich areas in aerobic bacteria. The vesicles may have incorporated some chromosomal (better: genophoric) DNA. According to Lewis (1970), the evolution of organelles has involved gradual differentiation in the cell. It may have been an advantage for the cell to maintain several distinct compartments, and partial transfer of DNA between evolving nucleus and organelles may have taken place. Opponents of the endosymbiotic hypothesis also include Klein and Cronquist (1967), Allsopp (1969) and Klein (1970). Klein assigns to the red algae, the only eukaryotic algae with phycobilins [*19g*], a special role in the transition from prokaryosis to eukaryosis. A detailed scheme for the direct evolution of eukaryotes from prokaryotes has recently been proposed by Uzzell and Spolsky (1974).

Raff and Mahler (1972, 1973) have likewise argued in considerable detail against the endosymbiotic hypothesis for the origin of mitochondria. They refer to what they call two particularly awkward aspects:

> The first is that the postulated protoeukaryote possessing many advanced cellular adaptations should have been so primitive and inefficient metabolically. In the face of competition from conventional prokaryotes possessing more efficient aerobic energy-yielding pathways already foreshadowing the patterns observed today, this should have left it at a considerable disadvantage. Second, the integration of the endosymbiont-protomitochondria required wholesale transfer of genes from the endosymbiont genome to an unrelated nuclear genome. A mechanism by which this may have been achieved is extremely difficult to conceive.

Raff and Mahler also feel that the resemblance of the protein synthesis system in mitochondria with that in prokaryotes has been overstated. This problem cannot be discussed here. The authors finally arrive at a model:

> We propose that the protoeukaryote was an advanced aerobic cell rather larger in size than is typical for prokaryotes. This trend to larger size necessitated . . . a large increase in respiratory membrane surface. This was achieved initially by invagination of the inner cell membrane, and later by formation of membrane-bound vesicles generated from the inner cell membrane. The respiratory particles thus generated were topologically closed objects surrounded by a membrane providing a selective permeability barrier between the respiratory elements and the cytoplasm . . . Because such constant turnover of this complex organelles was uneconomical, it would have been a considerable advantage for these cells to implant a system for protein synthesis on the inside of the organelle for organelle maintenance . . . We propose that the cell implanted a protein synthesis system into the respiratory organelle by simply incorporating a stable plasmid containing the appropriate genes for ribosomal components.

Raff and Mahler propose that the extranuclear genome is derived from extrachromosomal genomes (plasmids or episomes) that occur widely among prokaryotes (for discussion see also Edelman *et al.*, 1967; Mayer, 1973). The information content of plasmids may not necessarily be as restricted as has been thought up to now, but might also involve genes for RNA and various membrane proteins. Thus information did not flow from the proto-organelles to the nucleus, but in the opposite direction. However, a further discussion of this topic is beyond the scope of this book. For a rebuttal of the views of Raff and Mahler see Taylor (1974).

Cohen (1970) and Hall (1973b) as well as Raff and Mahler see a further difficulty in the existence of many aerobic biosynthetic pathways, e.g. for ascorbic acid, for oleic acid, and for sterols, in the cytoplasm of eukaryotes—outside the mitochondria [*13a*]. However, it is not clear why the host, after or even before the acquisition of mitochondria, should not have adapted, in one way or the other, his biosynthetic pathways to the presence of oxygen. Many aerobic bacteria have such adapted pathways. So perhaps this difficulty is not too serious.

In respect to the question, raised by the critics of the symbiotic hypothesis, what distinguished the original host or enabled it to combine with a protomitochondrion, Stanier (1970) has made a novel suggestion. He stressed the important and poorly understood fact that prokaryotes never harbour endosymbionts:

> A stable endosymbiosis in which the host is a prokaryote has never been described. The only stable association involving two prokaryotic partners are ectosymbioses . . . The impenetrability of the prokaryotic cytoplasmic membrane by any object of supramolecular dimensions effectively precludes the acquisition of endosymbionts. This may well be considered a fundamental biological difference between eukaryotes and prokaryotes.

It is this feature which is selected by Stanier as the starting-point for eukaryotes. Could not the ancestral prokaryotic cell qualify as host to symbionts precisely by learning endocytosis, even the uptake of whole bacteria? Selection among the emerging eukaryotes would have centred on the improvement of the efficiency for endocytosis, i.e. predation.

The cytological corollaries of this hypothesis were discussed by Stanier in some detail (see also Echlin, 1970b). In prokaryotes, the cell membrane may undergo deep invaginations, e.g. in the photosynthetic bacteria. But apparently it has not enough plasticity for complete involution and for the intake of a large external object. In contrast, the plasticity of the eukaryotic cell membrane is sufficient for various endocytotic and exocytotic activities which are observable under the light microscope. External cell walls were developed later by some eukaryotes according to need, while the walls of the endosymbionts vanished.

The opponents of the symbiotic hypothesis among the botanists derive the chloroplasts from one or other part of a prokaryotic cell. An amazing view is that of Bell (1970): "The transition from prokaryote to eukaryote . . . could also have taken place more or less instantaneously as a consequence of physical principles identical with those that probably led to the first cell . . . We should thus have reached in one step, purely as the result of physical forces, the eukaryotic condition." This view is not really shared by the present author.

20

ENERGY SUPPLY OF PROTOZOA AND FUNGI

(a) Protozoa

It has been said that aerobic organotrophy could develop after the atmosphere became oxygenic through the activity of the blue–greens. The eukaryotic protists [*18a*] include undifferentiated plants, protozoa and fungi. They all are basically aerobes; there is little doubt that the anaerobes among them have, in contrast to the strict anaerobes among the bacteria (clostridia and methane formers; *7 b*, *e*), adapted later to anoxygenic environments (see Hungate, 1967).

Reference is made elsewhere [*18g*, *21a*] to the notorious difficulty of dividing protozoa cleanly from plants. Here we shall disregard green "protozoa". Of the many species of colourless protozoa, some are free-living. Others are parasites (see von Brand, 1966, 1972; Levine, 1972). As in prokaryotic organotrophs, dissolved nutrients may in many species be absorbed anywhere through the cell surface ("osmotrophy"). There are, then, no specialized organs for the absorption of such nutrients. Protozoa that live as intracellular parasites may not only use the various kinds of nutrients already available in the host cell, but moreover they may, through complicated and specific mechanisms, induce the cell to assist actively in their nutrition. In the words of Trager (1974), the parasite forces his host, like a bandit that has cajoled his way in, to prepare a banquet for him. Other protozoa are "phagotrophs", i.e. eaters and even predators (Huxley, 1876; Lwoff, 1951; Hutner, 1961; Hall, 1967; Manwell, 1968). Of course, organized food must be digested within the protozoan cell by suitable mechanisms.

Whether protozoa are osmotrophs or phagotrophs, they utilize substances of low molecular weight by a bewildering variety of anaerobic and aerobic processes (see Lwoff, 1951; von Brand, 1966, 1972; Danforth, 1967; Honigberg, 1967; Ryley, 1967). Some of the bioenergetic processes are known elsewhere, but others look distinctly unfamiliar. The fuller evaluation of this rich experimental material is awaited. In particular, it would be important to extract the facts of evolutionary importance from the data. Unfortunately, many protozoa multiply slowly, and only a few kinds have so far been grown in the absence of contaminating organisms (axenically; see Hall, 1967). A much investigated organism is *Tetrahymena pyriformis*, which is both an osmotroph and a phagotroph (Hill, 1972).

Some parasitic protozoa that live in an anoxygenic environment (rumen) are reported to be strict anaerobes, e.g. *Trichomonas termopsidis* (Trager, 1934). Others, e.g. *Chilomonas paramecium*, live best in oxygen of low tension (Pace and Ireland, 1945). Many further parasites live in presence of a great deal of oxygen (in air or in blood) so that they can obtain much energy by respiration, e.g. *Trypanosoma cruzi* (Stoppani and DeBoiso, 1973). Apparently at some stage a need for improved oxygen transport and storage arose. Haemoglobin-like oxygen carriers have been found in protozoa (Sato and Tamiya, 1937; Keilin and Ryley, 1953; see Lascelles, 1964b). They have even been reported in free-living (prokaryotic!) bacteria of the genus *Rhizobium* (Appleby, 1969).

(b) Fungi

The nutritive mode and way of life of the fungi, of course, also differ from those of the plants. They are always osmotrophs (Fig. 20.1). Large mycelia grow from single spores (Fig. 20.2). According to Whittaker (1969):

> So far as is known the fungi have been wholly nonphotosynthetic from their origin from unicells to their present diversity of forms. Fungi characteristically live embedded in a food source or medium, in many cases excreting enzymes for external digestion, but in all cases feeding by absorption of organic food from the medium. Their organization, whether mycelial, chytrid, or the unicellular of yeast, is adapted to this mode of nutrition.

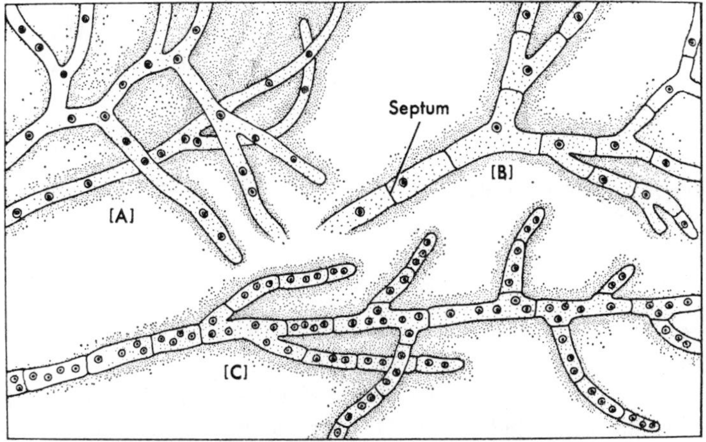

FIG. 20.1. Three types of fungal hyphae (Alexopoulos and Bold, 1967). A: Non-septate (coenocytic). B: Septate with uninuclear cells. C: Septate with multinuclear cells.

Many different organic substrates are utilized in fungal energy metabolism. The Embden–Meyerhof–Parnas, the pentose phosphate, and the Entner–Doudoroff pathways as well as other mechanisms for the anaerobic breakdown of carbohydrates occur (see Blumenthal, 1965; Burnett, 1968). The aerobic Krebs cycle usually operates, and oxidative phosphorylation takes place (see Lindenmayer, 1965). Haemoglobin-like oxygen transport proteins have been observed, as in protozoa [*20a*], in yeasts (Keilin, 1953) and moulds (Keilin and Tissières, 1953). A few kinds of strictly anaerobic fungi have been found (Emerson and Weston, 1967; Emerson and Held, 1969). No doubt these fungi have lost the capacity for aerobic life.

Various processes for the absorption of CO_2 exist (see Niederpruem, 1965). Species of *Fusarium* and *Cephalosporium* are reported to be capable of chemoautotrophic growth, with CO_2 as only source of carbon and with aerobic utilization of H_2 by a Knallgas reaction (*15d;* Mircha and DeVay, 1971).

For the time being, results with fungi appear to show that their energy metabolism is less diverse than that of protozoa. Maybe more research will change this view; on the other hand, a more standardized energy metabolism might indicate primary concern of fungi with

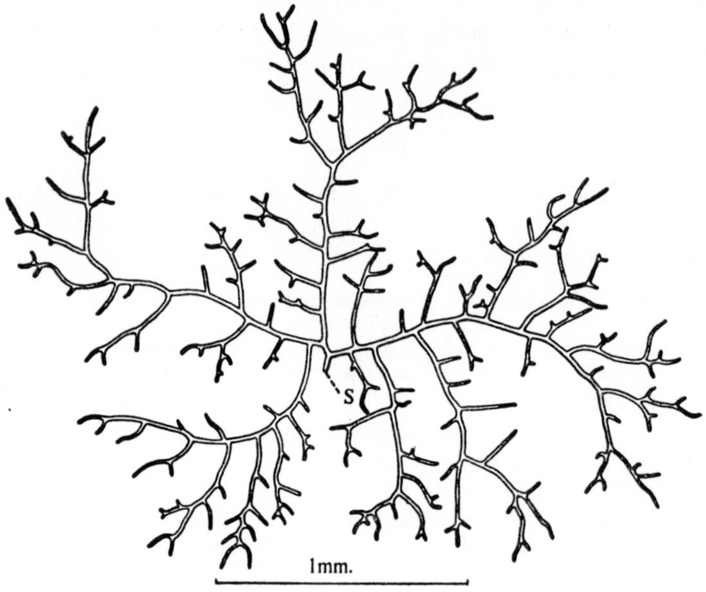

Fig. 20.2. Mycelium grown from a single spore of the primitive fungus *Phycomyces blakesleeanus* at 20°C in a day (after Ingold, 1969).

morphological and developmental aspects, as is the case with the higher plants (Van Niel, 1969).

Quite erroneously, the fungi were at times grouped with the bacteria, and this mistake is still reflected in the latters' name, schizomycetes (German (obsolete): *Spaltpilze*). In the confusion which prevailed before the advent of modern cytological and biochemical methods, the outstanding botanist von Wettstein (1935), for instance, came to the conclusion: ". . . ist es am zweckmäßigsten, die beiden Namen (of algae and fungi. E. B.) nicht im systematischen, sondern nur im biologischen Sinn zu gebrauchen, und mit dem Namen 'Algen' ganz allgemein die autotrophen, mit dem Namen 'Pilze' die heterotrophen Formen der ersten acht Stämme zu bezeichnen." Consequently, von Wettstein included among the "Pilze" not only the fungi, but also the bacteria. This has been the practice of some other authors, too, in a few cases quite recently.

True, most bacteria are, like fungi, heterotrophs and osmotrophs. Yet the eukaryotic fungi are clearly only most distantly related to the prokaryotic bacteria. But then, one may ask, why is it permissible, merely on the basis of identical energy metabolism, to group blue–green "algae" with eukaryotic, e.g. green or red, algae? Still, for the time being we are sticking to this terminology.

21

RELATIONSHIP OF THE BIOENERGETIC PROCESSES IN PLANTS, ANIMALS AND FUNGI

(a) Plants and Animals

Certainly transitions between plant and animal protists have taken place often. According to the Margulis version of the "symbiotic" hypothesis, eukaryotic organotrophs (protozoa) were formed first and eukaryotic plants later [*19e*]. The view is, of course, contrary to the old-established and still widely held idea that animals arose from plants. For instance, the order algae → protozoa has been accepted by Lwoff (1944, 1951), Pringsheim (1963 a, b, 1964), Stanier *et al.* (1966, 1970), Remane (1967), DeLey (1968), Klein (1970) and Stanier (1970). Also in a diagram drawn up by Cloud (1968b), incidentally a supporter of the endosymbiotic hypothesis and the ideas of Margulis, it appears that protozoa arose from algae by "leukosis".

The suggestion of a reversal, as proposed by Margulis, has become possible only through the conceptual separation of the prokaryotic plants (blue–green algae) from the other plants so that the former could provide the oxygenic atmosphere needed by the protozoa. Incidentally, the greater diversity of the "dark" bioenergetic processes among protozoa may be an argument for their precedence before plants.

The difficulty of a clean separation of protozoa from colourless or even coloured eukaryotic algae has been pointed out (*20a*; see Hutner and Provasoli, 1951; Lwoff, 1951; Whittaker, 1959; Pringsheim, 1963b). Thus, phycologists and protozoologists are warring for the possession of intermediate organisms, e.g. euglenids and trichomonads (Fott, 1959). Of course, many organisms are known that are clearly derived from plants, and yet have no chlorophyll. We still call these pale organisms plants. This is no doubt justified in the case of the higher plants with their elaborate morphology. But what is the position in respect to the unicellular eukaryotes?

In practice, many kinds of pale eukaryotic protists, obviously descended from plants, are also called plants. Other such organisms, where the kinship is less close, may be (and often are) called protozoa. In the Margulis form of the symbiotic hypothesis it is admitted that the ancestors of some, but by no means all, organisms known as protozoa temporarily owned (proto)chloroplasts as additional organelles. In her words "Although some fungal and protozoan genera probably did arise secondarily from nucleated algae that had lost their plastids, only these, and not all eukaryotes, as commonly thought, can be considered apochlorotic algae. The vast majority evolved directly from heterotrophs." There is, then, an alternative pathway for the appearance of protozoa.

Whether a given group ought, in fact, to be called colourless plants or protozoa is, to some extent, a matter of terminology. There are single-cell organisms that sometimes nourish themselves by photosynthesis, and at other times swim about digesting food particles (Margulis, 1970). For example, both the green *Euglena* and the pale *Astasia* (*18g*; see Jane, 1955; Buetow, 1968; Leedale, 1971), which are closely related, are often classed as protozoa. They have flagella, and they are surrounded by a proteinaceous pellicle rather than by a cell wall. Also some of the species, though not the green ones, are phagotrophs. The ambiguity of such organisms need not frighten us. On the basis of sound phylogeny such discipline breakers are understood. The problem should not be taken too seriously as long as the essence of the evolutionary position is clear.

Protozoa were, of course, the ancestors of the multicellular, higher, animals (metazoa). The algae, on the other hand, produced the higher plants (metaphyta). All higher plants appear to have in their chloroplasts similar accessory pigments, notably chlorophyll b, as have green algae in the narrow sense, the botanists' chlorophyta [*8c*]. On this and on other grounds it appears reasonable to derive the higher plants exclusively from the green algae (Fritsch, 1945; Banks, 1968, 1970; Swain, 1974). *Euglena* also has chlorophyll b, incidentally.

(b) Position of the Fungi

Formerly it was usual to derive the fungal protists, too, from algae, often polyphyletically. For references to the old views see Martin (1968). The idea is still maintained by some (Gäumann, 1964; Delevoryas, 1966; Klein and Cronquist, 1967; Mägdefrau, 1967; Fott, 1971; Metzner, 1973), but increasingly it is falling into disfavour (Martin, 1968; Swain, 1974). Alexopoulos and Bold (1967) took a somewhat cautious view:

> Structural resemblances, some of them superficial, between some fungi and some algae gave rise to the theory that the fungi originated from the algae by loss of chlorophyll. Present-day mycologists . . . however, are not convinced that this theory has any factual basis. On the contrary, the most popular current theory is that the fungi may have originated from a protozoan ancestor and may not, therefore, be related to green plants in any way. In the present state of our knowledge, there is, of course, no valid reason to assume that all fungi originated from a common ancestor . . . Some may have originated from a protozoan ancestor and others from a plant-like ancestor, perhaps some primitive algae which assumed a parasitic way of life and eventually lost its chlorophyll . . . The slime molds differ from . . . most fungi in their naked, amoeboid assimilative stages and in their holozoic mode of nutrition. They resemble the true fungi in producing walled spores, usually borne in characteristic fruiting bodies. Traditionally the slime molds are studied by the mycologists but they show at least as strong affinities with the protozoa as they do with the fungi.

Savile (1968) definitely invoked mere convergence to explain the similarities of algae and fungi:

> Most of the alga-fungus resemblances are such palpable examples of convergence that, even without the clues from physiology and biochemistry, they present no problem to the student of evolutionary mechanisms. Stagnant water, devoid of violent agitation, encourages the formation, in algae and fungi, of large coenocytic cells that would be prone to critical damage in many habitats.

Similarly, Hickman (1965) explains:

> . . . they [fungi. E. B.] have been regarded as being more like plants than like animals. Though they differ fundamentally from plants in being nonphotosynthetic, fungi have been grouped with them because of their, largely, non-motile habit, the presence of cell walls, and absorption of food materials in solution; and because of the theory, no longer generally held, that they have evolved from algae . . .

Many authors advocate derivation of fungi from protozoa, not always monophyletically (for references, see Martin, 1968). After emphasizing that algae and fungi do not intergrade,

Stanier *et al.* (1966) arrived at a diagram in which the fungi are derived from protozoa. They continued:

> It is essentially impossible to make a simple and clearcut separation between the protozoa and the fungi . . . Speaking very broadly . . . the protozoa are predominantly unicellular and motile, whereas the fungi are predominantly coenocytic and non-motile. However, there are exceptions in both groups.

Sequence studies with cytochrome c (McLaughlin and Dayhoff, 1973) confirm that higher fungi are more remote from higher plants than from higher animals. In the circumstances, there is no reason left to call fungi plants, and to assign mycology to botany departments. Fungi are neither animals nor plants (Kreisel, 1969). If it were true that fungi (and animals) had photosynthetic ancestors [*19g*] they would still, as the quotations show, be closer to the animals than to the plants. Consequently it also appears best to establish the higher fungi, descended from fungal protists, as a fifth kingdom on equal terms with the higher plants and the higher animals [*18a*].

While the bioenergetic processes of the fungi have not yet been correlated with their evolutionary position, it might be mentioned that biochemical features, notably cell wall composition (Bartnicki-Garcia, 1968, 1970; Léjohn, 1971) and pathways for lysine biosynthesis (Léjohn, 1971), are used as markers in evolutionary studies with fungi. The regrettable paucity of chemotaxonomic data about fungi has been pointed out by Sussman (1974) in an important survey.

Evolutionary trees were proposed for the higher fungi (Martin, 1968; Savile, 1968; Léjohn, 1971). According to Martin "the bulk of the fungi may be regarded . . . as constituting a very large, extremely variable, but on the whole remarkably coherent group of organisms, which may reasonably be treated as a discrete major taxonomic unit".

Addition in proof. R. Singer (personal communication; see also Singer, 1973, 1975) takes, in part, a different view:

> A first (needed. E.B.) step . . . is a redefinition of the word fungi. . . . A steadily growing number of competent mycologists restrict the fungi proper (Eumycetes) to Zygo-, Asco- and Basidiomycetes (including the corresponding imperfect and lichenized forms) and exclude the Myxo-"mycetes" and Mastigo-"mycetes" (Zoomycotina, Loquin, 1972), thus abandoning the now meaningless term "Phycomycetes". The evolutionary implication is that the excluded groups can be derived from protists close to protozoa, while the fungi proper contain at least one group (Laboulbeniales) whose close morphological and developmental similarities with some Rhodophyceae (red algae, E. B.) are difficult to ignore. Thus defined, the fungi proper, with or without the Laboulbeniales, certainly form a variable but homogeneous group with most probably a common ancestor which, with present methods and data, is difficult to identify. . . (A descent of fungi from algae is also proposed [Demoulin, 1974]. E.B.)

22

BIOENERGETICS OF TISSUES

(a) Differentiation and Energetics

Organisms that consist of well-differentiated cells are considered as higher organisms. Differentiation is the key to division of labour. Obviously, coherent assemblies of cells (tissues) organized for division of labour can tackle many problems in the struggle for existence better than a mere collection of individual, undifferentiated cells. Among these problems we may name movement, feeding, defence and propagation. The higher, tissue-forming, organisms include the higher animals (metazoa), the higher plants (metaphyta) and the higher fungi [*18a*]. Further, we find division of labour between several or many organisms, in families or communities, or, occasionally, among symbionts.

Following Haeckel, most biologists think that multicellular organisms evolved from colonies of protists. In contrast, some authors prefer the concept that the higher organisms arose through the formation of cell boundaries in multinucleate syncytical or plasmodial protists, followed by differentiation (Steinböck, 1937; Hadži, 1963; outlined by Cloud, 1968b).

In any case, the question is still open whether metazoa and metaphyta each arose monophyletically or polyphyletically. The latter possibility has been, for animals, sympathetically discussed by Cloud (1968b). For example, a separate origin from protozoa has been proposed for the mesozoa, extremely simple organisms consisting of few (20-30) cells only (Lapan and Morowitz, 1972). Another animal group, for which a separate descent has been considered (Cloud, 1968b), are the sponges (parazoa). These are relatively uncoordinated and poorly integrated.

However that may be, all existing higher organisms consist entirely of eukaryotic cells. Not impossibly at some time in the past prokaryotic cells embarked on more pronounced differentiation, and formed elaborate tissues. If so, such organisms have died out, and their remains have not yet been found. Eukaryotic cells have proved superior in their capacity to form efficient tissues.

Differentiation is, however, not an all-or-nothing principle. To a modest extent differentiation is observed already in bacteria (see Shapiro *et al.*, 1971). Another important example among prokaryotes is the formation of trichomes, which are physiological units, and of nitrogen-fixing heterocysts (see Fleming and Haselkorn, 1973) by blue-green algae. Among those groups of prokaryotes that are capable of a modest degree of differentiation and groups of eukaryotes remarkable morphological, physiological and ecological analogies have been found, just as exist between marsupial and placental mammals. Instances are the pairs actinomycetes–mycelial fungi, and the green bacterium *Pelodyctium*–the green alga *Hydrodyction* (Stanier, 1970). One of the most remarkable cases of convergence is provided by the (gliding) myxobacteria (prokaryotes!), on the one hand, and the acrasiales ("slime molds", probably best considered as amoebae; certainly eukaryotes!), on the other hand. Both of them may differentiate, in suitable conditions, to give fruiting bodies (see Cohen, 1967; Stanier *et al.*, 1970; Schlegel, 1972).

We have seen that all protists (according to Copeland-Whittaker [*18a*], by definition eukaryotes) have mitochondria, unless, in rare cases, they have lost them. The same is true of the cells of the higher organisms. Apparently a plentiful and reliable supply of oxygen and consequently of ATP made in respiration (oxidative phosphorylation) has been needed for the formation and the maintenance of tissues (Nursall, 1959; Gaffron, 1962b). Much information on the role of oxygen in higher organisms is found in the work edited by Dickens and Neil (1964). Warburg (1966) has called oxygen the "creator of differentiation". It has been remarked [*18c*] that normally for plants photophosphorylation is a more plentiful source of ATP than oxidative phosphorylation. Yet apparently by itself the source is not reliable enough to keep higher (or lower) eukaryotic plants going.

The need for oxygen in differentiated life is well illustrated in the slime molds. In their unicellular, vegetative forms they are facultative anaerobes, but for aggregation at the central collection points and for differentiation these organisms are obligate aerobes (see Smith and Galbraith, 1971).

At the basis of their capacity for respiration, all cells of all higher organisms still harbour the essentials of the ancient machinery for fermentation, notably glycolysis [*7c*]. The pentose phosphate pathway [*7f*] has also been maintained. Fermentation (by invertebrates) may in some cases yield other end products than lactate, notably succinate and alanine, and it has been contended that more ATP is in these processes generated than in lactic acid fermentation by substrate level phosphorylation (Hochachka and Mustafa, 1972).

The fundamental conservatism in the cellular bioenergetics of higher organisms is clearly due to the perfection that had been reached before differentiation began. But modifications in detail have probably been going on all along in respect to fermentation (as mentioned), to oxidative and to photophosphorylation. Increasingly, the powerful methods of modern biochemistry allow resolution of the differences in structure and function of the organelles for energy production between groups of higher organisms. But only rarely these differences can be correlated with the needs as they arise in the specific ways of life of the particular groups. Nevertheless, the analysis and comparison of the organelles, as found in the different groups, provide us with a fine additional tool for the construction of phylogenetic trees.

The eobionts and the organisms (cells) had to show their efficiency largely in the struggle for energy. They developed methods to get as much ATP as possible, first from the chemical energy of preformed organic substances, later by photosynthesis, and finally by the utilization of the products of photosynthesis in respiration. Higher organisms have concentrated on the methods for the efficient application of biochemical energy, admittedly often in the pursuit of more energy.

The more the division of labour was developed, the more important became intercellular and interorganismal communication and control. Hence for an understanding of more and more complicated systems, thermodynamics and kinetics must increasingly be supplemented by cybernetics, by applied systems analysis. Hormonal or neural mechanisms may be employed.

The present book is mainly devoted to cellular bioenergetics. Therefore it is not intended to deal *in extenso* with the effects of the division of labour within differentiated organisms on the localization and intensity of the energy supply processes among cells and tissues. Only a few examples will be given in this chapter to indicate the further lines of bioenergetic evolution among animals and plants. A very few data about the corresponding macrofossil record will be presented later [*23b*].

(b) Transition to Life on Land

Only rather late in evolution organisms colonized the land. Plants tend to go on land where the light intensity is far better than in water. One probable reason for the delay in the appearance of land plants was UV radiation. The difficulty presented to life in an anoxygenic biosphere by this radiation has often been pointed out. An atmosphere without free oxygen, and therefore without ozone, was largely transparent to short-wave UV.

Mainly because of its absorption by nucleic acid bases, this radiation is highly lethal to microorganisms, i.e. to organisms whose whole body is exposed to the rays. The unattenuated solar flux is enough to kill them within seconds (Sagan, 1957, 1961, 1965, 1973a; Rupert, 1964; Giese, 1964, etc.). Therefore, as long as the atmosphere was (almost) entirely anoxygenic, life could persist and evolve only when protected. Water absorbs short-wave UV fairly well (Hurlburt, 1928), and surely served as the shield (Sagan, 1957, 1973a; Wald, 1965). Suspended mud have helped to scatter and absorb harmful radiation. As Berkner and Marshall (1967) say:

> The environment must be protected by sufficient depth of water from lethal UV, but sufficiently shallow to maximize photosynthesis, with gentle convection to supply nutrients synthesized at the surface in the presence of UV radiation, but no violent convection that would dislodge and circulate primitive benthic organisms (having no advanced forms of control) toward the lethal surface. Thus primitive photosynthetic life must be restricted to shallow lakes or shallow protected seas.

As the oxygen content of the atmosphere increased, an ozone layer formed at a high altitude as a result of photochemical action [*25a.*] The ozone absorbed much of the UV, and so the thickness of water required for additional shielding decreased. Thus gradually large areas of shallow waters were opened up to life. Moreover, these waters, well exposed to the visible light needed in photosynthesis, were quite productive. Therefore the partial pressure of oxygen increased further rather rapidly [*25a*], and after a time, in Berkner and Marshall's view, organisms could renounce protection by a water layer altogether, and colonize the lands.

According to Fischer (1965), "soil plants" may have entered at first below-surface layers of the soil or other "shade oases", where UV did not reach well. Furthermore, not all organisms were equally sensitive to radiation. Some may (not only on land, but already in the water) have devised special mechanisms to protect themselves. One important way was to develop the remarkable mechanisms for the repair of DNA damaged by UV, which are so powerful even now (see Rupert, 1964; Hanawalt, 1968; Cook, 1970; Painter, 1970). (Incidentally, some of these repair mechanisms are helpful also against damage by ionizing radiation, although this radiation presumably was much less of a problem to life at any time during evolution. These are the "dark repair" processes, which act by "cutting and mending" of the damaged DNA strand along the second, healthy, strand as a template.) It has also been proposed (Sagan, 1973a) that some early organisms had a cover consisting of purines or pyrimidines as a shield against UV radiation.

Radiation damage was an extraneous obstacle to life on land. This is not to say that the factor was unimportant. But the transition from the waters to land must have been difficult in more fundamental ways, too, especially for the larger organisms, which are, incidentally, also less endangered by UV than microbes. As Keeton (1972) has well said, among the many problems faced by a land plant are the following:

1. Obtaining enough water when fluid no longer bathes the entire surface of the plant body.
2. Transporting water and dissolved substances from restricted areas of intake to other parts of the plant body, and transporting the products of photosynthesis to those parts of the plant that no longer carry out this process for themselves.
3. Preventing excessive loss of water by evaporation.
4. Maintaining a sufficiently extensive moist surface for gas exchange when the surrounding medium is air instead of liquid.
5. Supporting a large plant body against the pull of gravity when the buoyancy of an aqueous medium is no longer available.
6. Carrying out reproduction when there is little water through which flagellated sperms may swim and when the zygote and early embryo are in severe danger in desiccation.
7. Withstanding the extreme fluctuations in temperature, humidity, wind, light, and other environmental parameters to which terrestrial organisms are often subjected.

The land was surely invaded repeatedly by plants (see Chaloner, 1960; Chaloner and Allen, 1970). Probably primitive multicellular water plants had chances of success. They may have been at first sheet-like and encrusting, and later herbaceous (Schopf *et al.*, 1973). But only the vascular plants (tracheophytes) have really solved the problems, as the fossil record [*23b*] shows. They have, as pointed out by Keeton, evolved a host of adaptations that have enabled them to invade all but the most inhospitable land habitats.

The existing land plants may be descended from fresh water algae (Gessner, 1971), after they gained a foothold on wet lands. The argument is that land plants normally are not halophilic, and that they do not contain a great deal of sodium chloride. It is also clear that remaining NaCl, after evaporation of water, would have presented early land plants with most difficult problems. Schopf *et al.* (1973), in contrast, consider a marine origin for land plants. Their ancestors may have lived in a shelf-like setting and have developed anchoring devices to resist dislodgement by waves. However that may be, during the subsequent periods many kinds of vascular land plants have, of course, reentered the waters and have also learned to deal with NaCl. For instance, mangrove thrives in sea water. Such plants are vegetable equivalents of seals and dolphins.

The animals had their own physiological problems in the transition to life on land [*22e*]. But on top of these problems, they had to await the success of the land plants, which they clearly needed for food and also for shelter. From the beginning the fate of the metazoa must have been intertwined with that of the metaphyta (Schopf *et al.*, 1973). Again only highly differentiated organisms (animals) managed really well on land. Differentiated animals may at first have been pelagic suspension feeders, and subsequently lived in the "algal forest" (Schopf *et al.*, 1973). The success both of the metaphyta and the metazoa in the transition to life on land constitutes additional evidence for the superiority of organisms with extensive division of labour.

(c) Energetics of Plant Tissue

The arrangement of the energy-supplying organelles within the various organs of higher plants is described in the many treatises on plant physiology and anatomy. It might merely be emphasized here that in the metaphyta relatively thin leaf-like organs with large surfaces exposed to sunshine were a necessity. Such organs cannot be very strong mechanically, and they render movement of the whole plant difficult. This may be the main reason why higher plants, in contrast to animals, move only awkwardly, if at all. This stationary mode of life had the consequence that neither muscles nor nerves developed in them, a truly decisive fact of life.

For water conduction in large plants long tubes through dead tissue (xylem) are necessary, or else resistance to flow would be excessive (Gessner, 1971). The walls of these vessels had to be specially strengthened to resist turgor pressure by surrounding, living, cells: so lignin has been developed.

All eukaryotic plants have kept the mitochondria in evolution, although the plants with functional chloroplasts make most of their ATP by photophosphorylation [*18c*]. This also applies to plant tissues, i.e. to higher plants. Basically, structure and function of the mitochondria of higher plants are similar to those of animal mitochondria (see, for example, Beevers, 1961; Bonner, 1965; Hanson and Hodges, 1967; Lance and Bonner, 1968; Chance *et al.*, 1968; Packer *et al.*, 1970; Zelitch, 1971; Ikuma 1972; Storey, 1972). The P/O ratio for the "principal" pathway again seems to be 3, and the sites of phosphorylation also seem to correspond to those in animal mitochondria, to judge by the action of specific inhibitors. Like animal mitochondria, the plant mitochondria show respiratory control [*13b*].

In detail, however, differences are found in the principal pathway. For instance, the properties of the cytochromes and the flavoproteins differ somewhat from those in animal mitochondria. Moreover, additional pathways are observed in the plants and they may even surpass the principal pathway in quantitative importance. Above all, the mitochondrial respiration of plants has a cyanide (and CO)-insensitive component (see Bendall and Bonner, 1971). The importance of this component varies between species and tissues, and it also greatly depends on physiological conditions and development.

It appears that branching occurs in the region of the flavoproteins (Erecinska and Storey, 1970), and leads to a terminal oxidase different from cytochrome oxidase, and not affected by cyanide (Storey and Bahr, 1969). It is not yet clear to what extent phosphorylation takes place in this branch, if at all (Erecinska and Storey, 1970; Wilson, 1970). The difference between ATP production in this arm and in the ordinary pathway beyond the flavoproteins may make possible a role of the cyanide-insensitive pathway in metabolic control. The concentrations of ATP, AMP, etc., which are important modulators in enzyme systems [*6a*], and also heat production may be adjusted according to needs.

Haemoproteins considered as haemoglobins are found in higher plants, especially in the N_2 fixing tissue of legume root nodules (Bergersen, 1971; Bergersen *et al.*, 1972; see Ellfolk, 1972). Apparently they function as reversible oxygen carriers. Possibly these "leghaemoglobins" have evolved from the haemoglobin-like pigments in protists [*20 a, b*]. Alternatively they may have arisen anew; a polyphyletic origin for blood pigments (haemoglobin) in animals has been suggested [*22e*]. An evolutionary explanation will have to take account of similarities of the globin chains with those of myoglobin and the animal haemoglobins.

Quite possibly differences between the pathways of electron flow in chloroplasts from various higher plants, similar to the differences in mitochondria, can be found. Such differences have not yet been observed [*12b*], though morphologically the chloroplasts from different plants differ widely [*18g*]. But major functional differences certainly exist in respect to the assimilation and the loss of carbon during photosynthesis, the "dark" reactions. These differences greatly affect the efficiency of the plants for overall energy utilization. They may or may not show "photorespiration".

(d) Photorespiration

The reductive pentose phosphate (Calvin) cycle operates in the chloroplasts of all eukaryotic plants [*10e*]. Ideally, all carbon that enters the cycle is, ultimately as a consequence of the light reaction, converted to metabolites useful in body building or energy production. Some of the material is transferred sooner or later to the mitochondria, and serves there as substrate for ordinary "dark" respiration and thus for additional ATP production.

Dark respiration is possibly inhibited, and hardly promoted, by light (Marsh *et al.*, 1965; see Zelitch, 1971), though a preceding period of photosynthesis may be of help. In any case, mitochondrial respiration has relatively little importance in well-illuminated plants. However, in addition a process called "photorespiration" has fairly recently been found in prokaryotic and eukaryotic plants that functions only during illumination (see Zelitch, 1964, 1967, 1971; Goldsworthy, 1970; Jackson and Volk, 1970; Tolbert, 1971 a, b; Hatch *et al.*, 1971; Black, 1973; Björkman, 1973). In this curious process, a large part precisely of the newly assimilated carbon is lost by reoxidation to CO_2, possibly without generation of any ATP.

Photorespiration has been recognized first by a burst of CO_2 evolution after the end of illumination (Decker, 1955, 1959, 1970). Photorespiration is the reason why the compensation point of the plants affected is far higher than can be accounted for by dark respiration alone. The compensation point indicates the stationary (lowest) CO_2 concentration in the atmosphere above the plant that is reached during photosynthesis in strong light in a closed container. For example, the compensation point is 180 ppm at 35°C for wheat (see Zelitch, 1971). In contrast to dark respiration, which is practically unaffected by a reduction of atmospheric oxygen content to a small fraction of its normal value, photorespiration strongly decreases with decreasing oxygen pressure (Forrester *et al.*, 1966; Whittingham *et al.*, 1967; Björkman, 1971). The inhibition of net photosynthesis by oxygen had been discovered as early as 1920 by Warburg. Photorespiration also decreases with increasing CO_2 pressure. Photorespiration has a high temperature coefficient, and therefore is particularly important in warm climates.

The mechanism of photorespiration is still under discussion. It appears that part of the recent photosynthate is diverted, still within the chloroplasts, to glycolate (see Gibbs, 1971b). This may be derived from phosphoglycolate (see Richardson and Tolbert, 1961; Bassham and Kirk, 1973). In some algae, the glycolic acid is partly, depending on conditions, excreted as such (Tolbert and Zill, 1956; Bruin *et al.*, 1970; Tolbert *et al.*, 1971). To that extent, one cannot speak of photorespiration in a narrow sense, but the organism nevertheless suffers in net production of useful material. Another part of the glycolic acid is dehydrogenated in the algal cytoplasma, and later gives CO_2.

In higher plants, the glycolic acid is oxidized first (see Tolbert, 1963; Zelitch, 1964; Kisaki and Tolbert, 1969) to glyoxylic acid within peroxisomes [*18d*]. The glyoxylic acid is further partly converted, in a number of steps in the peroxisomes and elsewhere, to phosphoglycerate, which may enter the Calvin cycle in chloroplasts, while another part of the carbon is lost as CO_2 (Tolbert, 1963, 1971 a, b; Beevers, 1971; Gibbs, 1971b). This sequence of reactions is well represented, e.g. in wheat or spinach. Through the sequence, three-quarters of the carbon that went into glycolic acid are recovered.

The situation seems at first glance all the more puzzling as there are plants without photorespiration, and these plants thrive exceedingly well. In fact they include the most productive crop plants of warm climates, like maize and sugar cane; their high yields may well be due to the absence of photorespiration. These plants show no CO_2 burst after removal of light and have a compensation point close to zero (maize: 5 ppm; see Zelitch, 1971). However, it will be seen that at least some such plants nevertheless have photorespiration, though latently.

Interestingly, maize and sugar cane are also distinguished by their pathway of carbon assimilation (Kortschak *et al.*, 1965; Hatch and Slack, 1966, 1970; Hatch, 1971b; Downton, 1971). They fix CO_2 partly by reaction with phosphoenolpyruvate (PEP) rather than through the Calvin cycle, i.e. by reaction with ribulose diphosphate, RdP. The first fixation products are oxaloacetate and other C_4 compounds instead of the C_3 compound, phosphoglycerate. Hence the plants involved are known as "C_4 plants".

The C_4 plants have a characteristic structure. Two types of chloroplasts, unequally equipped with the Calvin and the Hatch-Slack enzymes, are found in different kinds of cells (Hatch and Slack, 1969; Laetsch, 1969, 1971, 1974; Black *et al.*, 1973). The relationship of the different mechanism may be elucidated through experiments in which the two kinds of cells and their chloroplasts are separated (Baldry *et al.*, 1968; Edwards and Black, 1971). It appears that photorespiration does take place (in the C_4 plants!) in the so-called bundle sheath cells, where also most peroxisomes are found (Newcomb and Frederick, 1971; Tolbert, 1971b). The bundle sheath is a single layer of dark green cells around the vascular bundle, rich in Calvin cycle enzymes. The mesophyll cells are interposed between the bundle sheath cells and the open air. They are rich in C_4 enzymes and poor in Calvin cycle enzymes, and have few peroxisomes only (Liu and Black, 1972). The light reaction is at least qualitatively similar in both kinds of cells and also similar to that in C_3 cells, though quantitative differences have been reported (see Boardman, 1971; Hatch *et al.*, 1971).

CO_2 from the air is mostly taken up into C_4 bodies in the mesophyll cells. Moreover, according to a working hypothesis the CO_2 set free in the bundle sheath by photorespiration is efficiently absorbed (recycled) by the mesophyll cells. The high affinity of phosphoenolpyruvate to CO_2 (see Walker, 1966) ensures efficient reabsorption. The C_4 compounds cannot, however, directly serve the biosynthesis of carbohydrates. They (e.g. malate) are transferred to the bundle sheath cells and converted according to $C_4 \rightarrow C_3 + CO_2$. The CO_2 enters the Calvin cycle, while C_3 compounds must return to the mesophyll for further duty (Hatch, 1971 a, b). A scheme is shown in Fig. 22.1. It implies, then, internal reversal of photorespiration within the plant. *Addition in proof*: According to Bahr and Jensen (1974), the velocity of fixation of CO_2 by RdP may, in contrast to previous views, be similar to that by PEP.

However, the reasons for the absence of net photorespiration may well be different in other C_4 plants; for instance, some plants have altogether few peroxisomes and only poor activity of the glycolate pathway enzymes (Osmond and Harris, 1971; Osmond, 1971). The diversity cannot surprise, as zero compensation has apparently arisen in warm countries, where photorespiration is particularly effective, polyphyletically (see Slatyer and Tolbert, 1971; Downton, 1971; Evans, 1971), i.e. convergent evolution obtains. Occasionally, some of the species of one and the same genus have predominantly fixation to C_4 compounds, and others to C_3 compounds, i.e. "tropical" or "temperate" characteristics (Downton *et al.*, 1969; Osmond and Harris, 1971).

The meaning of photorespiration is not yet clear (see Tolbert, 1971 a, b). It may serve regulation and control in some way. Naturally, photorespiration was non-existent as long as the Calvin cycle operated anaerobically. Nor did early plants, in an aquatic environment with little oxygen and a lot of CO_2, suffer greatly from photorespiration; only terrestrial

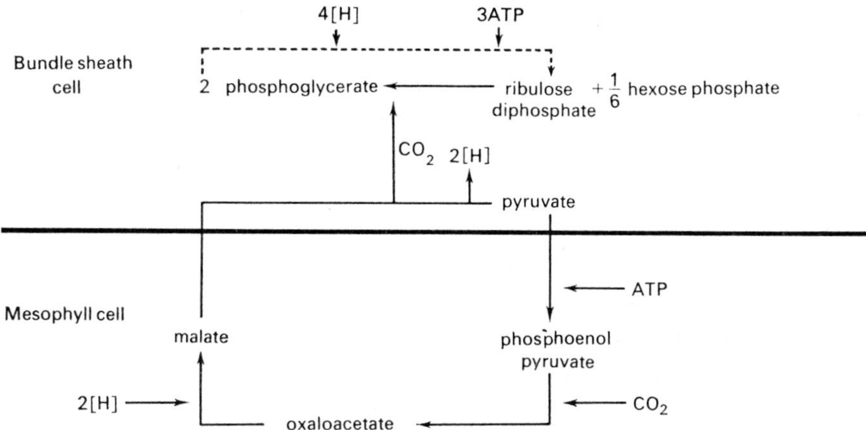

FIG. 22.1. Pathway of CO_2 assimilation in C_4 plants. Simplified representation of the ideas of the Australian school. In the bundle sheath, part of the carbon (not shown) is diverted as glycolate, and oxidized in peroxisomes. CO_2, one of the products of this process, is fixed again in mesophyll by phosphoenolpyruvate. In a parallel pathway (not shown), aspartate (also a C_4 compound) substitutes for malate. In the phosphorylation of pyruvate, ATP is converted to AMP.

plants, living in the recent atmosphere, rich in oxygen and poor in CO_2, and in warm climates, are strongly affected (Laetsch, 1969; Slatyer and Tolbert, 1971).

Goldsworthy (1969) has invoked interorganismic relationship to explain glycolate production. At early stages, excreted glycolate may have served to attract and feed heterotrophs that could supply useful metabolites. A related thought is that blue–greens, when they were in the process of being converted into chloroplasts, may have paid some of the entrance fee to their host in glycolate currency.

According to Lorimer and Andrews (1973), photorespiration is an inevitable side reaction in the C_3 pathway. Ribulose diphosphate, the CO_2 acceptor in the Calvin cycle, may be attacked by oxygen, and may therefore in part of the cases yield 1 molecule phosphoglycerate and 1 molecule phosphoglycolate instead of 2 molecules of phosphoglycerate. Both reactions of RdP may be catalysed by the same enzyme carboxydismutase. It is somewhat surprising that Nature should not have found it possible to modify the enzyme and in this way to avoid the useless side reaction.

Whatever the meaning (?) of photorespiration, the C_4 pathway may be a defence mechanism of relatively recent origin. While by itself it does not work as a cycle for assimilation, CO_2 from the bundle sheath is recaptured efficiently. Furthermore, capture of external CO_2 goes on even when the stomata of the leaves are, to save water, largely closed, and therefore admit CO_2 slowly only. The advantages of the C_4 pathway, in different conditions, have been discussed by Evans (1971). The change is not without cost, however. C_4 plants appear to need for the assimilation of each carbon atom, in addition to 2 NADPH, 5 ATP, while C_3 plants need 3 ATP only (Boardman, 1971).

The assimilation by C_4 plants shows similarities with that by "CAM" plants, i.e. by plants with "Crassulacean acid metabolism" (Walker, 1962, 1966; see Laetsch, 1969; Hatch *et al.*, 1971; Kluge, 1971; Ting, 1971; Zelitch, 1971; Black, 1973). Many succulents, e.g. pineapple, produce malate and other C_4 acids at night and accumulate it into the vacuole. The C_4 acids provide an internal carbon source in daytime when in a hot arid climate the stomata are closed. The malate is decarboxylated, and the carbon is used, with participation of NADPH and ATP obtained by photosynthesis, to make carbohydrate. It has been said that in a sense CAM plants separate in time what C_4 plants separate in space. Nothing is known about peroxisomes in CAM plants. For the polyphyletic origin of the CAM plants see Evans (1971).

(e) Energetics of Animal Tissues

Differentiation to produce metazoa required a fundamental revision of feeding methods, oxygen transfer and related anatomical features. Cloud (1968b) says: "In order to capitalize on the metabolic advantage of larger size combined with heterotrophic oxidative metabolism a metazoan characteristically requires a mouth and a digestive system, a circulatory system for oxygen distribution, and a nervous system to control these and related processes."

The mechanisms by which multicellular (invertebrate and vertebrate) animals ensure the efficient gas exchange needed for respiration have been explained in many treatises. Among them, Krogh's *Comparative Physiology of Respiration* (1941) is generally considered a classic. Recent books in the Krogh tradition, as already indicated in their titles, are those by Steen (1971) and by Jones (1972). These authors have been able to draw not only upon enormous new experimental material amassed by physiologists and zoologists, but also on results obtained by biochemists and molecular biologists on the oxygen carriers, notably haemoglobin and myoglobin. For the history of respiration research with higher organisms, Keilin (1966) and Florkin (1972) should again be consulted.

In every class of animals, the overall energy metabolism of different species will vary qualitatively and quantitatively. For instance, in mammals energy production per unit weight is inversely correlated with body size (see Kleiber, 1961). Moreover, each of the many different tissues within an animal has its own distinctive energy metabolism. Thus in the rat, the rates of respiration per milligram (Qo_2 values) vary by factors of 10 and more (see Aisenberg, 1961). Moreover, these tissues differ greatly (Aisenberg, 1961) in the relationship between fermentation and respiration, i.e. in their readiness for a Pasteur effect [*22f*].

Ultimately, in all animals known the intake of oxygen into cells occurs only by ordinary, passive diffusion, without energy-dependent (active) transport. But only very small animals, with diameters of the order of a millimetre, can cover their oxygen requirements without special organs. Many modern "diffusion animals" (Raff and Raff, 1970) are constructed as sheets with a thickness of less than a millimetre, e.g. flatworms (planaria) and, counting only metabolizing tissue, coelenterates, e.g. jellyfish and sea anemones. In early conditions of lower abundance of free oxygen, dimensions must have been smaller still. Any further increase in size necessitated the evolution of respiratory organs and a circulatory system. The first animals to have them were echinoderms (starfish, sea urchins).

These animals had, however, no circulating pigments as yet. Though pigments related to haemoglobin are observed in protozoa [*20a*], blood pigments are found first in annelids (segmented worms, e.g. earthworms). Higher animals generally have them and their quantity may be quite large; they may serve both transport and storage of oxygen. Thus rats contain about 250 times more haemoglobin than cytochrome (Drabkin, 1950). Haemoglobins appear, however, erratically, and may have arisen polyphyletically (Jones, 1972), though the precursors must have been related. Even fish without blood pigments exist, the antarctic icefishes (Ruud, 1954; see Steen, 1971). Surely they lost the pigments.

In molluscs and fishes, oxygen is taken in through gills. The large amounts of water that pass gills remove also a great deal of heat so that these animals are poikilothermic ("cold-blooded"). While some of the more advanced fish (tuna) do succeed, through use of a fine counter-current heat exchange mechanism, in maintaining relatively high temperatures in their muscles (Carey and Teal, 1966; Carey *et al.*, 1971), they are not truly warm-blooded.

A major change in the mechanisms for the uptake of oxygen was, of course, required in the transition to land. Pre-existing respiratory organs could be modified. Some of the mechanisms thus developed have proved their worth in limited groups of animals, some of which survive, but they have led these groups, from the evolutionist point of view, into blind alleys. Opportunity for important further evolution has been provided only by the introduction of the new organ, the lung. This fascinating story has been told well by Steen (1971). The exploitation of the lung principle in the end made possible the birth of truly warm-blooded animals, and ultimately of animals with efficient brains. According to Steen, a consequence is that man can write about fishes, and not vice versa.

Normally, respiratory electron flow in mitochondria serves the production of ATP, and hence uncoupling of phosphorylation from electron flow is considered as harmful. However, some kinds of mammalian tissue, namely, brown adipose tissue, have been adapted to serve as heaters ("thermogenic tissue"). Such tissue exists, e.g. in newly born mammals and also in grown-up rodents. The heat production of the brown fat tissue per unit weight may be 20 and more times larger than that of white tissue. It is regulated, as experiments with isolated mitochondria from the adipose tissue in different physiological conditions have shown, by way of a "loosely coupled state", and responds to biochemical stimuli (Skulachev *et al.*, 1973; Joel *et al.*, 1964; Brück, 1967; Prusiner and Poe, 1968; Christiansen *et al.*, 1969; Lindberg, 1970; Flatmark and Petersen, 1972, 1975; Hittelman *et al.*, 1974).

Uncoupling is, as may be noted in passing, rather a specialized mechanism for production of heat without work in warm-blooded animals. Shivering is, of course, another method for additional heating (see Hemingway, 1963). A further and very general and fundamental mechanism for the production of heat may consist in the stimulation by thyroid hormones of the pump for active Na extrusion, leading to additional ATP consumption. Relevant experiments were done on rat liver and muscle (Ismail-Beigi and Edelman, 1970). Increased pumping leads to increased back diffusion of Na, without resulting macroscopic work. It has been proposed that the process has evolved phylogenetically from the purposeful stimulation of muscle of lower vertebrates, namely, for the movement of the animal towards optimal external temperature (Stevens, 1973). One might say, then, that the pump was uncoupled from the organ it served, and the association of muscular activity with behavioural thermoregulation deleted. For a comparative physiology of heat production in various vertebrate groups see Whittow (1970, 1971).

Mechanisms to ensure a sufficient oxygen supply quite different from those of the vertebrates have been evolved by that immensely successful branch of the higher animals, the arthropods, including the insects (see Wigglesworth, 1972). The tracheal system, characteristic of them, has developed many times independently. There is no localized respiratory organ. Many small tubes, the tracheae, carry air directly to the cells, into which oxygen diffuses. Some, but not all, of the insects ventilate their tracheae by muscular contraction. This type of respiratory system did not allow insects to grow to very large sizes, for which we may be thankful (Keeton, 1972). The largest insects that ever existed were dragonflies of the Carboniferous period.

(f) Effects of Anaerobiosis

As the machinery for fermentation has been maintained in the higher organisms, the cells may in conditions of oxygen shortage return at least partly to fermentation. On this basis, many plants and animals that are normally aerobic resist oxygen shortage surprisingly well (see Siegel *et al.*, 1965). Of course, plants, supplying oxygen themselves, are expected to be less affected altogether, provided they obtain light.

It is a striking consequence of the capacity for fermentation that vertebrates can incur the well-known "oxygen debt" for maximum energy production during muscular exertions. The fast runner produces large amounts of lactate, without need for an immediate intake of an equivalent quantity of oxygen. Only subsequently the lactate is removed and largely reconverted to carbohydrate by means of energy obtained in respiration. But the fermentation disappears as soon as oxygen is abundant again. Closer investigation shows that fermentation is inhibited by oxygen. This has long been known to biochemists as the Pasteur effect (*13b*, *22e*; see Burk, 1939; Aisenberg, 1961; Racker, 1965, 1973; Krebs, 1972; Engelhardt, 1973) and has often been studied with simpler eukaryotes, namely, yeast. But it is already observed with facultative aerobes among prokaryotes, e.g. *E. coli*.

Many differentiated invertebrates that normally live at reduced oxygen pressure (mud dwellers and endoparasites, including annelids, helminths and molluscs) have come to like partial anaerobiosis; the preferred range of oxygen tension differs for various organisms (Fox and Taylor, 1955; von Brand, 1966, 1972). Thus fermentation has again, as with some protozoa [*20a*], permanently come to the foreground (Bueding, 1961; von Brand, 1966, 1972; Scheibel *et al.*, 1968; Fairbairn, 1970; Hochachka and Mustafa, 1972). But it has many times been shown that even "anaerobic" parasites maintain mitochondria, maybe modified, and remain capable of respiration. This applies to roundworms (Smith, 1969), to tapeworms (Cheah, 1971) and to the parasitic stage of blood flukes (Coles, 1972). It is still uncertain whether any of the parasites can, in the long run and at all developmental stages, survive in complete absence of oxygen (von Brand, 1966, 1972).

In any case, not all endoparasites are partial or complete anaerobes. Not only for protozoa [*20a*] blood provides an environment for aerobic life. Some parasitic worms are remarkably good at extracting oxygen from large volumes of blood (Tasker, 1961). Some endoparasites have become lazy. They excrete products of fermentation, even when supplied with ample oxygen. Evidently their Pasteur effect is deficient.

Various groups of vertebrates have adopted manifold measures to deal with prolonged shortage of oxygen in the environment, to store oxygen and to reduce consumption. Pioneers of the work in this field have been Irving (1939) and Scholander (1940). Of especial interest (to man!) are the diving vertebrates—for instance, turtles, seals and whales (Andersen, 1966, 1969; Lenfant, 1969; Steen, 1971; Jones, 1972). Most of these animals must from time to time breathe air. The most outstanding masters in submerged life are turtles, some of which can remain under water for many hours and more (Belkin, 1962, 1968). The turtle *Pseudemys scripta*, which can stay under water for weeks, deserves the rank of a vertebrate facultative anaerobe (Reeves, 1963; Storey and Hochachka, 1974). The turtles drastically reduce their aerobic energy metabolism and rely largely on fermentation, though the possibility of the uptake of some oxygen through the skin is not to be disregarded. During the long periods under water probably even brain and heart, known in other divers to get special allocations of air, go over to fermentation. As in other divers, the total rate of energy production in water is less than in air.

Addition in proof. A fine review of the metabolic consequences of diving in animals and man has now been published by Hochachka and Storey (1975).

23

PALAEONTOLOGICAL EVIDENCE

(a) Microfossils and Stromatolites

In the preceding chapters attempts have been made to deduce the path of bioenergetic evolution from the properties of extant organisms. This might be called the deductive approach. Obviously, it would be important to check the correctness of the results through induction, i.e. by unearthing, identifying and analysing fossils, and by determining the order in which they appeared. This would be particularly desirable for the prokaryotes, because they took the most important steps in bioenergetic evolution. A search for microfossils and other evidence in the most ancient sediments is needed. These results must be combined with those of radiometric determinations of absolute ages.

The investigation of the plentiful and diverse (macro-) fossils of higher organisms, of metazoa and metaphyta, has long been the task of palaeontology. In contrast, microfossils, though first reported nearly a hundred years ago, have not been more widely investigated until after the Second World War. Even now, only a few dozen useful sediments are known. References to the oldest papers, especially by J. J. H. Teall, by C. D. Walcott and by J. W. Gruner, are given by Schopf (1969, 1970b) and by Cloud and Licari (1972). The whole field of geomicrobiology, not only of the oldest sediments, has been surveyed by Moore (1969).

Most fruitful work began when Tyler and Barghoorn (1954, see Barghoorn and Tyler, 1965 a, b) found, by light microscopy, microfossils in the rich Middle Precambrian Gunflint formation in Ontario (age 2 Gy). Results obtained in this and in other formations, to be named below, have been reported, surveyed and discussed many times (Rutten, 1962; Glaessner, 1966; Welte, 1967; Schopf, 1967, 1969, 1970b, 1974; Cloud, 1968 a, b; 1969; Licari and Cloud, 1968; Swain, 1969a; Echlin, 1970a; Barghoorn, 1971; Lopuchin, 1975).

The fine preservation of the shapes in several formations (Fig. 23.1) is apparently due to the impregnation of the living cells by colloidal silica (see Schopf, 1969; Barghoorn, 1971) whereby the "chert" formed. Later the recrystallized silica preserved the microfossils without much structural alteration. Thus the silica is not of secondary origin. For conservation, the fact of this impregnation is most fortunate, but the conditions of life of the organisms may have been somewhat unusual (Schopf, 1970b), though Holland (1972) argues that generally sea water may have contained more silica before the advent of (eukaryotic) silica-using algae (see also Sillén, 1965, 1966, 1967). Not all species may have liked the silica. In model experiments, prokaryotic cells have been premineralized with silica, and the similarity of the products with natural microfossils has been noted (Oehler and Schopf, 1971).

Other sediments of the Early and Middle Precambrian bearing microfossils are the Soudan iron formation (Minnesota; >2.7 Gy). Still older are the sediments of the Fig

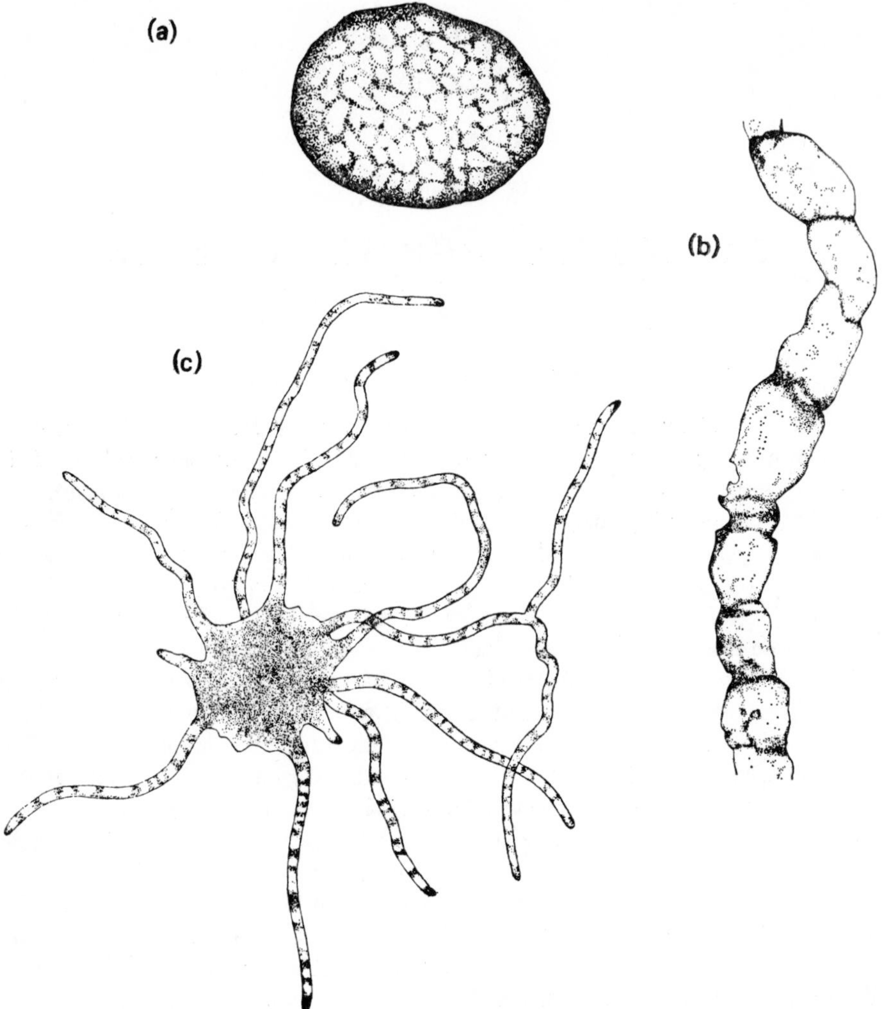

FIG. 23.1. Microfossils (Barghoorn, 1971). (A) is a spherical prokaryote and (B) a filamentous prokaryote from the Gunflint formation. They are called *Huroniosporia microreticulata* and *Gunflintia grandis* respectively. The filamentous *Eoastrion bifurcatum* (C) apparently has not left any descendants that look similar.

Tree series (3.2 Gy), the middle part of the Swaziland system in South Africa (Engel *et al.*, 1968). They are likewise fossiliferous, though the fossils are less well preserved and also less diverse than those of Gunflint (Barghoorn, 1971). Whether the spherical or, more often, cup-shaped bodies in the Onverwacht series that underlies the Fig Tree series are biogenic is still uncertain, though there is an increasing tendency to answer in the affirmative. The Onverwacht series is ascribed an age of 3.35 Gy (Hurley *et al.*, 1972; see Jahn and Shih, 1974) and is the oldest known sedimentary rock. The problems of the Swaziland system have been discussed recently by Barghoorn and Schopf (1966), Pflug (1967 a, b), Engel *et al.* (1968), Nagy and Nagy (1969), Scott *et al.* (1970), Brooks *et al.* (1973), Brooks and Shaw (1973) and Schopf (1974).

Similarities in form of fossil with extant cells are in many instances quite striking. In a number of suitable cases, the organisms have been classified by genera and species. Remarkable is the Precambrian genus *Kakabekia*, which first attracted attention among microfossils from Gunflint because of the curious shape of the cells, and only later was found alive in Wales and elsewhere (Siegel and Siegel, 1970). The living bacteria of this genus are facultative aerobes, presumably as a result of further evolution. Colonial prokaryotes have been reported in Amelia dolomite, North Australia, 1.6 Gy (Croxford *et al.*, 1973).

Some of the fossil bacteria have also been examined by electron microscopy (Cloud and Hagen, 1965; Schopf *et al.*, 1965; Barghoorn and Schopf, 1966; Oberlies and Prashnowsky, 1968; Schidlowski, 1970; Schopf 1970a). Nothing is known as yet about the biochemical processes in these bacteria, and especially we do not yet know which of them were fermenters or photosynthesizers.

From the point of view of bioenergetic evolution it is most interesting that not only bacteria, but also, to judge by morphology, blue–green algae are observed (see Schopf, 1970b; Barghoorn, 1971). This applies not only to Gunflint, but probably even to Fig Tree (Schopf and Barghoorn, 1967; Schopf, 1974). The blue–green algae have an unbroken record since their appearance, and their success in the long struggle for existence is expressed in their conservatism. Morphologically, fossil species often hardly differ from recent species. Filaments have been reported both from Gunflint and from Fig Tree (Schopf, 1974).

Thus the production of free oxygen can have begun earlier than 3 Gy ago. According to Schopf and Barghoorn (1969) the widespread distribution of blue–green algae as well as geological considerations (to which we shall turn below) point to the presence of a fairly high concentration of oxygen in the late Precambrian and make a direct correlation of the origin of multicellular organism with a rapid increase of the oxygen content at the end of the Precambrian questionable. From the presence of planktonic algae, which presumably floated near the surface, in Gunflint it has been concluded (Schopf, 1969) that an effective ozone shield existed as early as 2 Gy ago.

Whether any organisms in Gunflint were eukaryotes, is doubtful (Barghoorn, 1971; Schopf *et al.*, 1973). The first definite fossil eukaryotes, 1.2–1.4 Gy old, have been detected in Beck Springs, Eastern California (Cloud *et al.*, 1969). Further reports refer to the Belt series, U.S.A., aged more than 1 Gy (Pflug, 1965). Fine specimens of eukaryotes, 0.9–1 Gy old, have been found in the Bitter Springs area of Central Australia (Barghoorn and Schopf, 1965; Dunn *et al.*, 1966; Schopf, 1968, 1970b; Schopf and Blacic, 1971). Prokaryotes are found side by side with the eukaryotes, e.g. in Bitter Springs.

The eukaryotic cells have been identified as those of algae, some of them in mitosis, and, with less certainty, of aquatic fungi. No protozoon have as yet been found. Schopf (1970a) has applied electron microscopy to the eukaryotes; again, ultra-structural results are helpful in classification. The presence of the fundamentally aerobic eukaryotes in Beck Springs and Bitter Springs confirms that in those times the atmosphere was already oxygenic (Schopf, 1970b). This agrees with geological evidence, as mentioned. A tentative time-table of the appearance of the various classes of eukaryotic algae in the fossil record has been given by Schopf (1970b). Green algae were the first to appear. As reported before [*19g*], the green algae were probably also the ancestors of the metaphyta.

Further proof for the existence of prokaryotes in the Early and Middle Precambrian is provided by stromatolites, first described and so named by Kalkowski (1908). They are finely laminated sedimentary bodies, often many centimetres or even metres large (Cloud, 1968a). They still indicate in their structure the phototrophic response, and therefore are considered to have been built mainly by photosynthetic organisms. But traces of cells are left only rarely. Cherty stromatolites occur in the Gunflint formation, and calcareous stromatolites have been found, e.g. in the rock of Bulawayo, South Africa (MacGregor, 1940; Oberlies and Prashnowsky, 1968; Schopf, 1969, 1974; Schopf *et al.*, 1971). An age of the latter of > 2.9 Gy has been given (Bond *et al.*, 1973).

Most important components were probably blue–green algae that caught particles of sediment (silt) among their sticky filaments. Indeed the stromatolites resemble the "biohermal" deposits of modern blue–green algae. Well-preserved fossil remnants of filamentous, i.e. more advanced, blue–green algae have been observed in calcareous stromatolites from Transvaal (Nagy, 1974). Species of blue–green algae have also been identified in cherty stromatolites from Paradise Creek in Queensland (Licari and Cloud, 1972); no eukaryotes have been found in these structures, 1.6 Gy old.

Laminated mats, the raw material of stromatolites, have recently been shown to be formed also by *Chloroflexus* [*14g*]. If this filamentous organism really were a bacterium rather than a blue–green alga, this result would throw doubt on the assumption that the blue–greens are at least as old as the stromatolites (Walter *et al.*, 1972; Doemel and Brock, 1974). Of course, the evidence from the microfossils would remain valid.

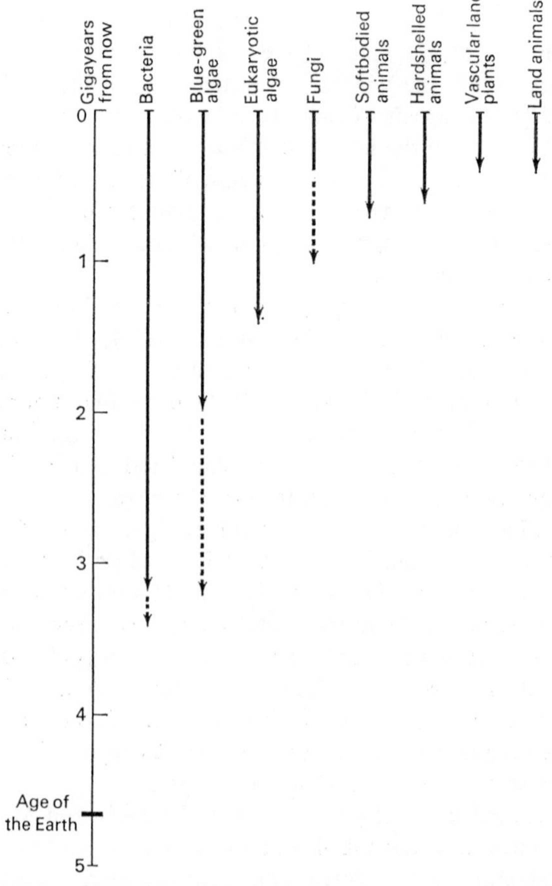

FIG. 23.2. Time of appearance of organisms, according to the palaeontological record. Broken lines indicate incertitude. No good information is available about protozoa.

The stromatolites changed in appearance during the Precambrian, probably owing to changes in the composition of the ecosystem. Therefore they are used, mainly in the Soviet Union, for stratigraphy; a critical survey of the literature has been given by Cloud and Semikhatov (1969). Stromatolites from the older times cover large areas (see Brock, 1973b). After the Ordovician, 0.45 Gy ago, they are found more seldom. This may be due to attack by grazing and burrowing animals (Garrett, 1970).

It used to be thought that reliable fossils date back to the base of the Cambrian only, 0.58 Gy ago. Thus seven-eighths of the past of the Earth were believed to be empty palaeontologically. Macrofossils have more recently been found in somewhat earlier sediments [*23b*]. The main thing is, however, that through the recent work on microfossils material traces of life can be followed back during a very much longer period. This period, previously considered as empty, is now sometimes called "Proterozoic", in contrast to the "Phanerozoic" period that comprises all time from which we have macrofossils. The emerging timetable (Fig. 23.2) gives grounds for surprise. The Fig Tree sediments are not much younger than the most ancient dated rocks on Earth [*3b*]. In particular, the great antiquity of the algae (Fig Tree: 3.2 Gy) is surprising. If the Earth as a whole is about 4.65 Gy old, at most $4.65 - 3.2 = 1.45$ Gy were available for the production of the secondary atmosphere and the primeval soup, for the evolution of the eobionts, of the non-photosynthetic and of the photosynthetic bacteria, including the photolithotrophs, and at last for the birth of the blue–green algae themselves.

(b) Macrofossils

The standard terminology for the periods of the geological past is shown in Fig. 23.3. It has often been suggested that metazoa did exist before the period now called Phanerozoic, but did not leave a distinct record (see Fischer, 1965). However, a thorough study of the problem led Cloud (1968b) to the double conclusion. (a) There are as yet no unequivocal records of metazoa in undoubtedly Proterozoic rocks, neither relics of organisms themselves, nor tracks, burrows, or after-death imprints. (b) There is no good reason why we should not find records of Proterozoic metazoa if they existed. Subsequently the many

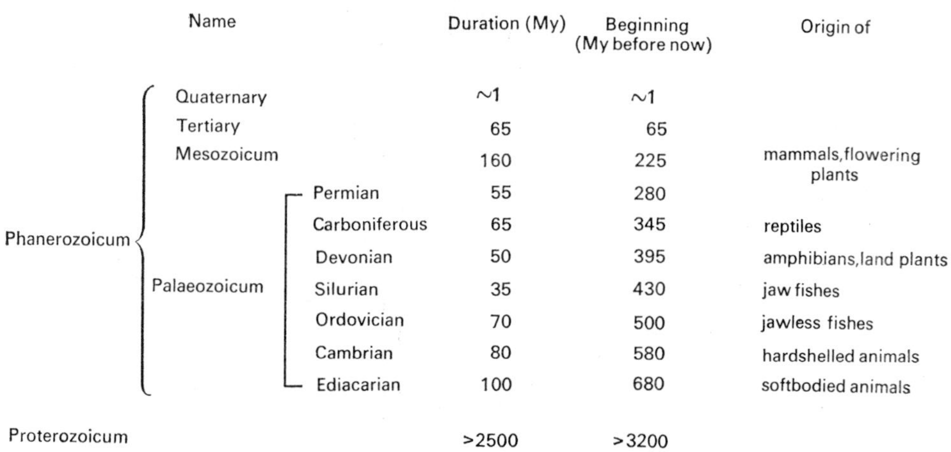

	Name		Duration (My)	Beginning (My before now)	Origin of
Phanerozoicum	Quaternary		~1	~1	
	Tertiary		65	65	
	Mesozoicum		160	225	mammals, flowering plants
	Palaeozoicum	Permian	55	280	
		Carboniferous	65	345	reptiles
		Devonian	50	395	amphibians, land plants
		Silurian	35	430	jaw fishes
		Ordovician	70	500	jawless fishes
		Cambrian	80	580	hardshelled animals
		Ediacarian	100	680	softbodied animals
Proterozoicum			>2500	>3200	

FIG. 23.3. Geological ages, as commonly dated.

reports of metazoan fossils from Proterozoic times were discussed (and discounted) individually by Cloud.

It has impressed observers that many different metazoan phyla appeared practically at the same time. The first period that left a rich macrofossil record was the early Cambrian, now thought to be 0.58 Gy old (Cloud, 1965, 1968 a, b, 1973; Fischer, 1965). The base of the Cambrian is defined by the lower limit of shelly fossils. An earlier assemblage of metazoa has fairly recently been found in the Ediacarian hills, Australia (age 0.68 Gy), and elsewhere (Sprigg, 1949; Glaessner, 1966, 1971; Cloud, 1968b, 1972, 1973; Pflug, 1971). It contains impressions of softbodied creatures: jellyfish, flatworms, corals, etc. The base of this system is now to be considered as the Proterozoic-Phanerozoic boundary. According to Cloud, the term Precambrian has become misleading, and should be abandoned. It does, however, linger on.

The metaphyta were already highly diversified at the beginning of the Cambrian (Schopf *et al.*, 1973). Among early fossil metaphyta, one finds bryophytes (mosses and liverworts), primitive plants without a vascular system. These plants colonized on land only moist places, they never developed to large sizes, and may be a side line (see Chaloner and Allen, 1970;

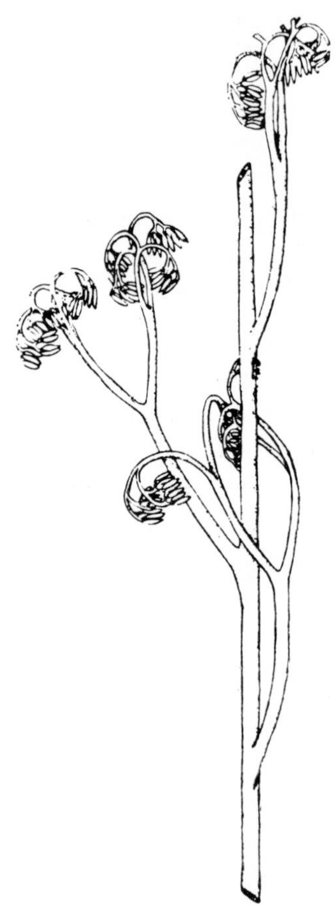

FIG. 23.4. Reconstruction of the early land plant, a fossil kind of *Psilophyton* (Banks, 1970). The leafless stem bore lateral branches that divided several times and terminated in pairs of elongated sporangia.

Gessner, 1971; Swain, 1974). The most successful line has been that from algae to the vascular plants [*22b*]. Some early vascular plants, including the psilophytes (Fig. 23.4), characterized by xylem, stomata, cuticle and waxy spores, appeared from the Silurian, 0.4 Gy ago (Banks, 1970b; Chaloner and Allen, 1970; Gessner, 1971; Swain, 1974; for a suggestion of much earlier land plants see Axelrod, 1959). In the Devonian (0.37 Gy) with its large forests, and in the Carboniferous period (0.3 Gy) further evolution of the land plants was rapid; lycopods, horsetails and ferns spread. The dry land apparently provided many opportunities for evolutionary radiation, once the basic methods for survival on land had evolved (see Ehrendorfer, 1972). In Swain (1974) useful tables on the evolution of higher plants are found.

The rapidity of evolution between Bitter Springs time and Ediacarian time has been attributed to the invention of sexuality by the eukaryotes (Schopf *et al.*, 1973, 1974). Some Cambrian animals had calcified exoskeletons; the great advantage of calcification at that stage has been emphasized by Schopf *et al.* (1973). Towe (1970) has suggested that increased ease of collagen formation, an aerobic process, in an atmosphere with increased oxygen content facilitated the enlargement of the musculature and the development of hard parts. But it is uncertain whether the oxygen content of the atmosphere did increase, and whether collagen synthesis was an aerobic process at that stage. However that may be, in the Ordovician, 0.45 Gy ago, all major invertebrates groups were already well represented. In the Silurian, fishes were common in the seas. Feeding on land plants, in the Silurian the first known land animals (scorpions) appeared, and it was in the Devonian that vertebrates (amphibia) also entered the land. Quite generally, the interconnection between the metaphyta and the metazoa during Phanerozoic times is remarkable, partly because of parallel evolution, and partly because of ecological relationships (Schopf *et al.*, 1973).

Microfossils identified with high probability as fungi have been found in the Bitter Springs sediments as mentioned, and fungi certainly existed in the Silurian (Schopf, 1970b). The fungi were varied and common in Lower Devonian time (Kidston and Lang, 1921; see Alexopoulos and Bold, 1967; Martin, 1968; Savile, 1968). Fairly early, fungi were abundant on land, often within the tissues of vascular plants (Banks, 1970). Also among the fungi a trend from moist towards moderately dry habitats has been observed.

24

GEOLOGICAL EVIDENCE

(a) Inorganic Palaeochemistry

Experimental (inductive) evidence about the oxidation state of rocks naturally must tie up with that obtained from estimates of the origin and development of the photophytotrophs.

Igneous rocks are not a good source of information about the atmosphere, as it was, as these rocks formed with only superficial contact or no contact at all with the gaseous envelope. We can learn about the composition of the atmosphere only from an analysis of erosion products and sediments (Rutten, 1962, 1969, 1971). Elements capable of different valencies, e.g. S, Fe and U, may occur there in reduced or in oxidized states. Though generally the most ancient sediments tend to be poor in oxygen (see Degens, 1968), as one would expect, some oxidized rocks have been reported even from early times (Rubey, 1955; Ramsay, 1963; Davidson, 1965; Rutten, 1970).

Really large amounts of partly or fully oxidized iron (typically magnetite, Fe_3O_4, and hematite, Fe_2O_3) appear in the curious and economically important "banded iron formations" (Lepp and Goldich, 1964; Govett, 1966). These are alternating layers of silica that are rich or poor in Fe. They are of marine origin and 3.2–1.9 Gy old (Cloud, 1965, 1968 a, b, 1969, 1972, 1974). Whether banded iron was deposited on a small scale in later periods is controversial (see Boyle and Davies, 1973; Holland, 1973). Microfossils are abundant in the banded iron of the Gunflint and even of the older Soudan formation (La Berge, 1967; Cloud and Licari, 1968). The oxygen for the banded iron was probably provided, somewhat patchily and irregularly at first, by the blue-green algae. The dissolved ferrous ion was gradually precipitated by the free oxygen in the form of ferric hydroxide, and in this way removed from the water. The rhythmic banding may result from a fluctuating balance between production and consumption of oxygen (Borchert, 1960; Cloud, 1968a; Holland, 1972). The removal of the oxygen may have prevented toxic effects of this gas on organisms that were not yet equipped with defence mechanisms, i.e. not yet prepared for it [*13a*]. On the other hand, the capture of the oxygen may also have delayed the elaboration of mechanism for respiration.

(The oxidation state of pyrite–uranium sands, contemporary with the banded iron, is lower (Ramdohr, 1958; Schidlowski, 1966). According to Rutten (1969, 1970), the reason is kinetic. Because of the specific place of these sands within the orogenic cycle, they were rapidly and largely mechanically eroded, and they also sedimented rapidly. Therefore they may have had less of a chance of being oxidized.)

"Red beds" (see Van Houten, 1973) are continental or marginal sediments, where fine grains of silica are coated with Fe oxides. The interpretation is that only after the dissolved iron was used up free oxygen could enter the atmosphere and oxidize iron compounds on the continents (Cloud, 1968 a, b, 1969, 1972, 1974). The end of large-scale banded iron does indeed coincide with the beginning of red beds, and the time of the transition from the anoxygenic to the oxygenic atmosphere has been put at 2.0–1.8 Gy ago. Older evidence about the transition time, obtained by Rankama, Ramdohr, Goldich and others, and consistent with Cloud's estimates, has been quoted by Rutten (1962).

It has been suggested *[11b]* that sulphate was formed by photosynthetic bacteria, presumably in the early Precambrian. Holland (1962) thinks that sulphide was gradually replaced, not necessarily as a result of bacterial activity, by sulphate when the Swaziland rocks were formed. Nevertheless, sulphate sediments are generally rare in the Precambrian (see also Urey, 1952 a, b; Hutchinson, 1954). The record of Precambrian sedimentary sulphates must be considered as both slim and moot (Cloud, 1968 a, b). Relatively large amounts of calcium sulphate are contained in the Grenville rocks in New York State from the Middle Precambrian (Engel and Engel, 1953; Silver, 1963, however, Cloud, 1968a).

(b) Organic Palaeochemistry

Much inductive evidence may be expected, in principle, from the investigation of organic compounds in sediments, i.e. from organic palaeochemistry (see Margulis, 1969). Whether the shapes of cells are still recognizable or not, biogenic compounds, either biomolecules or their metabolic products, may still be expected ("chemical fossils"). The tools of modern organic microchemistry are being applied to these problems: solvent extraction, molecular sieving, chromatography, various kinds of spectroscopy and mass spectrography. Surveys have been given by Hoering (1967), Calvin (1965, 1967, 1969), Eglinton and Murphy (1969), Albrecht and Ourisson (1971), Maxwell *et al.* (1971), Blumer (1973) and Eglinton (1973).

These investigations refer to organic matter in sedimentary inorganic rocks rather than to coals or oils or other organic bulk materials. It should not be forgotten, and is consistent with the newer views on early evolution, that hydrocarbons are found in Proterozoic strata, and are increasingly exploited there commercially (see Meinschein, 1969; Brooks and Shaw, 1973).

Numerous organic compounds, mostly water-insoluble and fat-soluble, have been identified in sediments of various ages (Eglinton *et al.*, 1964; Meinschein, 1965; Oró *et al.*, 1965; Oró and Nooner, 1967; see Calvin, 1965, 1967, 1969; Eglinton and Calvin, 1967; Eglinton, 1969; Maxwell *et al.*, 1971). Among these "geolipids" are alkanes, isoprenoids, aromatic compounds, pigments, including porphyrins (Baker, 1969; Kvenvolden and Hodgson, 1969) and carotenoids (Schwendinger, 1969). The isoprenoid alkanes pristane (2,6, 10,14-tetramethylpentadecane) and phytane (2,6,10,14-tetramethylhexadecane) presumably stem from the phytyl side chain of chlorophyll. They have often been found in Proterozoic sediments. There are also water-soluble substances, including carbohydrates (Swain, 1969b), aminoacids (Hare, 1969) and their degradation products.

It may be hoped that in the long run, with further refinement of the methods, criteria for the discrimination between different organisms will be developed on the basis of the chemico-fossil record: "palaeochemotaxonomy" or "molecular palaeontology" (see Chaloner and Allen, 1970; Calvin, quoted by Florkin, 1967). For example, among multicellular organisms lignin and flavonoids indicate vascular plants (Schopf, 1970b). Such compounds are biological "markers". Work with higher animals has been reported, e.g. by Abelson (1957), and a masterly survey of his own and other work given by Florkin (1971).

From our point of view, criteria for organisms using different bioenergetic processes are desired. For instance, if the hypotheses of this book are correct, porphyrins to their products ought not to appear syngenetically in sediments laid down before the advent of photosynthetic organisms. According to Schopf (1970b) there is chemical evidence for photosynthesis in Fig Tree.

Caveat: tremendous difficulties are ahead. One point is that abiogenic organic compounds are present within the crust [*4b*], as they are in some meteorites. Even petroleum might contain abiogenic components, as mentioned. Such compounds might be confused with products of living matter, but fortunately they will generally differ from them chemically. Moreover, the abiogenic material cannot be optically active.

The major difficulty is migration of biomolecules from one place to the other, especially from younger into older strata (see Schopf, 1970b). Soluble organic substance may move through pores and cracks. It may, then, not be syngenetic, and the presence of the markers may be misleading. The reality of this danger is strikingly demonstrated through the occurrence of certain aminoacids, e.g. serine, in very ancient sediments, in spite of their well-known chemical instability (Abelson, 1957; Hare and Abelson, 1968; Abelson and Hare, 1969; Schopf *et al.*, 1968; Kvenvolden *et al.*, 1969). Moreover, during the long periods in question the aminoacids ought to have been racemized, but sometimes they are still optically active (Abelson and Hare, 1969; Hare and Mitterer, 1969; Hare, 1969; Kvenvolden *et al.*, 1970b).

Indeed in samples from Gunflint and Bitter Springs far more biogenic material was found along surfaces than in the true interior (Smith *et al.*, 1970). Onverwacht chert (Nagy, 1970; Sanyal *et al.*, 1971; Brooks and Shaw, 1973) and other relevant rock (Smith *et al.*, 1970) are porous and permeable enough to account for the contents of soluble biogenic material through entry and transport.

So-called kerogen, the insoluble and non-volatile part of the organic material within the rock, is less likely to have been transported. Moreover, it often constitutes the major part of the organic material. The development of suitable methods for the reproducible degradation, analysis and interpretation of kerogen will certainly be of help (Robinson, 1969; Simmonds *et al.*, 1969; Scott *et al.*, 1970; Albrecht and Ourisson, 1971). Work on kerogen from the Onverwacht rock led Brooks and Shaw (1973) to consider it as derived from "sporopollenin", which is present in the outer coatings of pollens and arises by polymerization of fatty acid esters of carotenoids (Shaw, 1970). These authors therefore regard the existence of relics of higher plants in that rock as confirmed. The authors also consider the insoluble organic material in carbonaceous meteorites as chemically identical with terrestrial kerogen, and therefore as biogenic [*18a*]!

25

HISTORY OF ATMOSPHERIC OXYGEN

(a) Oxygen in the Secondary Atmosphere

Preplanetary matter with its abundance of hydrogen cannot, at temperatures low enough for chemical reactions, have contained free oxygen, i.e. it was anoxygenic. The secondary atmosphere of the Earth [*3c*] cannot have contained any free oxygen either as long as it was dominated by the reducing gases. We cannot be equally categorical about the oxygen content of the secondary atmosphere after the replacement of H_2, hydrocarbons and NH_3 by CO_2 and N_2. Nobody doubts that before the emergence of plant photosynthesis the free oxygen content was minute, compared to the present content. It is significant that no other planet has significant amounts of free oxygen. But how far from zero was the oxygen content in early times?

The only source of free, molecular, oxygen was the photolysis of water vapour in the high atmosphere by solar short-wave UV. The free hydrogen formed gradually escaped into space from the exosphere and left oxygen behind [*3c*]. But photolysis has an inbuilt brake, and is self-regulated (Urey, 1959). The radiation that produced free oxygen from water was also absorbed by the free oxygen so that the water was increasingly shielded from the rays. This effect was all the more pronounced as the partial pressure of the water, in contrast to that of oxygen, decreases with height more strongly than corresponds to the barometric equation. The water is frozen out by the cold trap in the top parts of the atmosphere.

On absorption of radiation by oxygen, species even more reactive than O_2 are produced, namely, atomic oxygen (O) and ozone (O_3). As long as there was little oxygen to begin with, the reactive species were formed near bottom level and reacted quickly with gases from volcanoes, at that time reducing [*3c*], and with reducing components of rocks, mostly with ferrous iron and with sulphides. So a steady state was set up.

Some authors have rather boldly estimated the amount of oxygen set free by photolysis or the stationary concentration of free oxygen, i.e. the concentration at the "Urey point" (Berkner and Marshall, 1964, 1965, 1967; Brinkmann, 1969; see van Valen, 1971). But there are so many unknown factors. It is hard to guess (1) how much water, largely as vapour, was accessible to the flux, the vapour pressure depending on temperature, and (2) which part of the hydrogen liberated escaped or else rereacted with oxygen (see Barth and Suess, 1960). But whatever the assumptions, the amount of photolytic oxygen was certainly far less than the amount of oxygen that is set free in photosynthesis now in an equal time.

For the time being, it will be better just to note that in the earliest sediments oxidized minerals, possibly to be ascribed to oxidation by free oxygen, are not common [*24a*]. For instance, ferric iron and sulphate are scarce. Moreover, blue–green algae, i.e. producers of free oxygen, have been identified in very old strata already [*23a*]. Indeed at present no time limit can be given for the first appearance of such organisms. Therefore even the rare examples of oxidized minerals in early sediments may have to be attributed at least in part to the action of phytotrophs. There is then no need at all to assume any appreciable concentration of free oxygen before plant photosynthesis arose.[1]

Furthermore, we can take as witnesses for an essentially anoxygenic past the organisms themselves. As Wald (1965; see also 1966) has said so well:

> That oxygen was absent during much of the Earth's early history is a crucial point, for it would be hard to imagine the accumulation of organic molecules over long ages of time if oxygen were present. . . . I think that living organisms themselves offer important testimony to the absence of oxygen from the atmosphere throughout their early history. It would be difficult otherwise to understand the ingenuity organisms displayed in performing almost all their oxidations (electron transfer, E. B.) anaerobically. It is as though organisms, having once learned ways of getting along without oxygen, persisted in these habits. Even now, with oxygen plentiful, the whole basic structure of metabolism, animal and plants, remains anaerobic.

It is meant that in the most important of all aerobic reaction sequences, the complete oxidation of metabolic hydrogen with coupled oxidative phosphorylation [*13d*], no step but one (the last) requires free oxygen rather than only water. This applies to all variants of the respiratory chain, in prokaryotes and eukaryotes.

The replacement of the anoxygenic by an oxygenic atmosphere had far-reaching consequences in respect to the relative supply of electrons and "holes" for electrons. Originally the organisms had to hunt for electron acceptors; witness, the fermenting anaerobes. Nowadays, in contrast, we are used to the thought that "fuel" (electron donors) is limiting, and precious. In other words, in the biosphere the need for oxidants has been replaced by the need for reductants.

(b) The Rise of Photosynthesis and of Respiration

According to Berkner and Marshall (1964, 1965, 1967), only living matter could break the steady state and overcome the Urey point. True, the additional oxygen from photosynthesis absorbed more and more of the UV radiation. But because of self-shielding the fraction of the existing free oxygen affected decreased. For the same reason, the altitude where the more reactive species were formed from molecular oxygen also tended to increase with increasing oxygen pressure, and the opportunity of these species to react with rocks diminished. The overall effect was that the oxygen pressure rose more and more rapidly as plant photosynthesis developed.

Berkner and Marshall have introduced the further concept of the "Pasteur point". This is the partial pressure of free oxygen at which it was worth while for organisms to switch over from fermentation to the more efficient energy generating process of aerobic respiration, i.e. where the Pasteur effect [*22f*] takes place. Berkner and Marshall assume that the Pasteur point was reached at a partial pressure of oxygen of the order of 0.2% (0.01 present atmospheric level—"PAL").

It is uncertain how soon after arrival at the Pasteur point aerobic respiration actually developed. But quite possibly no particularly long periods were needed for the evolution of respiration on a large scale. Conversion of the machinery for photosynthesis into that for respiration [*14e*] took place many times independently, not only among the blue–green algae, but also among the most diverse groups of photosynthetic bacteria, photoorganotrophic and photolithotrophic. So apparently the difficulties were not excessive, and the rate of conversion was fairly fast.

Berkner and Marshall have suggested that the further rise of oxygen pressure may precisely have been accelerated by respiration. Eukaryotic protists and, later, higher organisms need the vastly improved energy supply offered by respiration, which is secure also in the shade, at night and in winter [*21a*]. Respirers were better equipped to deal with the difficulties of life in inhospitable parts of the sea and ultimately on land [*22b*]. Now the oceans could be colonized further, and subsequently also terrestrial plants could greatly add to oxygen production.

(Rutten (1970) has, on the contrary, ascribed the Pasteur effect a stabilizing influence on the composition of the atmosphere. Excess oxygen, produced in photosynthesis, was now consumed not only by dead matter, inorganic and organic, but also in respiration. But apparently stabilization near the Pasteur point by this mechanism, if it existed, was by no means complete, as shown by the continuing rise of oxygen pressure.)

According to Berkner and Marshall (1967), the partial pressure of oxygen reached 10% of its present value at the boundary between Ordovician and Silurian, and may have overshot it during the Upper Carboniferous. Somewhat similar, but generally somewhat higher, values were offered by Rutten (1970). In contrast, Broecker (1970a) argues that no major change in oxygen pressure took place in Phanerozoic times, i.e. during the period that left a record of macrofossils (0.7 Gy). It is amusing to recall that Lord Kelvin thought a few dozen million years sufficient for all terrestrial evolution. His time scale was altogether a bit short [*3b*].

(c) The Stationary Concentration of Oxygen and Carbon Dioxide

The question arises how the present stationary composition of the atmosphere (21% oxygen) has been reached and is being maintained. One has the broad impression that the rate of oxygen consumption somehow caught up with the rate of oxygen production a fairly long time ago. The nature of the control mechanisms in the oceans and therefore in the atmosphere has been considered by Broecker (1970a) and by Holland (1972), to whom we must refer.

Oxygen consumption through any mechanism must increase with oxygen pressure. After most of the land had been covered by plants, organic matter, living or dead, has played the main role in oxygen consumption. When the cover possible with existing plants, limited by CO_2 pressure, light flux, water supply, temperature, etc., reached a ceiling, respiration by the plants themselves and by other organisms could catch up with photosynthesis. Respiration is not aided substantially by oxidation of rocks although new surface is exposed all the time in crustal processes.

The CO_2 content in the biosphere cannot fall below the compensation point [*22d*]. On the contrary, because most plants live at light levels far below saturation, the stationary content must be more than corresponds to the compensation point, probably much more. In fact, it is of the order of 300 ppm, far higher than the compensation point of the photorespiring plants (the majority of the plants) at prevailing temperatures. Compared with the rapid processes in living matter, the reaction of CO_2 with silicates, tending towards the Urey equilibrium [*3c*], cannot count for much in present conditions. Thus in respect to both O_2 and CO_2 absorption, life processes have the lion's share.

As far as we can limit ourselves to the activities of life and neglect permanent addition of O_2 and CO_2 to the biosphere, or permanent loss, the biosphere constitutes a closed system in respect to these substances. Their steady concentrations, not corresponding to chemical equilibria, are maintained dynamically throught input of free energy [*1f, 3c*].

(d) Carbon and Oxygen Inventory in the Biosphere

The amount of carbon in the biosphere in various forms has been estimated. The sediments are included, as they should be. They are not static, and take part in the processes of the biosphere. For instance, the half-life of shales by erosion may be 0.6 Gy (Garrels and Mackenzie, 1969), but new shales are laid down all the time.

According to Table 25.1, after Revelle and Fairbridge, where the carbon is measured in trillion (10^{18}) grams, the total amount of reduced carbon in the biosphere is 6.8×10^{21} g, or about 5.5×10^{20} moles. Degens (1968) has 3.8×10^{15} tons as "organic matter", probably corresponding to $(2 - 3) \times 10^{20}$ moles of elementary carbon, and Broecker (1970a) has 6×10^{20} moles.

The bulk of the reduced carbon of the biosphere is contained in sediments. Gehman (1962) found in limestone (Cambrian to late Tertiary) on an average organic matter corresponding to 0.2% carbon, and in shales from the same period a 5 times higher content. The silicates may retain organic matter by adsorption better than limestones; in recent sediments, the difference between limestone and silicate is much less. But in any case the bulk of the shales is 10 times more than that of the carbonate rock, i.e. limestone and dolomite (Correns, 1948; Borchert, 1951). The content of reduced carbon in shales may not have varied greatly during Phanerozoic times (Ronov, 1968).

TABLE 25.1. THE CARBON IN THE BIOSPHERE

Form	Amount of carbon (10^{18}g) (recalculated from Revelle and Fairbridge 1957)
Carbonate in sediment	18,000†,‡
Organic carbon in sediment	6800†,§
CO_2 in the atmosphere	0.65
Living organic matter on land	0.08
Dead organic matter on land	0.7
CO_2 as H_2CO_3 in ocean	0.22
CO_2 as HCO_3^- in ocean	31.3
CO_2 as CO_3^{--} in ocean	3.9
Living organic matter in ocean	0.008
Dead organic matter in ocean	2.7

† According to Ronov (1968), only 18% of the sedimentary carbon are in organic form.

‡ From data of Hoefs (1965), a value of 6000 is computed.

§ For different values by other authors, and by himself, see Broecker (1970a).

The amount of sedimentary carbonate is enormous, and most of it is found in younger sediments. An explanation (Rubey, 1951, 1955; Cloud, 1968a) is that the CO_2 arrived only gradually on the surface of the Earth through outgassing, and could thereafter be converted to carbonate through the Urey equilibrium [*3c*].

Attempts have also been made to correlate the carbonate of the sediments with that of the "igneous" rock, which customarily includes metamorphic rock. This "igneous" rock contains often 0.02 %, the metamorphic rock 0.1 % as carbon of carbonate, in any case much less than corresponds to the average value in sediments, all counted (1.3 %). Yet it should not be argued that the difference has been contributed by juvenile CO_2, as a large and unknown part of the "igneous" rock consists of sediments that were remelted and lost carbon in the process. Incidentally the content of reduced carbon (graphite, carbides, etc.) in the "igneous" rock (typically 0.02 %; Hoefs, 1965) is quite appreciable, though much less than in sediments (0.55 %). This latter figure has been computed from the mass of the sediments (Borchert, 1951; see below) and the amount of reduced carbon in them (Table 25.1). An unknown part of the reduced carbon of the "igneous" rocks, too, may be derived from sediments. Another part is no doubt original, just as there is reduced carbon (0.04 %) in stony meteorites (Noddack, 1937).

However that may be, about one-half by order of magnitude of all the carbon of the crust is held by "igneous" rock. According to Borchert (1951) the crust to a depth of 16 km contains 1.2×10^{18} t of sediments, but 21×10^{18} t of "igneous" rock. The greater bulk of the latter about compensates for the greater CO_2 content of the former (Borchert, 1951). Schidlowski (1971), also quoting Wedepohl and Ronov, has rather similar values.

An estimate of the turnover time of carbon in photosynthesis is possible. According to Table 25.1 the amount of carbon available for photosynthesis in the short run (C as CO_2, bicarbonate and carbonate in air and water—"pool of available carbon") is 36×10^{18} g. Values for the annual yield of photosynthesis scatter widely. Thus Bolin (1970) has 7.5×10^{16}, and Fairbridge (1972) 13.5×10^{16} g C (compare also Steemann-Nielsen, 1960; Broecker, 1970b; Whittaker, 1970). With Bolin's value the mean residence or turnover time of the element in the pool of available CO_2 would be of the order of 500 years only, with Fairbridge's value correspondingly less.

Incidentally, the presumed share of marine plants in global photosynthesis has gone down from 85 % (Hutchinson, 1954; Revelle and Fairbridge, 1957) to 53 % (Bolin, 1970) or even to 30 % (Whittaker, 1970) or 22 % (Fairbridge, 1972). More estimates of marine productivity are due to Steemann-Nielsen (1960), Ryther (1969) and Vishniac (1971), and a survey to Morris (1974).

The smallness of the turnover times imply, of course, that the annual contribution of volcanism to biospheric CO_2 in recent times has been enormously less than that of bio-processes. If the total amount of biospheric carbon, including sediments, i.e. the amount due to degassing, were uniformly distributed over the time of 4.66 Gy [*3b*], the annual addition of new carbon would work out as 0.0005×10^{16} g only. In reality, this addition must be much less now than in earlier periods.

Be it noted that with an inventory of reduced carbon in the biosphere, including sediments, of 6.8×10^{21} g and a yearly rate of photosynthesis of 7.5×10^{16} g the total reduced carbon in the biosphere corresponds to no more than the production or consumption during about 90,000 years, not much compared with the age of the Earth! Even the carbon laid down as carbonate in sediments (which need not ever have been part of organisms) does not correspond to more than 240,000 years, if the ratio of carbonate and organic carbon given by Revelle and Fairbridge is taken as basis. In other words, the present rate of carbon reduction and oxidation by living matter is extremely fast in comparison with the resources of the biosphere.

If past history were very crudely, to get the order of magnitude, approximated by a constant rate of photosynthesis (Bolin's value) through 1 Gy, this would mean that only 1 part in 10,000 of plant matter was converted into fossil carbon in sediments, and preserved.

Only a fraction of this is, of course, sufficiently rich to be classed as fuel; figures will be given below.

By comparing Bolin's annual rate of photosynthesis (2×10^{17} g O_2) with the atmospheric reservoir of free oxygen (1.2×10^{21} g), a mean turnover time of oxygen in the air of 6000 years is found. The residence time of oxygen in available "carbonate" is again, of course, 500 years, as is that of carbon.

Directly, the oxygen liberated in photosynthesis is derived from water rather than from CO_2 [*12a*]. Therefore, the water takes part in the photosynthesis—respiration cycle, and it makes sense to compute, on the basis of the present rate of photosynthesis, the average residence time of the oxygen atoms within the water of the biosphere, the amount of which is 1.4×10^{24} g. Taking Bolin's rate of photosynthesis, the surprisingly short residence time of about 2×10^7 years is obtained. From this simplified model of three uniform pools for mobile oxygen it is concluded that oxygen spends on an average 0.0025% of its time as CO_2, 0.03% as O_2, and all the rest of its time as water.

Attempts have been made to correlate the amount of reduced carbon in the biosphere, including the sediments, with the amount of free oxygen. At first glance, the quantity of the former (5.7×10^{20} moles) far exceeds that of the latter (4×10^{19} moles). The "missing oxygen" could be accounted for through the oxygen content of oxidized sediments (Hutchinson, 1954; Holland 1962, 1964; Broecker, 1970a; Cloud, 1972). Unfortunately, any agreement must be looked at as accidental. Above all, the crust, including the secondary atmosphere produced by outgassing, initially contained a lot of reduced carbon [*3c*]. No oxygen can correlate with this carbon.

In so far as oxygen derived from plant photosynthesis is ultimately taken up by rocks, the overall action of the plants can be considered as a light-powered oxidation of rock by CO_2. Typically, we may write for standard conditions

$$4FeO + CO_2 + H_2O = 2Fe_2O_3 + (CH_2O);\ \Delta G_0 = 6 \text{ kcal.} \qquad (25.1)$$

and

$$0.5H_2S + CO_2 + H_2O = H^+ + 0.5SO_4^{2-} + (CH_2O);\ \Delta G_0 = 28 \text{ kcal.} \qquad (11.5 = 25.2)$$

These processes are not necessarily irreversible, as many organisms can reduce ferric iron [*13a*] and sulphate [*16a*].

(e) Influence of Man

The quantity of minable coal in the world is estimated at only 7.6×10^{18} g (Averitt, 1971), and that of recoverable oil (three-fifths of it in tar sands and oil shales) at 0.6×10^{18} g (Hubbert, 1971). These figures include reserves still to be discovered. Up to 1970 the cumulative world production of coal has been 0.14×10^{18} g, of oil 0.04×10^{18} g, corresponding to about 2 and 6.5%, respectively, of the economic reserves, known and anticipated (Hubbert, 1971).

These amounts are trifling compared with an estimated amount of 6.8×10^{21} g of reduced carbon in sediments. But they are not small compared with the amounts involved in photosynthesis and respiration. This is a consequence of the fact that the pool size of available carbon is so much smaller than the store of reduced carbon in sediments. The minable or recoverable fossil carbon corresponds to the yield in photosynthesis in only about 100 years. Of course, it also corresponds to the total energy of the sunlight that comes to the Earth in a very much shorter time—a few weeks.

What is the influence of civilization on the composition of the atmosphere? Man consumes fossil carbon at the rate of 3×10^9 tons $= 3 \times 10^{15}$ g per year. If consumption continued at this rate and there were no compensating influences, the pool of available carbon in the biosphere (36×10^{18} g) would double in about 10,000 years; at a rate expanding as now, in a very much shorter time (Revelle and Suess, 1957; Zimen and Altenhein, 1973). Because of a time lag in the transition of atmospheric CO_2 to the ocean and into organic matter (see Revelle and Suess, 1957; Münnich, 1963; Zimen and Altenheim, 1973) CO_2 pressure must rise more rapidly than corresponds to the whole pool of available carbon. On the other hand, increasing photosynthesis must act as a brake. Excess CO_2 will, of course, greatly affect any ecosystem.

The actually observed changes in the atmosphere due to man have been discussed by Sawyer (1972). The increase in the CO_2 content is unmistakable already now. No decrease in oxygen content has been observed (Machta and Hughes, 1970), or is anticipated for the near future, as the base-line is so much higher than with CO_2 (see Ryther, 1970; Broecker, 1970b). If all fuel reserves (8.2×10^{18} g) were burnt, they would still consume only 1.8 % of the atmospheric oxygen (1.2×10^{21} g). Broecker (1970b) computed that even the oxygen stores in the broad expanses of open ocean are not threatened by Man's activities. How increased CO_2 pressure will affect the climate is another question (see Sawyer, 1972). The "greenhouse" effect will warm up the biosphere, but there will also be, acting in the opposite direction, increased evaporation of water, changed cloud cover, and larger albedo.

A few words might be added here about the utilization of solar energy in agriculture. A theoretical limit to harvests is given by the quantum requirement in photosynthesis. With a minimum requirement of 9 quanta per atom of carbon [*12e*], the optimum energy yield in the biosynthesis of carbohydrate, namely, with algal suspensions, is for green light (500 nm) about 20 %, if 1 "mole" of (CH_2O) is worth 115 kcal. For practical farming, optimum values of the order of 2 % have been given, but referred to light in the useful part of the solar spectrum and during the vegetation period only (see Talling, 1961; Gates, 1971; Loomis *et al.*, 1971).

Rabinowitch and Govindjee (1969) estimate the overall average photosynthetic yield for the production of reduced carbon on the whole Earth, again compared to the radiation energy that is potentially useful (excluding the 50 % contributed by infrared) and is actually absorbed by plants, as 1 %. A very much smaller value is computed for the used fraction of the total solar energy hitting the Earth. With Bolin's (1970) value for the annual production of reduced carbon (7.5×10^{16} g), and a solar energy flux of 1.7×10^{14} kW (solar constant 1.92 gcal/min $cm^2 = 1340$ W/m^2; maximum cross-section of the Earth 1.27×10^{18} cm^2), a yield of 0.05 % is obtained. The difference must be due to reflection into space, to absorption by non-plant matter and to the share of the infrared in energy.

Hopes for improved utilization of solar energy by means of plants centre on advances in social organization, farming practice and plant breeding. Naturally, the production of plants without photorespiration [*22d*] is a goal of more promise. Wheat, rice and soy bean would be the most important cases.

Almost the only energy exploited by man that is not of solar origin is nuclear energy; it might be called a kind of cosmic energy. The record of the nuclear energy industry is certainly impressive. However, in the long run and with ever-increasing energy requirements, it may become difficult to manage the nuclear reactors, the factories for the regeneration of nuclear fuel and the facilities for the storage of radioactive wastes. Especially the administration of the long-lived radioactive wastes and of the even more long-lived plutonium require a stability of the institutions of human civilization over enormous periods that is unheard of (Weinberg, 1972). Plutonium is the strongest radioactive poison of practical importance, and also a raw material for nuclear weapons.

Great concentration of technology on the tremendous untapped possibilities of solar power appears indicated. It is a real scandal that so little effort and finance, compared with nuclear energy, is spent on the development of solar energy.

Most experts on solar energy think in terms of heat, or they work on the semiconductor cells, which are, and probably will remain, immensely expensive. It would be more in keeping with the traditions of mankind and of our ancestors in evolution to explore the possibilities of the photolysis of water (see, for example, Daniels, 1972; Broda, 1974; Calvin, 1974).

After all, it has been the achievement of the phytotrophs, the green plants, to reduce the hydrogen of water to the electrochemical level of free hydrogen. This is then contained in the reduced ferredoxin. Only secondarily the hydrogen is used by the plant to assimilate CO_2. It is quite conceivable that biogenic (enzyme-containing) systems could be built up that release the hydrogen instead of applying it as a reactant. Steps in that direction were indeed taken by several authors, e.g. lately, by Benemann *et al.* (1973), who noted that some hydrogen is evolved on illumination of a mixture of spinach chloroplasts with hydrogenase from *Clostridium kluyveri* and with ferredoxin. But in the long run it may be better to learn rather than to borrow from Nature and to construct hydrogen-evolving systems in a well-considered and planned way, i.e., partly or entirely from man-made components.

The essence of the man-made system would be, then, that by means of the membrane principle, evolved by the modest photosynthetic bacteria more than 3 gigayears ago, reductant (hydrogen) and oxidant (oxygen) are set free in separate compartments. In this way, the back reaction of primary photochemical products *in statu nascendi* would be prevented.

The hydrogen obtained should then be used as a technical fuel and as a metallurgical and chemical reductant. Some of the hydrogen could also be employed for the direct production of food.

Food production could be achieved in abiotic industrial processes. Alternatively, the hydrogen could be fed to hydrogen bacteria [*15d*]. This would be a biotic carbon reduction process in competition with plant photosynthesis. The reduction would, however, not be directly coupled with the photosynthetic reaction. The resulting increased flexibility constitutes an important technical advantage. The reduction and assimilation of the carbon could be carried out at times and in places where no sunlight, or not much, is available.

The continued existence of some uranium 235 on Earth, the only ultimate source of fission energy, is a cosmic accident. If the half-life of that nuclide were by chance considerably smaller than it is (7×10^8 years) no fission power plants could be built at all any more. In contrast, electromagnetic radiation from the stars, including our Sun, is the essence of the world we know.

REFERENCES

AARONSON, H. and HUTNER, S. H. (1966) *Quart. Rev. Biol.* **41,** 13.
ABELSON, P. A. (1953) *Carnegie Inst. Yearbook* no. 53, Washington.
ABELSON, P. A. (1957) *Ann. N.Y. Acad. Sci.* **69, 276.**
ABELSON, P. A. (1966) *Proc. Nat. Acad. Sci., Wash.* **55,** 1365.
ABELSON, P. A., Ed. (1967) *Researches in Geochemistry*, New York, Vol. 2.
ABELSON, P. A. and HARE, P. E. (1969) *Carnegie Inst. Yearbook* 1967–68.
ACKRELL, B. A. C. and JONES, C. W. (1971) *Eur. J. Biochem.* **20,** 22, 29.
ADAIR, F. W. (1966) *J. Bact.* **92,** 899.
AGENO, M. (1972) *J. Theor. Biol.* **37,** 187
AHMADJIAN, V. (1963) *Scient. Amer.* **208** (2), 122.
AHMADJIAN, V. (1967) *The Lichen Symbiosis*, New York.
AHMADJIAN, V. (1970) in: Dobzhansky (1967), Vol. 4.
AHMADJIAN, V. and HALE, M. H. Eds. (1973) *The Lichens*, New York.
AHRENS, L. H., PRESS, F., RANKAMA, K. and RUNCORN, S. K. (1957) *Physics and Chemistry of the Earth.*
AHRENS, W. and SCHLEGEL, H. G. (1972) *Arch. Mikrobiol.* **85,** 142.
AINSWORTH, G. C. and SUSSMAN, A. S., Eds. (1965) *The Fungi*, New York, Vol. 1.
AISENBERG, A. C. (1961) *The Glycolysis and Respiration of Tumors*, New York.
AKABORI, S. (1957) in: Oparin *et al.* (1957).
AKAGI, J. M. (1964) *J. Bact.* **88,** 813.
AKAGI, J. M. (1967) *J. Biol. Chem.* **242,** 2478.
ALBERTY, R. A. (1969) *J. Chem. Educ.* **46,** 713.
ALBRECHT, P. and OURISSON, G. (1971) *Angew. Chem.* **83,** 221.
ALEEM, M. I. H. (1965) *J. Bact.* **90,** 95.
ALEEM, M. I. H. (1966a) *Biochim. Biophys. Acta.* **113,** 216.
ALEEM, M. I. H. (1966b) *J. Bact.* **91,** 729.
ALEEM, M. I. H. (1966c) *Biochim. Biophys. Acta.* **128,** 1.
ALEEM, M. I. H. (1968) *Biochim. Biophys. Acta.* **162,** 338.
ALEEM, M. I. H. (1970) *Ann. Rev. Plant Physiol.* **21,** 67.
ALEEM, M. I. H. (1972) *J. Gen. Microbiol.* **69,** IV.
ALEEM, M. I. H., LEES H. and NICHOLAS, D. J. D. (1963) *Nature*, **200,** 759.
ALEEM, M. I. H. and NASON, A. (1960) *Proc. Nat. Acad. Sci., Wash.* **46,** 763.
ALEXANDER, M. (1971) *Microbial Ecology*, New York.
ALEXOPOULOS, C. J. and BOLD, H. C. (1967) *Algae and Fungi*, New York.
ALFVÉN, H. and ARRHENIUS, G. (1972) *The Moon*, **5,** 210.
ALLEN, G. (1957) *Amer. Naturalist*, **91,** 65.
ALLEN, J. E., GOODMAN, D. B., BESARAB, A. and RASMUSSEN, H. (1973) *Biochim. Biophys. Acta*, **320,** 708.
ALLEN, J. M., Ed. (1967) *Molecular Organization and Biological Function*, New York.
ALLSOPP, A. (1969) *New Phytol.* **68,** 591.
ALSTON, R. E. (1966) in: Swain (1966).
ALSTON, R. E. (1967) in: Dobzhansky, Hecht and Steere (1967), Vol. 1.
ALTMANN, R. (1894) *Die Elementarorganismen und ihre Beziehungen zu den Zellen*, Leipzig.
ALVING, R. and LAKI, K. (1972) *J. Theoret. Biol.* **34,** 199.
AMESZ, J. (1963) *Biochim. Biophys. Acta*, **66,** 22.
AMINUDDIN, M. and NICHOLAS, D. J. D. (1974) *J. Gen. Microbiol.* **82,** 103, 115.
ANDERS, E. (1962) *Rev. Mod. Phys.* **34,** 287.
ANDERS, E. (1963) in: Middlehurst and Kuiper (1963).
ANDERS, E. (1964) *Space Science Rev.* **3,** 583.
ANDERS, E. (1969) *Naturwiss.* **56,** 180.
ANDERS, E. (1970) *Science*, **169,** 1309.
ANDERS, E. (1971) *Ann. Rev. Astron. Astrophys.* **9,** 1.
ANDERS, E., DUFRESNE, E. R., HAYATSU, R., DUFRESNE, A., CAVAILLÉ, A. and FITCH, F. W. (1964) *Science*, **146,** 1161.

ANDERS, E., HAYATSU, R. and STUDIER, M. H. (1973) *Science*, **182,** 781.
ANDERS, E., HAYATSU, R. and STUDIER, M. H. (1974) *Origin of Life*, **5,** 57.
ANDERSEN, B. and USSING, H. H. (1960) in: Florkin and Mason (1960), Vol. 2.
ANDERSEN, H. T. (1966) *Physiol. Rev.* **46,** 212.
ANDERSEN, H. T., Ed. (1969) *The Biology of Marine Mammals*, New York.
ANDERSON, C. A. and HINTHORPE, J. R. (1973) *Geochim. Cosmochim. Acta.* **37,** 745.
ANDERSON, D. L., SAMMIS, C. and JORDAN, T. (1971) *Science*, **171,** 1103.
ANDERSON, E. (1967) in: Chen (1967), Vol. 1.
ANDERSON, E. S. (1966) *Nature*, **209,** 637.
ANDERSON, L. and FULLER, R. C. (1967) *Plant Physiol.* **42,** 487, 491, 497.
ANDERSON, R. L. and WOOD, W. A. (1969) *Ann. Rev. Microbiol.* **23,** 569.
ANDREESEN, J. R., ELGHAZZAWI, E. and GOTTSCHALK, G. (1974) *Arch. Mikrobiol.* **96,** 103.
ANDREESEN, J. R. and GOTTSCHALK, G. (1969) *Arch. Mikrobiol.* **69,** 160.
ANDREESEN, J. R., GOTTSCHALK, G. and SCHLEGEL, H. G. (1970) *Arch. Mikrobiol.* **72,** 154.
ANDREESEN, J. R., SCHAUPP, A., NEURAUTER, C., BROWN, A. and LJUNGDAHL, L. G. (1973) *J. Bact.* **114,** 743.
ANDREWES, C. H. (1965) *Bact. Rev.* **29,** 1.
ANDREWS, I. G. and MORRIS, J. G. (1965) *Biochim. Biophys. Acta.* **97,** 176.
ANFINSEN, C. B. (1959) *The Molecular Basis of Evolution*, New York.
ANTHONY, C. (1975) *Biochem. J.* **146,** 289.
APPLEBY, C. A. (1969) *Biochim. Biophys. Acta*, **172,** 88.
ARCHAKOV, A. I., DEVICHENSKY, V. M. and KARJAKIN, A. V. (1975) *Arch. Biochem. Biophys.* **166,** 295, 308, 313.
ARNON, D. I. (1956). *Ann. Rev. Plant Physiol.* **7,** 325.
ARNON, D. I. (1959) *Nature*, **184,** 10.
ARNON, D. I. (1961) *Bull. Torrey Bot. Club*, **88,** 215.
ARNON, D. I. (1965) *Science*, **149,** 1460.
ARNON, D. I. (1966) *Experientia*, **22,** 1.
ARNON, D. I. (1967a) *Physiol. Rev.* **47,** 317.
ARNON, D. I. (1967b) in: Goodwin (1967).
ARNON, D. I. (1967c) *Nature*, **214,** 562.
ARNON, D. I. (1969a) *Naturwiss.* **56,** 295.
ARNON, D. I. (1969b) in: Metzner (1969), Vol. 3.
ARNON, D. I. (1971) *Proc. Nat. Acad. Sci. Wash.* **68,** 2883.
ARNON, D. I., ALLEN, M. B. and WHATLEY, F. R. (1954) *Nature*, **174,** 394.
ARNON, D. I., KNAFF, D. B., MCSWAIN, D. B., CHAIN, R. K. and TSUJIMOTO, H. Y. (1971) *Photochem. Photobiol.* **4,** 397.
ARNON, D. I., LOSADA, M., NOZAKI, M. and TAGAWA, K. (1961a) *Nature*, **190,** 601.
ARNON, D. I., LOSADA, M., WHATLEY, F. R., TSUJIMOTO, H. Y., HALL, D. O. and HORTON, A. A. (1961b) *Proc. Nat. Acad. Sci., Wash.* **47,** 1314.
ARNON, D. I., TSUJIMOTO, H. Y. and MCSWAIN, B. D. (1964) *Proc. Nat. Acad. Sci., Wash.* **51,** 1274.
ARNON, D. I., TSUJIMOTO, H. Y. and MCSWAIN, B. D. (1965) *Nature*, **207,** 1367.
ARNON, D. I., WHATLEY, F. R. and ALLEN, M. B. (1958) *Science*, **127,** 1026.
ARRHENIUS, G. (1974) quoted by Cloud (1974b).
ASANO, A. and BRODIE, A. F. (1965) *J. Biol. Chem.* **240,** 4002.
ASANO, A., COHEN, N. S., BAKER, F. R. and BRODIE, A. F. (1973) *J. Biol. Chem.* **248,** 3386.
ASHWELL, M. and WORK, T. S. (1970) *Ann. Rev. Biochem.* **39,** 251.
ASTON, F. W. (1924) *Nature*, **114,** 786.
ATKINSON, D. E. (1969) *Ann. Rev. Microbiol.* **23,** 47.
ATKINSON, D. E. (1971) in: Greenberg (1967), Vol. 5 (Ed. Vogel, H. J.).
ATKINSON, M. R. and MORTON, R. K. (1960) in: Florkin and Mason, (1960), Vol. 2.
AUBERT, J. P., MILHAUD, G. and MILLET, J. (1957) *Ann. Inst. Pasteur*, **92,** 516.
AVERITT, P. (1971), quoted by Hubbert (1971).
AVRON, M. (1967) *Curr. Topics Bioenergetics*, **2,** 1.
AVRON, M. (1971) in: Gibbs (1971).
AVRON, M., KROGMANN, D. W. and JAGENDORF, A. T. (1958) *Biochim. Biophys. Acta*, **30,** 144.
AVRON, M. and NEUMANN, J. (1968) *Ann. Rev. Plant Physiol.* **19,** 137.
AXELROD, B. (1967) in: Greenberg (1967), Vol. 1.
AXELROD, D. I. (1959) *Evolution*, **13,** 264.
AYALA, F. J. (1974) *Amer. Scient.* **62,** 692.
BAAK, J. M. and POSTMA, P. W. (1971) *FEBS Letters*, **19,** 189.
BAALSRUD, K. (1954) *Symp. Soc. Gen. Microbiol.* **4,** 54.
BAALSRUD, K. and BAALSRUD, K. S. (1954) *Arch. Mikrobiol.* **20,** 34.

BAARS, J. K. (1930) Thesis, Delft, quoted by Kluyver and Van Niel (1956).
BAAS-BECKING, L. G. M. (1925) *Ann. Bot.* **39**, 319.
BAAS-BECKING, L. G. M. and PARKS, G. S. (1927) *Physiol. Rev.* **7**, 85.
BACHOFEN, R., BUCHANAN, B. B. and ARNON, D. I. (1964) *Proc. Nat. Acad. Sci., Wash.* **51**, 690.
BADASH, L. (1968) *Proc. Amer. Phil. Soc.* **112**, 157.
BAHR, J. T. and JENSEN, R. G. (1974) *Plant Physiol.* **53**, 39.
BAILLIE, R. D., HOU, C. and BRAGG, P. D. (1971) *Biochim. Biophys. Acta*, **234**, 46.
BAKER, B. L. (1971) *Space Life Sciences*, **2**, 472.
BAKER, E. W. (1969) in: Eglinton and Murphy (1969).
BALDRY, C. W., COOMBS, J. and GROSS, D. (1968) *Z. Pflanzenphysiol.* **60**, 78.
BALDWIN, E. (1937) *Introduction to Comparative Biochemistry*, Cambridge.
BALL, G. H. (1969) in: Chen (1967), Vol. 3.
BALTSCHEFFSKY, H. (1967a) *Acta Scand.* **21**, 1973.
BALTSCHEFFSKY, H. (1967b) in: Goodwin (1967).
BALTSCHEFFSKY, H. (1969a) Federation of European Biochemical Societies Meeting, Madrid.
BALTSCHEFFSKY, H. (1974a) *Origin of Life*, **5**, 387.
BALTSCHEFFSKY, H. (1974b) in: Dose *et al.* (1974).
BALTSCHEFFSKY, H. and ARVIDSON, B. (1962) *Biochim. Biophys. Acta*, **65**, 425.
BALTSCHEFFSKY, H., BALTSCHEFFSKY, M. and THORE, A. (1971) *Current Topics Bioenerget.* **4**, 273.
BALTSCHEFFSKY, M. (1967) *Nature*, **216**, 241.
BALTSCHEFFSKY, M. (1968) *BBA Libr.* **11**, 277.
BALTSCHEFFSKY, M. (1969) *Arch. Biochem. Biophys.* **133**, 46.
BALTSCHEFFSKY, M. and BALTSCHEFFSKY, H. (1963) in: Gest *et al.* (1963).
BALTSCHEFFSKY, M. BALTSCHEFFSKY, H. and VON STEDINGK, L. V. (1966a), *Brookhaven Symp. Biol.* **19**, 246.
BALTSCHEFFSKY, M., VON STEDINGK, L. V., HELDT, H. W. and KLINGENBERG, M. (1966b) *Science*, **153**, 1120.
BANDURSKI, R. S. (1965) in: Bonner and Varner (1965).
BANKS, B. E. C. (1969) *Chem. in Britain*, **5**, 514.
BANKS, B. E. C. and VERNON, C. A. (1970) *J. Theor. Biol.* **29**, 301.
BANKS, H. P. (1968) in: Drake (1968).
BANKS, H. P. (1970) *Evolution and Plants of the Past*, Belmont, Calif.
BARGHOORN, E. S. (1971) *Scient. Amer.* **224** (5), 30.
BARGHOORN, E. S., MEINSCHEIN W. G. and SCHOPF, J. W. (1965) *Science*, **148**, 461.
BARGHOORN, E. S. and SCHOPF, J. W. (1965) *Science*, **150**, 337.
BARGHOORN, E. S. and SCHOPF, J. W. (1966) *Science*, **152**, 758.
BARGHOORN, E. S. and TYLER, S. A. (1965a) *Science*, **147**, 563.
BARGHOORN, E. S. and TYLER, S. A. (1965b) in: Mamikunian and Briggs (1965).
BARKER, H. A. (1935) *Cell. Comp. Physiol.* **7**, 73.
BARKER, H. A. (1936) *Cell Comp. Physiol.* **8**, 231.
BARKER, H. A. (1956) in: *Bacterial Fermentations*, New York.
BARKER, H. A. (1967) *Biochem. J.* **105**, 1.
BARKER, H. A., RUBEN, S. and KAMEN, M. D. (1940) *Proc. Nat. Acad. Sci., Wash.* **26**, 426.
BAR-NUN, A. (1975) *Origin of Life*, **6**, 109.
BAR-NUN, A., BAR-NUN, N., BAUER, S. H. and SAGAN, C. (1970) *Science*, **168**, 470; **170**, 1001.
BAR-NUN, A., BAR-NUN, N., BAUER, S. H. and SAGAN, C. (1971) in: Buvet and Ponnamperuma (1971).
BAR-NUN, A. and TAUBER, M. E. (1972) *Space Life Sciences*, **3**, 254.
BARTH, C. A. and SUESS, H. E. (1960) *Z. Phys.* **158**, 85.
BARTHOLOMEW, J. W. and WITTWER, T. (1952) *Bact. Rev.* **16**, 1.
BARTLEY, W., KORNBERG, H. L. and QUAYLE, J. R., Eds. (1970) *Essays in Cell Metabolism*, London.
BARTNICKI-GARCIA, S. (1968) *Ann. Rev. Microbiol.* **22**, 87.
BARTNICKI-GARCIA, S. (1970) in: Harborne (1970).
BARTSCH, R. G. (1968) *Ann. Rev. Microbiol.* **22**, 181.
BASSHAM, J. A. (1963) *J. Theor. Biol.* **4**, 52.
BASSHAM, J. A. (1964) *Ann. Rev. Plant Physiol.* **15**, 101.
BASSHAM, J. A. (1971a) *Science*, **172**, 526.
BASSHAM, J. A. (1971b) *Proc. Nat. Acad. Sci., Wash.* **58**, 2877.
BASSHAM, J. A. and CALVIN, M. (1957) *The Path of Carbon in Photosynthesis*, Englewood Cliffs.
BASSHAM, J. A. and CALVIN, M. (1961) 5th Int. Congr. Biochem., Moscow.
BASSHAM, J. A. and JENSEN, R. G. (1967) in: San Pietro *et al.* (1967).
BASSHAM, J. A. and KIRK, M. (1962) *Biochem. Biophys. Res. Commun.* **9**, 376.
BASSHAM, J. A. and KIRK, M. (1973) *Plant Physiol.* **52**, 407.
BASSHAM, J. A. and KRAUSE, G. H. (1969) *Biochim. Biophys. Acta*, **189**, 207.
BATRA, P. (1969) in: Giese (1964), Vol. 3.

BAUCHOP, T. and ELSDEN, S. R. (1960) *J. Gen. Microbiol.* **23,** 457.
BAUDHUIN, P. (1968) *Ann. N.Y. Acad. Sci.* **168,** 214.
BAUMGARTEN, J., REH, M. and SCHLEGEL, H. G. (1974) *Arch. Microbiol.* **100,** 207.
BAXTER, R. (1971) in: Reinert and Ursprung (1971).
BEADLE, G. W. (1945) *Physiol. Rev.* **25,** 643.
BEALE, G. H., JURAUD, A. and PREER, J. R. (1969) *J. Cell Sci.* **5,** 65.
BEARDEN, A. J. and MALKIN, R. (1974) *Quart. Rev. Biophys.* **7,** 131.
BECK, A., LOHRMANN, R. and ORGEL, L. E. (1967) *Science,* **157,** 952.
BECK, A. and ORGEL, L. E. (1965) *Proc. Nat. Acad. Sci., Wash.* **54,** 664.
BEEVERS, H. (1961) *Respiratory Metabolism in Plants,* New York.
BEEVERS, H. (1969) *Ann. N.Y. Sci.* **168,** 313.
BEEVERS, H. (1971) in: Hatch *et al.* (1971).
BELITSER, V. A. and TSYBAKOVA, E. T. (1939) *Biokhimiya,* **4,** 516.
BELKIN, D. A. (1962) *The Physiologist,* **5,** 105.
BELKIN, D. A. (1968) *Respir. Physiol,* **4,** 1.
BELL, R. P. (1970) in: Miller (1970).
BELLAMY, D. (1968) *New Scientist,* **37,** 532.
BENDALL, D. S. and BONNER, W. D. (1971) *Plant Physiol.* **47,** 236.
BENDALL, D. S. and HILL, R. (1968) *Ann. Rev. Plant Physiol.* **19,** 167.
BENEMANN, J. R., BERENSON, J. A., KAPLAN, N. O. and KAMEN, M. D. (1973) *Proc. Nat. Acad. Sci., Wash.* **70,** 2317.
BENEMANN, J. R. and VALENTINE, R. C. (1971) *Adv. Microb. Physiol.* **5,** 135.
BENEMANN, J. R. and VALENTINE, R. C. (1972) *Adv. Microb. Physiol.* **8,** 59.
BENEMANN, J. R. and WEARE, N. M. (1974) *Science,* **184,** 174.
BENNETT, R., RIGOPOULOS, N. and FULLER, R. C. (1964) *Proc. Nat. Acad. Sci., Wash.* **52,** 762.
BEREZNEV, R., FUNK, L. K. and CRANE, F. L. (1970) *Biochim. Biophys. Acta,* **223,** 61.
BERGERON, J. A. (1963) *Nat. Acad. Sci. Nat. Res. Council,* Publ. 1145, 527.
BERGERSEN, F. J. (1971) *Ann. Rev. Plant Physiol.* **22,** 121.
BERGERSEN, F. J., TURNER, G. L. and APPLEBY, C. A. (1972) *Biochim. Biophys. Acta,* **292,** 272.
BERKNER, L. V. and MARSHALL, L. C. (1964) *Disc. Faraday Soc.* **37,** 122.
BERKNER, L. V. and MARSHALL, L. C. (1965), *Proc. Nat. Acad. Sci., Wash.* **53,** 1215.
BERKNER, L. V. and MARSHALL, L. C. (1967) *Adv. Geophysics,* **12,** 309.
BERNAL, J. D. (1951) *The Physical Basis of Life,* London.
BERNAL, J. D. (1967) *The Origin of Life,* London.
BERNARD, U., PROBST, I. and SCHLEGEL, H. G. (1974) *Arch. Microbiol.* **95,** 29.
BERNARD, U. and SCHLEGEL, H. G. (1974) *Arch. Microbiol.* **95,** 39.
BETZ, J. L., BROWN, P. K., SMYTH, M. J. and CLARKE, P. H. (1974) *Nature,* **247,** 261.
BEUSCHER, N. and GOTTSCHALK, G. (1972) *Z. Naturforsch.* **27b,** 967.
BIGGINS, J. (1969) *J. Bact.* **99,** 570.
BIGGINS, J. and DIETRICH, W. E. (1968) *Arch. Biochem. Biophys.* **128,** 40.
BIRCH, F. (1965) *Bull. Geol. Soc. Amer.* **76,** 133.
BISHOP, N. I. (1962) *Nature,* **195,** 55.
BISHOP, N. I. (1966) *Ann. Rev. Plant Physiol.* **17,** 185.
BISHOP, N. I. and GAFFRON, H. (1962) *Biochem. Biophys. Res. Commun.* **8,** 471.
BISSET, K. A. (1962) *Symp. Soc. Gen. Microbiol.* **12,** 361.
BISSET, K. (1973) *New Scientist,* 8th Feb., p. 296.
BISSET, K. A. and GRACE, J. B. (1954) *Symp. Soc. Gen. Microbiol.* **4,** 28.
BJÖRKMAN, O. (1971) in: Hatch *et al.* (1971).
BJÖRKMAN, O. (1973) in: Giese (1964), Vol. 8.
BLACK, C. C. (1966) *Biochim. Biophys. Acta,* **120,** 332.
BLACK, C. C. (1973) *Ann. Rev. Plant Physiol.* **24,** 253.
BLACK, C. C., CAMPBELL, W. H., CHEN, T. M. and DITTRICH, P. (1973) *Quart. Rev. Biol.* **48,** 299.
BLACKMORE, M. A., QUAYLE, J. R. and WALKER, I. O. (1968) *Biochem. J.* **107,** 699.
BLAUROCK, A. and STOECKENIUS, W. (1971) *Nature New Biol.* **233,** 152.
BLINKS, L. R. (1954) *Symp. Soc. Gen. Microbiol.* **4,** 224.
BLINKS, L. R. (1964) in: Giese (1964), Vol. 1.
BLINKS, L. R. and VAN NIEL, C. B. (1963) in: *Studies on Microalgae and Photosynthetic Bacteria* (Special Issue of *Plant and Cell Physiology,* Tokyo).
BLUMENTHAL, H. J. (1965) in: Ainsworth and Sussman (1965).
BLUMER, M. (1973) in: Swain (1973).
BOARDMAN, N. K. (1968) *Adv. Enzymol.* **30,** 1.
BOARDMAN, N. K. (1970) *Ann. Rev. Plant Physiol.* **21,** 115.

Boardman, N. K. (1971) in: Hatch *et al.* (1971).
Boardman, N. K. (1972) *Biochim. Biophys. Acta*, **283**, 469.
Böhme, H. and Trebst, A. (1969) *Biochim. Biophys. Acta*, **180**, 137.
Böhmeke, H. (1939) *Arch. Mikrobiol.* **10**, 385.
Bogorad, L. (1967) in: Allen (1967).
Bolin, B. (1970) *Scient. Amer.* **223** (3), 124.
Boltzmann, L. (1886) quoted by Broda (1955).
Boltzmann, L. (1904) quoted by Broda (1955).
Bond, G., Wilson, J. F. and Winnall, N. J. (1973) *Nature*, **244**, 275.
Bone, D. H., Bernstein, S. and Vishniac, W. (1963) *Biochim. Biophys. Acta*, **67**, 581.
Bonner, D. M., deMoss, J. A. and Mills, S. E. (1965) in : Bryson and Vogel (1965).
Bonner, W. D. (1965) in: Bonner and Varner (1965).
Bonner, J. and Varner, J. E. (1965) *Plant Biochemistry*, New York.
Booij, H. J. and Bungenberg de Jong, H. G. (1956) *Biocolloids and Their Interactions, Protoplasmatologia*, Vol. 1, Vienna (Eds. Heilbrunn, L. V. and Weber, F.).
Borchert, H. (1951) *Geochim. Cosmochim. Acta*, **2**, 62.
Borchert, H. (1960) *Trans. Inst. Min. Metall.* **69**, 261.
Borisov, A. Yu. and Godik, V. I. (1973) *Biochim. Biophys. Acta*, **301**, 227.
Borst, P. (1970) in: Miller (1970).
Borst, P. and Kroon, A. M. (1969) *Int. Rev. Cytol.* **26**, 108.
Bose, S. K. and Gest, H. (1963) *Proc. Nat. Acad. Sci., Wash.* **49**, 337.
Bosshard-Heer, E. and Bachofen, R. (1969) *Arch. Microbiol.* **65**, 61.
Boulter, D. (1970) *Science Prog.* **60**, 217.
Boulter, D. (1973) in: Swain (1973).
Boulter, D., Laycock, M. V., Ramshaw, J. and Thompson, E. W. (1970) in: Harborne (1970a).
Boulter, D., Ramshaw, J., Thompson, E. W., Richardson, M. and Brown, R. H. (1972) *Proc. Roy. Soc. Lond.* B, **181**, 441.
Boulter, D., Thompson, E. W., Ramshaw, J. and Richardson, M. (1970b) *Nature*, **228**, 55.
Bovarnick, J. G., Chang, S. W., Schiff, J. A. and Schwartzbach, S. D. (1974a) *J. Gen. Microbiol.* **83**, 51.
Bovarnick, J. G., Schiff, J. A., Freedman, Z. and Egan, J. M. (1974b) *J. Gen. Microbiol.* **83**, 63.
Bowen, E. J., Ed. (1965) *Recent Progress in Photobiology*, Oxford.
Boyd, D. B. and Lipscomb, W. N. (1969) *J. Theor. Biol.* **25**, 403.
Boyer, P. D. (1965) in: King, T. E., Mason, H. S. and Morrison, M., Eds. (1965) *Oxidases and Related Redox Systems*, New York.
Boyer, P. D., Cross, R. L. and Momsen, W. (1973) *Proc. Nat. Acad. Sci., Wash.* **70**, 2873.
Boyer, P. D., Lardy, H. and Myrbäck, K., Eds. (1959, etc.) *The Enzymes*, New York.
Boyle, R. W. and Davies, J. L. (1973) *Geochim. Cosmochim. Acta*, **37**, 1389.
Brachet, J. and Mirsky, A. E., Eds. (1961, etc.) *The Cell*, New York.
Brancazio, P. J. and Cameron, A. G. W., Eds. (1964) *The Origin and Evolution of Atmospheres and Oceans*, New York.
Brand, T. von (1966) *Biochemistry of Parasites*, New York.
Brand, T. von (1972) *Parasitenphysiologie*, Stuttgart.
Branton, D. (1968) in: Giese (1964), Vol. 3.
Breidenbach, R. W. and Beevers, H. (1967) *Biochem. Biophys. Res. Commun.* **27**, 462.
Breuer, H. D. (1974) *Angew. Chem.* **86**, 401.
Brillouin, L. (1949) *Amer. Scient.* **37**, 554.
Brinkmann, R. T. (1969) *J. Geophys. Res.* **74**, 5355.
Britten, R. and Davidson, E. (1971) *Quart. Rev. Biol.* **46**, 111.
Brock, T. D. (1973a) in: Carr and Whitton (1973).
Brock, T. D. (1973b) *Science* **179**, 480.
Brockmann, H. and Knobloch, G. (1972) *Arch. Mikrobiol.* **85**, 123.
Broda, E. (1955) *Ludwig Boltzmann, Mensch-Physiker-Philosoph*, Vienna.
Broda, E. (1959) *Naturwiss. Rundsch.* **12**, 331.
Broda, E. (1960) *Radioactive Isotopes in Biochemistry*, Amsterdam.
Broda, E. (1968) in: Marie Sklodowska-Curie Centenary Lectures, International Atomic Energy Agency, Vienna.
Broda, E. (1970) *Prog. Biophys. Molec. Biol.* **21**, 146.
Broda, E. (1971) in: Buvet and Ponnamperuma (1971).
Broda, E. (1974a) *Origin of Life*, **6**, 247.
Broda, E. (1974b) *24th Pugwash Conference on Science and World Affairs*, Baden.
Broecker, W. S. (1970a) *J. Geophys. Res.* **75**, 3533.
Broecker, W. S. (1970b) *Science*, **168**, 1537.

BROOKS, J., MUIR, M. D. and SHAW, G. (1973) *Nature*, **244**, 215.
BROOKS, J. and SHAW, G. (1973) *Origin and Development of Living Systems*, London.
BROWN, A. H. and WEBSTER, C. G.(1953) *Amer. J. Bot.* **40**, 753.
BROWN, H. (1949) in: Kuiper, G. P., Ed., *The Atmospheres of the Earth and of the Planets*, Chicago.
BROWN, H. (1952) in: Kuiper (1952).
BROWN, H. (1964) *Science*, **145**, 1177.
BRÜCK, K. (1967) *Naturwiss.* **54**, 156.
BRUEMMER, J. H., WILSON, P. W., GLENN, J. L. and CRANE, F. L. (1957) *J. Bact.* **73**, 113.
BRUIN, W. J., NELSON, E. B. and TOLBERT, N. E. (1970) *Plant Physiol.* **46**, 386.
BRUNO, G. (1584) *Del infinito universo e mondi*, Venice.
BRUSH, S. (1969) *Phys. Teacher*, **7**, 271.
BRYAN-JONES, D. G. and WHITTENBURY, R. (1969) *J. Gen. Microbiol.* **58**, 247.
BRYANT, M. P., WOLIN, E. A., WOLIN, M. J. and WOLFE, R. S. (1967) *Arch. Mikrobiol.* **59**, 20.
BRYSON, V. and VOGEL, H. J., Eds. (1965) *Evolving Genes and Proteins*, New York.
BUCHANAN, B. B. (1966) *Structure and Bonding*, Berlin, Vol. 1.
BUCHANAN, B. B. and ARNON, D. I. (1965) *Biochim. Biophys. Res. Commun.* **20**, 163.
BUCHANAN B. B. and ARNON, D. I. (1970) *Adv. Enzymol.* **33**, 119.
BUCHANAN, B. B., BACHOFEN, R. and ARNON, D. I. (1964) *Proc. Nat. Acad. Sci., Wash.* **52**, 389.
BUCHANAN, B. B. and EVANS, M. C. W. (1965) *Proc. Nat. Acad. Sci., Wash.* **54**, 1212.
BUCHANAN, B. B. and EVANS, M. C. W. (1969) *Biochim. Biophys. Acta*, **180**, 123.
BUCHANAN, B. B., EVANS, M. C. W. and ARNON, D. I. (1967) *Arch. Mikrobiol.* **59**, 32.
BUCHANAN, B. B., MATSUBARA, H. and EVANS, M. C. W. (1969) *Biochim. Biophys. Acta*, **189**, 46.
BUCHANAN, B. B., SCHÜRMANN, P. and SHANMUGAM, K. T. (1972) *Biochim. Biophys. Acta*, **283**, 136.
BUCHNER, P. (1966) *Endosymbiosis of Animals with Plant Microorganisms*, New York.
BUEDING, E. (1961) 5th Int. Congr. Biochem., Moscow.
BUETOW, D. E., Ed. (1968) *The Biology of Euglena*, New York.
BUHL, D. (1971) *Nature*, **234**, 332.
BUHL, D. (1974) *Origin of Life*, **5**, 29.
BUHL, D. and PONNAMPERUMA, C. (1971) *Space Life Sciences*, **3**, 157.
BURK, D. (1939) *Cold Spring Harbor Symp. Quant. Biol.* **7**, 420.
BURNETT, J. H. (1968) *Fundamentals of Mycology*, London.
BURRIS, R. H. (1961) 5th Int. Congr. Biochem., Moscow.
BURRIS, R. H. (1966) *Ann. Rev. Plant Physiol.* **17**, 155.
BURTON, K. (1957) in: Krebs and Kornberg (1957).
BUTLEROW, A. (1861) *Ann. Chem.* **120**, 295.
BUTLIN, K. R. (1953) *Research* **6**, 184.
BUTLIN, K. R. and POSTGATE, J. R. (1954) *Symp. Soc. Gen. Microbiol.* **4**, 271.
BUTT, V. S. and PEEL, M. (1963) *Biochem. J.* **88**, 31 P
BUVET, R. and LEPORT, L. (1973) *Space Life Sciences*, **4**, 434.
BUVET, R. and PONNAMPERUMA, C. (1971) *Chemical Evolution and the Origin of Life* (*Molecular Evolution* I), Amsterdam.
CAIN, A. J. (1962) *Symp. Soc. Gen. Microbiol.* **12**, 1.
CAIRNS, J., STENT, G. and WATSON, J. D., Eds. (1966) *Phage and the Origins of Molecular Biology.*
CALDWELL, P. C (1969) *Curr. Topics Bioenerget.* **3**, 251.
CALVIN, M. (1957) in: Oparin *et al.* (1957).
CALVIN, M. (1959) *Science*, **130**, 1170.
CALVIN, M. (1963) quoted by Baltscheffsky (1967a).
CALVIN, M. (1965) *Proc. Roy. Soc., Lond.* A, **288**, 441.
CALVIN, M. (1967) in: Dobzhansky *et al.* (1967), Vol. 1.
CALVIN, M. (1969) *Chemical Evolution*, Oxford.
CALVIN, M. (1974) *Science*, **184**, 375.
CAMERON, A. G. W. (1970) *Trans. Amer. Geophys. Union*, **51**, 628.
CAMMACK, R., HALL, D. and RAO, K. (1971) *New Scientist*, 23rd Sept.
CAMPBELL, J. J., GRONLUND, A. F. and DUNCAN, M. G. (1963) *Ann. N.Y. Acad. Sci.* **102**, 669.
CANALE-PAROLA, E., UDRIS, Z. and MANDEL, M. (1968) *Arch. Mikrobiol.* **63**, 385.
CANOVAS, J. L., ORNSTON, L. N. and STANIER, R. Y. (1967) *Science*, **156**, 1695.
CAREY, F. G. and TEAL, J. M. (1966) *Proc. Nat. Acad. Sci., Wash.* **56**, 1464.
CAREY, F. G., TEAL, J. M., KANWISHER, J. W., LAWSON, K. V. and BECKETT, J. S. (1971) *Amer. Zool.* **11**, 135.
CARLETON, N. P. and TRAUB, W. A. (1972) *Science*, **177**, 988.
CARLILE, M. J. and SKEHEL, J. J. Eds. (1974) *Symp. Soc. Gen. Microbiol.* **24**.
CARR, N. G. and CRAIG, I. W. (1970) in: Harborne (1970).
CARR, N. G. and WHITTON, B. A., Eds. (1973) *The Biology of Blue–Green Algae*, Oxford.

Caskey, C. T. (1970) *Quart. Rev. Biophys.* **3,** 296.
Cassio, D. and Waller, J. P. (1971) *Eur. J. Biochem.* **20,** 283.
Castenholz, R. W. (1973) in: Carr and Whitton (1973).
Cavalier-Smith, T. (1970) *Nature,* **228,** 333.
Chaloner, W. G. (1960) *Science Prog.* **48,** 524.
Chaloner, W. G. and Allen, K. (1970) in: Harborne (1970).
Chamberlin, T. C. (1897) *J. Geol.* **5,** 653.
Chamberlin, T. C, and Salisbury, R. D. (1905) *Geology,* New York.
Chance, B. (1961a) *Nature,* **189,** 719.
Chance, B. (1961b) *J. Biol. Chem.* **236,** 1569.
Chance, B. and Baltscheffsky, M. (1958) *Biochem. J.* **68,** 283.
Chance, B., Bonner, W. D. and Storey, B. T. (1968) *Ann. Rev. Plant Physiol.* **19,** 295.
Chance, B. and Hollunger, G. (1960) *Nature,* **185,** 666.
Chance, B. and Hollunger, G. (1961) *J. Biol. Chem.* **236,** 1534.
Chance, B., Lees, H. and Postgate, J. R. (1972) *Nature,* **238,** 330.
Chance, B. and Olson, J. M. (1960) *Arch. Biochem. Biophys.* **88,** 54.
Chance, B. and Williams, G. R. (1956) *Adv. Enzymol.* **13,** 65.
Chang, S., Williams, J. A., Ponnamperuma, C. and Rabinowitz, J. (1970) *Space Life Sciences,* **2,** 144.
Chargaff, E. (1955) in: Chargaff, E. and Davidson, J. N., Eds. (1955, etc.) *The Nucleic Acids,* Vol. 1. New York.
Charles, H. P. and Knight, B. C., Eds. (1970) *Organization and Control in Prokaryotic and Eukaryotic Cells,* Symp. Soc. Gen. Microbiol., Vol. 20, Cambridge.
Chatton, E. (1937) *Titres et Trouvaux Scientifiques,* Sète.
Cheah, K. S. (1971) *Biochim. Biophys. Acta,* **253,** 1.
Chen, T. T., Ed. (1967) *Research in Protozoology,* Oxford.
Cheniae, G. M. (1970) *Ann. Rev. Plant Physiol.* **21,** 467.
Chibnall, A. C. (1939) *Protein Metabolism in the Plant,* New Haven.
Christiansen, E. N., Pedersen, J. I. and Grav, H. J. (1969) *Nature,* **222,** 857.
Clark-Walker, G. D. and Linnane, A. W. (1967) *J. Cell Biol.* **34,** 1.
Clarke, P. H. (1974) in: Carlile and Skehel (1974).
Claus, G. (1968) *Ann. N.Y. Acad. Sci.* **147,** 363.
Claus, G. and Madri, P. P. (1972) *Ann. N.Y. Acad. Sci.* **196,** 385.
Clayton, R. K. (1962) *Bact. Rev.* **26,** 151.
Clayton, R. K. (1966) *Brookhaven Symp. Biol.* **19,** 62.
Clayton, R. K. (1972) *Proc. Nat. Acad. Sci., Wash.* **69,** 44.
Clayton, R. K. (1973) *Ann. Rev. Biophys. Bioeng.* **2,** 131.
Cloud, P. E. (1965) *Science,* **148,** 27.
Cloud, P. E. (1968a) *Science,* **160,** 729.
Cloud, P. E. (1968b) in: Drake (1968).
Cloud, P. E. (1972) *Amer. J. Sci.* **272,** 537.
Cloud, P. E. (1973) *Amer. J. Sci.* **273,** 193.
Cloud, P. E. (1974a) *Amer. Sci.* **62,** 54.
Cloud, P. E. (1974b) *Science,* **183,** 739.
Cloud, P. E. and Hagen, H. (1965) *Proc. Nat. Acad. Sci., Wash.* **54,** 1.
Cloud, P. E. and Licari, G. R. (1968) *Proc. Nat. Acad. Sci., Wash.* **61,** 779.
Cloud, P. E. and Licari, G. R. (1972) *Amer. J. Sci.* **272,** 138.
Cloud, P. E., Licari, G. R., Wright, L. A. and Troxel, B. W. (1969) *Proc. Nat. Acad. Sci., Wash.* **62,** 623.
Cloud, P. E. and Semikhatov, M. A. (1969) *Amer. J. Sci.* **267,** 1017.
Clowes, F. A. L. and Juniper, B. E. (1968) *Plant Cells,* Oxford.
Cohen, A. L. (1967) *Arch. Mikrobiol.* **59,** 59.
Cohen, S. S. (1970) *Amer. Scient.* **58,** 281.
Cohen, S. S. (1973) *Amer. Scient.* **61,** 437.
Cohen-Bazire, G. and Kunisawa, R. (1960) *Proc. Nat. Acad. Sci., Wash.* **46,** 1543.
Cohen-Bazire, G. and Kunisawa, R. (1963) *J. Cell Biol.* **16,** 410.
Cohen-Bazire, G., Pfennig, N. and Kunisawa, R. (1964) *J. Cell. Biol.* **22,** 207.
Cohen-Bazire, G. and Sistrom, W. R. (1966) in: Vernon and Seely (1966).
Cohen-Bazire, G. and Stanier, R. Y. (1958) *Nature,* **181,** 250.
Cohn, F. (1853) *Nova act. Acad. Leop. Carol.* **24,** 103, quoted by Pringsheim (1949).
Cohn, F. (1874) *Beitr. Biol. Pflanz.* **3,** 141, quoted by Stanier (1964).
Cole, A., Ed. (1968, etc.) *Theoretical and Experimental Biophysics,* New York.
Cole, J. A. and Hughes, D. E. (1965) *J. Gen. Microbiol.* **38,** 65.
Cole, J. S. and Aleem, M. I. H. (1970) *Biochem. Biophys. Res. Commun.* **38,** 736.

COLES, G. C. (1972) *Nature*, **240**, 488.
CONN, E. E. (1960) in: Florkin and Mason (1960), Vol. 1.
CONOVER, T. E. (1967) *Current Topics Bioenergetics*, **2**, 235.
CONOVER, T. E. (1970) *Arch. Biochem. Biophys.* **136**, 541.
COOK, J. S. (1970) in: Giese (1964), Vol. 5.
COOPER, T. G. and WOOD, H. G. (1971) *J. Biol. Chem.* **246**, 5488.
COPELAND, H. F. (1956) *The Classification of Lower Organismus*, Palo Alto.
CORLISS, J. A. (1962) *Symp. Soc. Gen. Microbiol.* **12**, 44.
CORRENS, C. W. (1948) *Naturwiss.* **35**, 7.
CORYELL, C. D. (1941) Personal communication to Kalckar (1941).
COX, G. B. and GIBSON, F. (1974) *Biochim. Biophys. Acta*, **346**, 1.
CRICK, F. H. C. (1958) Symp. Soc. Exptl. Biol. **12**, 138.
CRICK, F. H. C. (1968) *J. Molec. Biol.* **38**, 367.
CRICK, F. H. C. (1970) *Nature*, **227**, 322.
CRICK, F. H. C. and WATSON, J. (1953) *Nature*, **171**, 737.
CRIDDLE, R. S. and SCHATZ, G. (1969) *Biochemistry*, **8**, 322.
CROXFORD, N. J. W., JANECEK, J., MUIR, M. D. and PLUMB, K. A. (1973) *Nature*, **245**, 28.
CRUDEN, D. L., COHEN-BAZIRE, G. and STANIER, R. Y. (1970) *Nature*, **228**, 1345.
CRUDEN, D. L. and STANIER, R. Y. (1970) *Arch. Mikrobiol.* **72**, 115.
CURTIS, V. A., SIEDOW, J. N. and SAN PIETRO, A. (1973) *Arch. Biochem. Biophys.* **158**, 898.
CUSANOVICH, M. A., BARTSCH, R. B. and KAMEN, M. D. (1968) *Biochim. Biophys. Acta*, **153**, 397.
DAMON, P. E. and KULP, J. L. (1958) *Geochim. Cosmochim. Acta*, **13**, 280.
DANFORTH, W. F. (1962) in: Lewin (1962).
DANFORTH, W. F. (1967) in: Chen (1967), Vol. 1.
DANIEL, R. M. and ERICKSON, S. K. (1969) *Biochim. Biophys. Acta*, **180**, 63.
DANIELS, F. (1972) *Biophys. J.* **12**, 723.
DANON, A. and STOECKENIUS, W. (1974) *Proc. Nat. Acad. Sci., Wash.* **71**, 1234.
DARWIN, C. (1871) quoted among others, by Fox (1960) and by Bernal (1967).
DAVIDSON, C. F. (1965) *Proc. Nat. Acad. Sci., Wash.* **53**, 1194.
DAVIDSON, E., HOUGH, B. R., AMENSON, C. and BRITTEN, R. (1973) *J. Molec. Biol.* **77**, 1.
DAVIES, R. E. and KREBS, H. A. (1952) *Biochem. Soc. Symp.* **8**, 77.
DAVIES, S. L. and WHITTENBURY, R. (1970) *J. Gen. Microbiol.* **61**, 227.
DAVIS, B. D. (1961) *Cold Spring Harb. Symp. Quant. Biol.* **26**, 1.
DAWES, E. A. and RIBBONS, D. W. (1964) *Bact. Rev.* **28**, 126.
DAWES, E. A. and SENIOR, P. J. (1973) *Adv. Microb. Physiol.* **10**, 136.
DAYHOFF, M. O. (1969a) *Scient. Amer.* **221** (1), 87.
DAYHOFF, M. O. (1969b) *Atlas of Protein Sequences and Structure*, Silver Springs, Md. Also later editions.
DAYHOFF, M. O. (1971) in: Buvet and Ponnamperuma (1971).
DAYHOFF, M. O. (1972) in: Ponnamperuma (1972b).
DAYHOFF, M. O., BARKER, W. C. and MCLAUGHLIN, P. J. (1974) *Origin of Life* **5**, 311.
DAYHOFF, M. O. and ECK, R. V. (1969) in: Eglinton and Murphy (1969).
DAYHOFF, M. O., LIPPINCOTT, E. R. and ECK, R. V. (1964) *Science*, **146**, 1461.
DEAMER, D. W. (1969) *J. Chem. Educ.* **46**, 201.
DEAN, A. C. R. and HINSHELWOOD, C. (1966) *Growth, Function and Regulation in Bacterial Cells*, Oxford.
DECKER, J. P. (1955) *Plant Physiol.* **30**, 82.
DECKER, J. P. (1959) *Plant Physiol.* **34**, 100.
DECKER, J. P. (1970) *Early History of Photorespiration*. Bull. 10, Engng. Res. Center, Arizona State Univ., Temple, quoted after Zelitch (1971).
DECKER, K., JUNGERMANN, K. and THAUER, R. K. (1970) *Angew. Chem.* **82**, 153.
DE DUVE, C. (1969a) *Proc. Roy. Soc. Lond.* B, **173**, 71.
DEDUVE, C. (1969b) *Ann. N.Y. Acad. Sci.* **168**, 369.
DEDUVE, C. (1973) *Science*, **182**, 85.
DE DUVE, C. and BAUDHOUIN, P. (1966) *Physiol. Rev.* **46**, 323.
DEGANI, C. and HALMANN, M. (1967) *Nature*, **216**, 1207.
DEGENS, E. T. (1968) *Geochemie der Sedimente*, Stuttgart.
DEGENS, E. T. and MATHEJA, J. (1971) in: Kimball and Oró (1971).
DELEVORYAS, T. (1966) *Plant Diversification*, New York.
DE LEY, J. (1962) *Symp. Soc. Gen. Microbiol.* **12**, 164.
DE LEY, J. (1964) *Ann. Rev. Microbiol.* **18**, 17.
DE LEY, J. (1968a) in: Dobzhansky *et al.* (1967), Vol. 2.
DE LEY, J. (1968b) in: Prauser, H., Ed. (1968) *The Actinomycetales*, Jena.
DE LEY, J. (1971) in: Schoffeniels (1971).

DELSEMME, A. H. (1973) *Space Science Rev.* **15,** 103.
DELWICHE, C. C. (1956) in: McElroy and Glass (1956).
DELWICHE, C. C. (1970) *Scient. Amer.* **223** (3) 136.
DEMOULIN, V. (1974) *Bot. Rev.* **40,** 263.
DICKENS, F. and NEIL, Eds. (1964) *Oxygen in the Animal Organism*, Oxford.
DICKERSON, R. E. (1971) *J. Molec. Evol.* **1,** 26.
DICKERSON, R. E. (1972) *Scient. Amer.* **226** (4), 58.
DIENES, M., Ed. (1967) *Energy Conversion by the Photosynthetic Apparatus*, Springfield.
DIERS, L. (1970) in: Miller (1970).
DILLON, L. S. (1974) *Bot. Rev.* **39,** 301.
DOBZHANSKY, T., HECHT, M. K. and STEERE, W. C., Eds. (1967, etc.) *Evolutionary Biology*, Amsterdam.
DODGE, J. D. (1974) *Sci. Prog.* **61,** 257.
DOEMEL, W. N. and BROCK, T. D. (1974) *Science* **184,** 1083.
DOLE, M. (1948) *Science*, **109,** 77.
DOLIN, M. I. (1955) *Arch. Biochem. Biophys.* **55,** 415.
DOLIN, M. I. (1961) in: Gunsalus and Stanier (1960), Vol. 2.
DONAWA, A. L., ISHAQUE, M. and ALEEM, M. I. H. (1971) *Eur. J. Biochem.* **21,** 292.
DOSE, K. (1974) *Origin of Life* **5,** 239.
DOSE, K., FOX, S. W., DEBORIN, G. A. and PAVLOVSKAYA, T. E. (1974) *The Origin of Life and Evolutionary Biochemistry*, New York.
DOSE, K. and RAJEWSKY, B. (1957) *Biochim. Biophys. Acta*, **25,** 225.
DOSE, K. and ZAKI, L. (1971) in: Buvet and Ponnamperuma (1971).
DOTY, P., MARMUR, J., EIGNER, J. and SCHILDKRAUT, C. (1960) *Proc. Nat. Acad. Sci., Wash.* **46,** 461.
DOUCE, R., HOLTZ, R. B. and BENSON, A. A. (1973) *J. Biol. Chem.* **248,** 7215.
DOUDOROFF, M. (1966) in: Kaplan and Kennedy (1966).
DOUDOROFF, M. and STANIER, R. Y. (1959) *Nature*, **183,** 1440.
DOWNTON, W. J. S. (1971) in: Hatch *et al.* (1971).
DOWNTON, J., BERRY, J. and TREGUNNA, E. B. (1969) *Science*, **163,** 78.
DRABKIN, D. (1950) *J. Biol. Chem.* **182,** 317.
DRAKE, E. T., Ed. (1968) *Evolution and Environment*, New Haven.
DRAKE, J. W. (1974) in: Carlile and Skehel (1974).
DREWS, G. (1965) *Naturwiss. Rundsch.* **18,** 274.
DREWS, G. (1973) in: Carr and Whitton (1973).
DREWS, G., LAMPE, H. H. and LADWIG, R. (1969) *Arch. Mikrobiol.* **65,** 12.
DREYSEL, J., THIELE, O. W. and HERMANN, D. (1970) *Experientia*, **26,** 1304.
DUFRESNE, E. R. and ANDERS, E. (1962) *Geochim. Cosmochim. Acta*, **26,** 1085.
DUFRESNE, E. R. and ANDERS, E. (1963) in: Middlehurst and Kuiper (1961), Vol. 4.
DUNN, P. P., PLUMB, K. A. and ROBERTS, H. G. (1966) *J. Geol. Soc. Australia* **13,** 595.
DUYSENS, L. N. M. (1951) *Nature*, **168,** 548.
DUYSENS, L. N. M. (1952) Thesis, Leiden; quoted after Knox (1969).
DUYSENS, L. N. M. (1953) *Nature*, **173,** 692.
DUYSENS, L. N. M. (1959) *Brookhaven Symp. Biol.* **11** (1958), 10.
DUYSENS, L. N. M. (1962) *Plant. Physiol.* **37,** 407.
DUYSENS, L. N. M. (1964) *Prog. Biophys. Molec. Biol.* **14,** 1.
DUYSENS, L. N. M. (1967) *Brookhaven Symp. Biol.* **19** (1966), 71.
DUYSENS, L. N. M. and AMESZ, J. (1962) *Biochim. Biophys. Acta*, **64,** 243.
DUYSENS, L. N. M. and AMESZ, J. (1967) in: Florkin and Stotz (1962, etc.) Vol. 27.
DUYSENS, L. N. M., AMESZ, J. and KAMP, B. M. (1967). *Nature*, **190,** 510.
EBERHARDT, U. (1969) *Arch. Mikrobiol.* **66,** 91.
ECHLIN, P. (1964) *Arch. Mikrobiol.* **49,** 267.
ECHLIN, P. (1966a) *Science J.* **2** (4), 42.
ECHLIN, P. (1966b) *Scient. Amer.* **214** (6), 74.
ECHLIN, P. (1967) *Brit. Phycol. Bull.* **3,** 225.
ECHLIN, P. (1970a) in: Harborne (1970).
ECHLIN, P. (1970b) in: Charles and Knight (1970).
ECHLIN, P. and MORRIS, I. (1965) *Biol. Rev.* **40,** 143.
ECK, R. V. and DAYHOFF, M. O. (1966) *Science*, **152,** 363.
ECK, R. V., LIPPINCOTT, E. R., DAYHOFF, M. O. and PRATT, Y. T. (1966) *Science*, **153,** 628.
EDELMAN, M., SWINTON, D., SCHIFF, J. A., EPSTEIN, H. T. and ZELDIN, B. (1967) *Bact. Rev.* **31,** 315.
EDWARDS, G. E. and BLACK, C. C. (1971) in: Hatch *et al.* (1971).
EGAMI, F. (1957) *Svensk Kem. Tidskr.* **69,** 652.
EGAMI, F. (1973) *Z. Allg. Mikrobiol.* **13,** 177.

EGAMI, F. (1974) *Origin of Life* **5**, 405.
EGGLESTON, L. V. and KREBS, A. H. (1974) *Biochem. J.* **138**, 321.
EGLINTON, G. (1969) in: Eglinton and Murphy (1969).
EGLINTON, G. (1973) in: Swain (1973).
EGLINTON, G. and CALVIN, M. (1967) *Scient. Amer.* **216** (1), 32.
EGLINTON, G. and MURPHY, M. T. J., Eds. (1969) *Organic Geochemistry*, Berlin.
EGLINTON, G., SCOTT, P. M., BELSKY, T., BURLINGAME, A. L., RICHTER, W. and CALVIN, M. (1964) *Science*, **145**, 263.
EHRENDORFER, F. (1972) *Symp. Biol. Hung.* **12**, 227.
EIGEN, M. (1971a) *Quart. Rev. Biophys.* **4**, 149.
EIGEN, M. (1971b) *Naturwiss.* **58**, 465.
EISMA, E. and JUNG, J. W. (1969) in: Eglinton and Murphy (1969).
ELGHAZZAWI, E. (1967) *Arch. Mikrobiol.* **57**, 1.
ELIAS, W. E. (1972) *J. Chem. Educ.* **49**, 448.
ELLFOLK, N. (1972) *Endeavour*, **31**, 139.
ELSDEN, S. R. (1954) *Symp. Soc. Gen. Microbiol.* **4**, 202.
ELSDEN, S. R. (1962) in: Gunsalus and Stanier (1960), Vol. 3.
ELSDEN, S. R., KAMEN, M. D. and VERNON, L. P. (1953) *J. Amer. Chem. Soc.* **75**, 6347.
EMERSON, R. and ARNOLD, W. (1932) *J. Gen. Physiol.* **15**, 391.
EMERSON, R., CHALMERS, R. and CEDERSTRAND, C. (1957) *Proc. Nat. Acad. Sci., Wash.* **43**, 133.
EMERSON, R. and RABINOWITCH, E. (1960) *Plant Physiol.* **35**, 477.
EMERSON, R. and HELD, A. A. (1970) *Amer. J. Bot.* **56**, quoted by Stanier (1970).
EMERSON, R. and LEWIS, C. M. (1943) *Amer. J. Bot.* **30**, 165.
EMERSON, R., STAUFFER, J. F. and UMBREIT, F. F. (1944) *Amer. J. Bot.* **31**, 107.
EMERSON, R. and WESTON, W. H. (1967) *Amer. J. Bot.* **54**, 702.
ENGEL, A. E. J. and ENGEL, C. G. (1953) *Geol. Soc. Amer. Bull.* **64**, 1013.
ENGEL, A. E. J., NAGY, B., NAGY, L. A., ENGEL, C. C., KREMP, G. O. W. and DREW, C. M. (1968) *Science*, **161**, 1005.
ENGELHARDT, W. A. (1930) *Biochem. Z.* **227**, 16.
ENGELHARDT, W. A. (1932) *Biochem. Z.* **251**, 343.
ENGELHARDT, W. A. (1973) *Conference on the Historical Development of Bioenergetics, American Academy of Arts and Sciences.*
ENGELMANN, T. W. (1883) *Pflügers Arch. ges. Physiol.* **30**, 95.
ENGELMANN, T. W. (1884) *Botan. Z.* **42**, 81, 97.
ENGELMANN, T. W. (1887) *Botan. Z.* **46**, 393, 409, 425, 441, 457.
EPHRUSSI, B. (1953) *Nucleo-Cytoplasmic Relations in Microorganisms*, Oxford.
EPHRUSSI, B. and SLONIMSKI, P. P. (1950) *C.R. Acad, Sci., Paris*, **230**, 685.
EPHRUSSI, B. and SLONIMSKI, P. P. (1955) *Nature*, **176**, 1207.
ERECINSKA, M. and STOREY, B. T. (1970) *Plant Physiol.* **46**, 618.
ERICKSON, S. K. (1971) *Biochim. Biophys. Acta*, **245**, 63.
ERICKSON, S. K. and PARKER, G. L. (1969) *Biochim. Biophys. Acta*, **180**, 56.
ERNSTER, L. and LEE, C. H. (1964) *Ann. Rev. Biochem.* **33**, 729.
ERNSTER, L. and LUFT, R. (1963) *Exptl. Cell Res.* **32**, 26.
ERWIN, I. and BLOCH, K. (1964) *Science*, **143**, 1006.
EVANS, L. T. (1974) in: Hatch *et al.* (1971).
EVANS, M. C. W. and BUCHANAN, B. B. (1965) *Proc. Nat. Acad. Sci., Wash.* **53**, 420.
EVANS, M. C. W., BUCHANAN, B. B. and ARNON, D. I. (1966) *Proc. Nat. Acad. Aci., Wash.* **55**, 928.
EVANS, M. C. W. and WHATLEY, F. R. (1970) in: Charles and Knight (1970).
FAIRBAIRN, D. (1970) *Biol. Rev.* **45**, 29.
FAIRBRIDGE, R. W., Ed. (1972) *The Encyclopedia of Geochemistry and Environmental Science*, New York.
FERNANDEZ-ALONSO, J. I. (1964) *Adv. Chem. Phys.* **7**, 3.
FERRIS, J. P. (1968) *Science*, **161**, 53.
FERRIS, J. P. and NICODEM, D. E. (1972) *Nature*, **238**, 268.
FERRIS, J. P., SANCHEZ, R. A. and ORGEL, L. E. (1968) *J. Molec. Biol.* **33**, 693.
FERRIS, J. P., WOS, J. D., RYAN, T. J., LOBO, A. P. and DONNER, D. B. (1974) *Origin of Life* **5**, 153.
FICK, A. (1882) *Mechanische Arbeit und Wärmeentwicklung bei der Muskeltätigkeit*, Leipzig, quoted by Höber (1931).
FISCHER, A. G. (1965) *Proc. Nat. Acad. Sci., Wash.* **53**, 1205.
FISCHER-HJALMARS, I. (1969) *Quart. Rev. Biophys.* **1**, 311.
FITCH, W. M. and MARGOLIASH, E. (1967) *Science*, **155**, 279.
FITCH, W. M. and MARGOLIASH, E. (1970) in: Dobzhansky (1967), Vol. 4.
FLATMARK, T. and PEDERSEN, J. I. (1972) *Biochim. Biophys. Acta*, **292**, 64.

FLATMARK, T. and PEDERSEN, J. I. (1975) *Biochim. Biophys. Acta*, **416,** 53.
FLAVELL, R. B. (1971) *Nature*, **230,** 504.
FLEISCHER, R. L., PRICE, P. B. and WALKER, R. M. (1965a) *J. Geophys. Res.* **70,** 2703.
FLEISCHER, R. L., PRICE, P. B. and WALKER, R. M. (1965b) *Ann. Rev. Nucl. Sci.* **15,** 1.
FLEMING, H. and HASELKORN, R. (1973) *Proc. Nat. Acad. Sci., Wash.* **70,** 2727.
FLORKIN, M. (1944) *L'évolution biochimique*, Paris,
FLORKIN, M. (1966) *A Molecular Approach to Phylogeny*, Amsterdam.
FLORKIN, M. (1971) in: Buvet and Ponnamperuma (1971).
FLORKIN, M. (1972) in: Florkin and Stotz (1962), Vol. 30.
FLORKIN M. (1974) in: Florkin and Stotz (1962) Vol. 29A.
FLORKIN, M. and MASON, H., Eds. (1960, etc.) *Comparative Biochemistry*, New York.
FLORKIN, M. and MASON, H. (1960) in: Florkin and Mason (1960), Vol. 1.
FLORKIN, M. and SCHEER, B. T., Eds. (1967) *Chemical Zoology*, New York.
FLORKIN, M. and STOTZ, E. H., Eds. (1962, etc.) *Comprehensive Biochemistry*, Amsterdam.
FÖRSTER, TH. (1967) in: Florkin and Stotz (1962, etc.), Vol. 22.
FOGG, G. E., STEWART, W. D. P., FAY, P. and WALSBY, A. E. (1973) *The Blue–Green Algae*, London.
FORK, D. C, and AMESZ, J. (1969) *Ann. Rev. Plant Physiol.* **20,** 305.
FORK, D. C. and AMESZ, J. (1970) in: Giese (1964), Vol. 5.
FORREST, W. W. (1969) *Symp. Soc. Gen. Microbiol.* **19,** 65.
FORREST, W. W. and WALKER, D. J. (1971) *Adv. Microbiol. Physiol.* **5,** 213.
FORRESTER, M. L., KROTKOV, G. and NELSON, C. D. (1966) *Plant Physiol.* **41,** 422.
FOSTER, J. W. (1940) *J. Gen. Physiol.* **24,** 123.
FOTT, B. (1959) *Algenkunde*, Jena; 2nd ed., Stuttgart, 1971.
FOWLER, C. F. and KOK, B. (1974) *Biochim. Biophys. Acta*, **357,** 299.
FOWLER, C. F. and SYBESMA, C. (1970) *Biochim. Biophys. Acta*, **197,** 276.
FOWLER, W. A., GREENSTEIN, J. L. and HOYLE, F. (1961) *Amer. J. Phys.* **29,** 293.
FOX, H. M. and TAYLOR, A. E. R. (1955) *Proc. Roy. Soc., Lond.* B, **143,** 214.
FOX, S. W. (1960) *Science*, **132,** 200.
FOX, S. W., Ed. (1965a) *The Origin of Prebiological Systems and of Their Molecular Matrices*, New York.
FOX, S. W. (1965b) *Nature*, **205,** 328.
FOX, S. W. (1969) *Naturwiss.* **56,** 1.
FOX, S. W. (1971a) *Chem. Engng. News*, **49,** 46.
FOX, S. W. (1971b) in: Kimball and Oró (1971).
FOX, S. W. (1973) *Naturwiss.* **60,** 359.
FOX, S. W. (1974) *Mol. Cell. Biochem.* **3,** 129.
FOX, S. W. and DOSE, K. (1972) *Molecular Evolution and the Origin of Life*, San Francisco.
FOX, S. W., HARADA, K., KRAMPITZ, G. and MUELLER, G. (1970) *Chem. Engng. News*, **48,** 80.
FOX, S. W., JUNGCK, J. R. and NAKASHIMA, T. (1974) *Origin of Life* **5,** 227.
FRANCK, J. and LOOMIS, W. E., Eds. (1949) *Photosynthesis in Plants*, Ames.
FRANK, H., LEFORT, M. and MARTIN, H. H. (1962) *Z. Naturforsch.* **17b,** 262.
FRENKEL, A. W. (1954) *J. Amer. Chem. Soc.* **76,** 5568.
FRENKEL, A. W. (1970) *Biol. Rev.* **45,** 569.
FRENKEL, A. W. and RIEGER, C. (1951) *Nature*, **167,** 1030.
FREUNDLICH, M. M. and WAGNER, B. M., Eds. (1969) *Exobiology*, Tarzana, Calif.
FREY-WYSSLING, A. and MÜHLETHALER, K. (1965) *Ultrastructural Plant Cytology*, Amsterdam.
FRIDOVICH, I. (1975) *Amer. Scient.* **63,** 54.
FRIEDMANN, N., HAVERLAND, W. J. and MILLER, S. L. (1971) in: Buvet and Ponnamperuma (1971).
FRITSCH, E. F. (1945) *Ann. Botany*, **9,** 1.
FRITSCH, E. F. (1965) *The Structure and Reproduction of the Algae*, Cambridge.
FROMAGEOT, C. and SENEZ, J. C. (1960) in: Florkin and Mason (1960), Vol. 1.
FUHS, W. G. (1969) *The Nuclear Structure of Prokaryotic Organisms*, Vienna.
FUJITA, A. and KODAMA, T. (1934) *Biochem. Z.* **293,** 186.
FULLER, R. C. (1969) in: Metzner (1969), Vol. 3.
FULLER, R. C. (1971) in: Schoffeniels (1971).
FULLER, R. C. and NUGENT, N. A. (1969) *Proc. Nat. Acad. Sci., Wash.* **63,** 1311.
FULLER, R. C., SMILLIE, R. M., SISLER, E. C. and KORNBERG, H. L. (1961) *J. Biol. Chem.* **236,** 2140.
GABEL, N. W. and PONNAMPERUMA, C. (1967) *Nature*, **216,** 453.
GÄUMANN, E. (1964) *Die Pilze*, Basel.
GAFFRON, H. (1933) *Biochem. Z.* **260,** 11.
GAFFRON, H. (1935) *Biochem. Z.* **279,** 1.
GAFFRON, H. (1939) *Nature*, **143,** 204.
GAFFRON, H. (1942) *J. Gen. Physiol.* **26,** 241.

GAFFRON, H. (1944) *Biol. Rev. Cambr. Phil. Soc.* **19,** 1.
GAFFRON, H. (1946) in: Green (1946).
GAFFRON, H. (1960) in: Steward (1960), Vol. 1B.
GAFFRON, H. (1961) 5th Int. Congr. Biochem., Moscow.
GAFFRON, H. (1962a) in: Kasha and Pullman (1962).
GAFFRON, H. (1962b) *Comp. Biochem. Physiol.* **4,** 205.
GAFFRON, H. (1965) in: Fox (1965).
GAFFRON, H. and RUBIN, J. (1942) *J. Gen. Physiol.* **26,** 219.
GAFFRON, H. and WOHL, K. (1936) *Naturwiss.* **24,** 81.
GARCIA, A. F., DREWS, G. and KAMEN, M. D. (1974) *Proc. Nat. Acad. Sci., Wash.* **71,** 4213.
GARFINKEL, D. (1958) *Arch. Biochem. Biophys.* **77,** 493.
GARRELS, M. G. and MACKENZIE, F. T. (1969) *Science,* **163,** 570.
GARRETT, P. (1970) *Science,* **169,** 171.
GATES, D. M. (1966) in: Brode, W. R., Ed., *Science in Progress,* Vol. 15, New Haven.
GATES, D. M. (1971) *Scient. Amer.* **224** (3), 88.
GAUTHIER, D. K., CLARK-WALKER, G. D., GARRARD, W. T. and LASCELLES, J. (1970) *J. Bact.* **102,** 797.
GEHMAN, H. M. (1962) *Geochim. Cosmochim. Acta,* **26,** 885.
GEITLER, L. (1923) *Arch. Protistenkunde,* **47,** 1.
GEITLER, L. (1960) *Schizophyceen,* Berlin.
GELMAN, N. S., LUKOYANOVA, M. A. and OSTROVSKII, D. H. (1967) *Respiration and Phosphorylation of Bacteria,* New York.
GEORGE, P. (1964) in: King *et al.* (1964).
GEORGE, P. and RUTMAN, R. J. (1960) *Prog. Biophys. Chem.* **10,** 1.
GEORGE, P., WITONSKY, R. J., TRACHTMAN, M., WU, C., DORWART, W., RICHMAN, L., RICHMAN, W., SHURAYH, F. and LENTZ, B. (1970) *Biochim. Biophys. Acta,* **223,** 1.
GERHARDT, B. (1974) *Biol. i.u. Zeit.* **4,** 169.
GERLING, E. K. (1942) *Dokl. Akad. Nauk. SSSR,* **34,** 259.
GERLING, E. K. and PAVLOVA, T. G. (1951) *Dokl. Akad. Nauk SSSR,* **77,** 85.
GERSCHMAN, R. (1964) in: Dickens and Neil (1964).
GESSNER, F. (1971) *Umschau,* **71,** 377.
GEST, H. (1951) *Bact. Rev.* **15,** 183.
GEST, H. (1954) *Bact. Rev.* **18,** 43.
GEST, H. (1963) in: Gest *et al.* (1963).
GEST, H. (1966) *Nature,* **209,** 879.
GEST, H. (1972) *Adv. Microb. Physiol.* **7,** 243.
GEST, H. and KAMEN, M. D. (1949) *J. Bact.* **48,** 239.
GEST, H. and KAMEN, M. D. (1960) in: Ruhland (1960), Vol. V/2.
GEST, H., SAN PIETRO, A. and VERNON, L. P., Eds. (1963) *Bacterial Photosynthesis,* Yellow Springs.
GIBBS, M. (1962) in: Lewin (1962).
GIBBS, M. (1967) *Ann. Rev. Biochem.* **36,** 757.
GIBBS, M., Ed. (1971) *Structure and Function of Chloroplasts,* Berlin.
GIBBS, M. (1971a) in: Gibbs (1971).
GIBBS, M. (1971b) in: Hatch *et al.* (1971).
GIBBS, M., LATZKO, E., HARVEY, M. J., PLAUT, Z. and SHAIN, Y. (1970) *Ann. N.Y. Acad. Sci.* **175,** 541.
GIBBS, M. and SCHIFF, J. A. (1960) in: Steward (1960), Vol. 1B.
GIBBS, S. P. (1970) *Ann. N.Y. Acad. Sci.* **175,** 454.
GIBOR, A. (1967a) *Symp. Int. Soc. Cell Biol.* **6,** 305.
GIBOR, A. (1967b) in: Warren (1967).
GIBOR, A. and GRANICK, S. (1964) *Science,* **145,** 890.
GIBSON, J. (1967) *Arch. Mikrobiol.* **59,** 104.
GIESE, A. C. (1964, etc.) *Photophysiology,* New York.
GILES, K. L. and SARAFIS, V. (1971) *Nature New Biology,* **236,** 56.
GINGRAS, G. (1966) in: Goedheer, J. C. and THOMAS, J. B., Eds. (1966) *Currents in Photosynthesis,* Amsterdam.
GINGRAS, G., GOLDSBY, R. A. and CALVIN, M. (1963) *Arch. Biochem. Biophys.* **100,** 178.
GLAESSNER, M. F. (1966) *Earth Science Rev.* **1,** 29.
GLAESSNER, M. F. (1971) *Geol. Soc. Amer. Bull.* **82,** 509.
GLASSTONE, S. (1968) *The Book of Mars,* NASA SP-179, Washington, D.C.
GLOCK, G. E. and MCLEAN, P. (1955) *Biochem. J.* **61,** 388.
GLOE, A. and PFENNIG, N. (1974) *Arch. Mikrobiol.* **96,** 93.
GLOVER, J., KAMEN, M. D. and VAN GENDEREN, H. (1952) *Arch. Biochem. Biophys.* **35,** 384.
GOEDHEER, J. C. (1959) *Biochim. Biophys. Acta,* **35,** 1.

GOEDHEER, J. C. (1969) *Biochim. Biophys. Acta*, **172,** 252.
GOKSÖYR, J. (1967) *Nature*, **214,** 1161.
GOLDFINE, H. and BLOCH, K. (1963) in: Wright (1963).
GOLDICH, S. S. and HEDGE, C. E. (1974) *Nature*, **252,** 467.
GOLDICH, S. S., HEDGE, C. E. and STERN, T. W. (1970) *Geol. Soc. Amer. Bull.* **81,** 3671.
GOLDSCHMIDT, V. M. (1938) *Skrifter Norske Videnskaps-akad., Mat.-naturv. Kl.*, no. 4.
GOLDSWORTHY, A. (1969) *Nature*, **224,** 501.
GOLDSWORTHY, A. (1970) *Botan. Rev.* **36,** 321.
GOODWIN, T. W. (1961) 5th Int. Biochem. Congr., Moscow.
GOODWIN, T. W., Ed. (1967) *Biochemistry of Chloroplasts*, London.
GOODWIN, T. W., Ed. (1968) *Citrate in the Regulation of Energy Metabolism*, New York.
GOODWIN, T. W. (1971a) in: Gibbs (1971).
GOODWIN, T. W. (1971b) in: Kimball and Oró (1971).
GOTTLIEB, D. and SHAW, P. D., Eds. (1967) *Antibiotics*, New York.
GOTTSCHALK, G. (1968) *Eur. J. Biochem.* **5,** 346.
GOTTSCHALK, G. and BARKER, H. A. (1967) *Biochemistry*, **6,** 1027.
GOTTSCHALK, G. and CHOWDHURY, A. A. (1969) *FEBS Letters*, **2,** 342.
GOULD, J. M. and IZAWA, S. (1973) *Biochim. Biophys. Acta*, **314,** 212.
GOULD, J. M. and IZAWA, S. (1974) *Biochim. Biophys. Acta*, **333,** 509.
GOULD, S. J. (1974) *Science*, **183,** 739.
GOVETT, G. J. S. (1966) *Bull. Geol. Soc, Amer.* **77,** 1191.
GOVINDJEE, R., RABINOWITCH, E. and GOVINDJEE (1968) *Biochim. Biophys. Acta*, **162,** 539.
GOVINDJEE, R. and SYBESMA, C. (1970) *Biochim. Biophys. Acta*, **223,** 251.
GRABE, B. (1959) *Ark. Fysik.* **15,** 207.
GRANICK, S. (1951) *Ann. Rev. Plant Physiol.* **2,** 115.
GRANICK, S. (1957) *Ann. N.Y. Acad. Sci.* **69,** 292.
GRANICK, S. (1961) in: Brachet and Mirsky (1961), Vol. 2.
GRANICK, S. (1965) in: Bryson and Vogel (1965).
GRANICK, S. (1967) in: Goodwin (1967).
GRANICK, S. and GIBOR, A. (1967) *Prog. Nucl. Acid Res.* **6,** 143.
GRAY, C. T. and GEST, H. (1965) *Science*, **148,** 186.
GRAY, C. T., WIMPENNY, J. W. T. and MOSSMAN, M. R. (1966) *Biochim. Biophys. Acta*, **117,** 33.
GREEN, D. E., Ed. (1946) *Currents in Biochemical Research*, New York.
GREEN, D. E. (1961) Plenary Lecture, 5th Int. Congr., Biochem., Moscow.
GREEN, D. E. (1974) *Biochim. Biophys. Acta*, **346,** 27.
GREEN, D. E., ASAI, J., HARRIS, R. A. and PENNISTON, J. T. (1968) *Arch. Biochem. Biophys.* **125,** 684.
GREEN, D. E. and JI, S. (1972a) *Proc. Nat. Acad. Sci., Wash.* **69,** 726.
GREEN, D. E. and JI, S. (1972b) *J. Bioenerget.* **3,** 159.
GREENBERG, D. M., Ed. (1960, etc.) *Metabolic Pathways*, New York.
GREENBERG, D. M., Ed. (1967, etc.) *Metabolic Pathways*, 3rd ed., New York.
GRELL, K. (1973) *Protozoology*, New York.
GREVILLE, G. D. (1969) *Current Topics Bioenergetics*, **3,** 1.
GRIFFITHS, D. E. (1965) *Essays Bioenergetics* **1,** 91.
GRIFFITHS, M., SISTROM, W. R., COHEN-BAZIRE, G. and STANIER, R. Y. (1955) *Nature*, **176,** 1211.
GROSSENBACHER, K. A. and KNIGHT, C. A. (1965) in: Fox (1965a).
GROSSMAN, L. (1975) *Scient. Amer.* **232**(2), 30.
GROTH, W. and SUESS, H. (1938) *Naturwiss.* **26,** 77.
GROTH, W. and WEYSSENHOFF, H.v. (1960) *Planet. Space Sci.* **2,** 79.
GUILLORY, R. J. and FISHER, R. R. (1972) *Biochem. J.* **129,** 471.
GUILLIERMOND, A., MANGENOT, G. and PLANTEFOL, L. (1933) *Traité de Cytologie Végétale*, Paris.
GUNSALUS, I. C. and STANIER, R. Y., Eds. (1960, etc.) *The Bacteria*, New York.
GUNSALUS, I. C. and SHUSTER, C. W. (1961) in: Gunsalus and Stanier (1960), Vol. 2.
GUNSALUS, I. C. and UMBREIT, W. W. (1945) *J. Bact.* **49,** 347.
HACKETT, D. P., RICE, B. and SCHMID, C. (1960) *J. Biol. Chem.* **235,** 2140.
HADZI, J. (1963) *The Evolution of the Metazoa*, New York.
HAECKEL, E. (1866) *Generelle Morphologie der Organismen*, Berlin.
HAECKEL, E. (1894) *Systematische Phylogenie*, Berlin.
HAHN, F. E., Ed. (1969) *Prog. Molec. Subcell. Biol.*, Berlin.
HALDANE, J. B. S. (1929) *The Rationalist Annual*, reprinted in Bernal (1967).
HALDAR, D., FREEMAN, K. and WORK, T. S. (1966) *Nature*, **211,** 9.
HALL, D. O., CAMMACK, R. and RAO, K. K. (1971a) *Nature*, **233,** 136.
HALL, D. O., CAMMACK, R. and RAO, K. K. (1973a) *Space Life Sciences*, **4,** 455.

HALL, D. O., CAMMACK, R. and RAO, K. K. (1973b) in: Swain (1973).
HALL, D. O., CAMMACK, R. and RAO, K. K. (1974) *Origin of Life*, **5**, 363.
HALL, D. O., CAMMACK, R. and RAO, K. K. (1975) *Science Prog.* **62**, 285.
HALL, D. O. and EVANS, M. C. W. (1969) *Nature*, **223**, 1342.
HALL, D. O., REEVES, S. C. and BALTSCHEFFSKY, H. (1971b) *Biochem. Biophys. Res. Commun.* **43**, 359.
HALL, J. B. (1971) *J. Theoret. Biol.* **30**, 429.
HALL, J. B. (1973a) *J. Theoret. Biol.* **38**, 413.
HALL, J. B. (1973b) *Space Life Sciences*, **4**, 204.
HALL, R. P. (1967) in: Chen (1967), Vol. 1.
HALL, W. T. and CLAUS, G. (1963) *J. Cell Biol.* **19**, 551.
HALL, W. T. and CLAUS, G. (1967) *J. Phycol.* **3**, 37.
HALLAM, A. (1974) *Nature*, **251**, 568.
HAMILTON, E. I. (1965) *Applied Geochronology*, London.
HAMMOND, A. L. (1973) *Science*, **179**, 463.
HANAWALT, P. C. (1968) in: Giese (1964), Vol. 4.
HANSEN, T. A. and VAN GEMERDEN, H. (1972) *Arch. Mikrobiol.* **86**, 49.
HANSEN, T. A. and VELDKAMP, H. (1973) *Arch. Mikrobiol.* **92**, 45.
HANSON, E. D. (1966) *Quart. Rev. Biol.* **41**, 1.
HANSON, J. B. and HODGES, T. K. (1967) *Current Topics Bioenergetics*, **2**, 65.
HARADA, K. and FOX, S. W. (1964) *Nature*, **201**, 335.
HARBORNE, J. B., Ed. (1970) *Phytochemical Phylogeny*, London.
HARE, P. E. (1969) in: Eglinton and Murphy (1969).
HARE, P. E. and ABELSON, P. H. (1968) *Carnegie Inst. Wash. Yearbook*, **66**, 526.
HARE, P. E. and MITTERER, R. M. (1969) *Carnegie Inst. Yearbook* 1967–68, 205.
HAROLD, F. M. (1966) *Bact. Rev.* **30**, 772.
HAROLD, F. M. (1970) *Adv. Microb. Physiol.* **4**, 46.
HAROLD, F. M. (1972) *Bact. Rev.* **36**, 172.
HAROLD, R. and STANIER, R. Y. (1955) *Bact. Rev.* **19**, 49.
HARRISON, K. (1960) *A Guide-Book to Biochemistry*, Cambridge.
HARTECK, P. and JENSEN, J. H. D. (1948) *Z. Naturforsch.* **3a**, 591.
HARTLEY, B. S. (1974) in: Carlile and Skehel (1974).
HASTINGS, J. W. (1968) *Ann. Rev. Biochem.* **37**, 597.
HATCH, M. D. (1971a) *Biochem. J.* **125**, 425.
HATCH, M. D. (1971b) in: Hatch *et al.* (1971).
HATCH, M. D., OSMOND, C. B. and SLATYER, R. O., Eds. (1971) *Photosynthesis and Photorespiration*, New York.
HATCH, M. D. and SLACK, C. R. (1966) *Biochem. J.* **101**, 103.
HATCH, M. D. and SLACK, C. R. (1969) *Biochem. Biophys. Res. Commun.* **43**, 589.
HATCH, M. D. and SLACK, C. R. (1970) *Ann. Rev. Plant Physiol.* **21**, 141.
HAUGAARD, N. (1968) *Physiol. Rev.* **48**, 311.
HAUKELI, A. D. and LIE, S. (1971) *J. Gen. Microbiol.* **69**, 135.
HAUSKA, G., REIMER, S. and TREBST, A. (1974) *Biochim. Biophys. Acta*, **357**, 1.
HAYASHI, O. (1962) *Ann. Rev. Biochem.* **31**, 25.
HAYASHI, O., Ed. (1974) *Molecular Mechanisms of Oxygen Activation*, New York.
HAYASHI, O. and NOZAKI, M. (1969) *Science*, **164**, 389.
HAYATSU, R., STUDIER, M. H., MATSUOKA, S. and ANDERS, E. (1972) *Geochim. Cosmochim. Acta*, **36**, 555.
HAYES, J. M. (1967) *Geochim. Cosmochim. Acta*, **31**, 1395.
HEBER, U. (1973) *Biochim. Biophys. Acta*, **305**, 140.
HEBERER, G., Ed. (1967) *Die Evolution der Organismen*, Stuttgart, Vol. 1.
HEDGES, R. W. (1972) *Heredity*, **28**, 39.
HEGEMAN, G. D. and ROSENBERG, S. L. (1970) *Ann. Rev. Microbiol.* **24**, 429.
HEGNAUER, R., Ed. (1962, etc.) *Chemotaxonomie der Pflanzen*, Basel.
HEIDT, L. J. (1966) *J. Chem. Educ.* **43**, 623.
HELDT, H. W. and SAUER, F. (1971) *Biochim. Biophys. Acta*, **234**, 83.
HELMHOLTZ, H. VON (1869) *Über das Ziel und die Fortschritte der Naturwissenschaft*, reprinted in: Hörz and Wollgast (1971).
HEMPFLING, W. P. (1970) *Biochim. Biophys, Acta*, **205**, 169.
HEMPFLING, W. P. and VISHNIAC, W. (1965) *Biochem. Z.* **342**, 272.
HEMINGWAY, A. (1963) *Physiol. Rev.* **43**, 397.
HENDLEY, D. D. (1955) *J. Bact.* **70**, 625.
HENDRIE, M. S., HOTCHKISS, W. and SHEWAN, J. M. (1970) *J. Gen. Microbiol.* **64**, 151.
HERRARA, A. L. (1942) *Science*, **96**, 14.

HESLOP-HARRISON, J. (1962) *Symp. Soc. Gen. Microbiol.* **12,** 14.
HEVESY, G. VON (1948) *Radioactive Indicators*, New York.
HEYWOOD, V. H. (1966) in: Swain (1966).
HICKMAN, C. J. (1965) in: Ainsworth and Sussman (1965).
HICKMAN, M. H. (1974) *Nature*, **251,** 295.
HILL, D. L. (1972) *The Biochemistry and Physiology of Tetrahymena*, New York.
HILL, R. (1939) *Proc. Roy Soc.* B, **127,** 192.
HILL, R. (1951) *Symp. Soc. Exp. Biol.* **5,** 223.
HILL, R. (1956) *Proc. 3rd Int. Congr. Biochem. Brussels.*
HILL, R. (1965) *Essays Bioenergetics*, **1,** 121.
HILL, R. and BENDALL, F. (1960) *Nature*, **186,** 136.
HILL, T. L. and MORALES, M. F. (1951) *J. Amer. Chem. Soc.* **73,** 1656.
HIND, G. and OLSON, J. M. (1968) *Ann. Rev. Plant Physiol.* **19,** 249.
HINSHELWOOD, C. (1961) *Chem. and Ind.* 1050.
HIRSCH, P. (1968) *Nature*, **217,** 555.
HITTELMAN, K. J., BERTIN, R. and BUTCHER, R. W. (1973) *Biochim. Biophys. Acta* **338,** 398.
HOCH, G. and KNOX, R. S. (1968) in: Giese (1964), Vol. 3.
HOCH, G. and KOK, B. (1961) *Ann. Rev. Plant Physiol.* **12,** 155.
HOCHACHKA, P. W. and MUSTAFA, T. (1972) *Science* **178,** 1056.
HOCHACHKA, P. W. and STOREY, K. B. (1975) *Science* **187,** 613.
HOCHSTIM, A. R. (1963) *Proc. Nat. Acad. Sci., Wash.* **50,** 200.
HOCHSTIM, A. R. (1971) in: Buvet and Ponnamperuma (1971).
HODGSON, G. W. and BAKER, B. L. (1969) *Geochim. Cosmochim. Acta*, **33,** 943.
HODGSON, G. W. and PONNAMPERUMA, C. (1968) *Proc. Nat. Acad. Sci., Wash.* **59,** 22.
HÖBER, R. (1931) *Lehrbuch der Physiologie des Menschen*, 6. Aufl., p. 350, Berlin.
HOEFS, J. (1965) *Geochim. Cosmochim. Acta*, **29,** 399.
HÖPNER, T. and TRAUTWEIN, A. (1971) *Arch. Mikrobiol.* **77,** 26.
HOERING, T. C. (1967) in: Abelson (1967).
HÖRZ, H. and WOLLGAST, S. (1971) *Hermann von Helmholtz, Naturphilosophische Vorträge und Aufsätze*, Berlin.
HOFFMANN-OSTENHOF, O. and WEIGERT, W. (1952) *Naturwiss.* **38,** 303.
HOGG, J. F. (1969) *Ann. N.Y. Acad. Sci.* **168,** 281.
HOGNESS, D. S., COHN, M. and MONOD, J. (1955) *Biochim. Biophys. Acta*, **16,** 99.
HOLLAND, H. D. (1962) in: Engel, A. E. J. *et al.*, Eds., *Petrologic Studies: A Volume in Honor of A. F. Buddington*, Geol. Soc. Amer., New York.
HOLLAND, H. D. (1964) in: Brancazio and Cameron (1964).
HOLLAND, H. D. (1965) *Proc. Nat. Acad. Sci., Wash.* **53,** 1173.
HOLLAND, H. D. (1972) *Geochim. Cosmochim. Acta*, **36,** 637.
HOLLAND, H. D. (1973) *Geochim. Cosmochim. Acta*, **37,** 1390.
HOLM-HANSEN, O. (1968) *Ann. Rev. Microbiol.* **22,** 47.
HOLMES, A. (1946) *Nature*, **157,** 680.
HONIGBERG, B. M. (1967) in: Florkin and Scheer (1967), Vol. 1.
HORECKER, B. L. (1961) 5th Int. Congr. Biochem., Moscow.
HORECKER, B. L. (1962) *The Harvey Lectures*, Series 57, New York.
HORECKER, B. L. (1965) *J. Chem. Education*, **42,** 244.
HORIO, T. and KAMEN, M. D. (1962) *Biochemistry*, **1,** 1141.
HORIO, T. and KAMEN, M. D. (1970) *Ann. Rev. Microbiol.* **24,** 399.
HOROWITZ, N. H. (1945) *Proc. Nat. Acad. Sci., Wash.* **31,** 153.
HOROWITZ, N. H. (1965) in: Bryson and Vogel (1965).
HOROWITZ, N. H. and MILLER, S. L. (1962) *Fortsch. Chem. organ. Naturst.* **20,** 423.
HORTON, A. A. and HALL, D. O. (1968) *Nature*, **218,** 386.
HORVATH, R. S. (1974) *J. Theor. Biol.* **48,** 361.
HOUTERMANS, F. G. (1946) *Naturwiss.* **33,** 185.
HOUTERMANS, F. G. (1953) *Nuovo Cimento*, **10,** 1623.
HOYER, B. H., MCCARTHY, B. J. and BOLTON, E. T. (1964) *Science*, **144,** 959.
HRUBAN, Z. and RECHCIGL, M. (1969) *Microbodies and Related Particles*, New York.
HUBBERT, K. M. (1971) *Scient. Amer.* **224** (3), 60.
HUBLEY, J. H., MITTON, J. R. and WILKINSON, J. F. (1974) *Arch. Mikrobiol.* **95,** 365.
HUENNEKENS, F. M. and WHITELEY, H. R. (1960) in: Florkin and Mason (1960), Vol. 1.
HUGHES, D. E. and WIMPENNY, J. W. T. (1969) *Adv. Microb. Physiol.* **3,** 197.
HUNGATE, R. E. (1967) *The Rumen and its Microbes*, London.
HUNT, G. E. (1974) *Endeavour* ,**33,** 23.

HUNT, G. E, and BARTLETT, J. T. (1973) *Endeavour*, **32,** 39.
HURLBURT, E. O. (1928) *J. Opt. Soc. Amer.* **17,** 15.
HURLEY, P. M., PINSON, W. H., NAGY B. and TESKA, T. M. (1972) *Earth Planet. Sci. Lett.* **14,** 360.
HUTCHINSON, G. E. (1954) in: Kuiper (1954).
HUTNER, S. H. (1961) *Symp. Soc. Gen. Microbiol.* **11,** 1.
HUTNER, S. H. and PROVASOLI, L. (1951) in: Lwoff (1951).
HUXLEY, A. F. (1970) *Chem. in Brit.* **6,** 477.
HUXLEY, T. H. (1876), quoted by Hutner (1961).
IBANEZ, J., KIMBALL, A. P. and ORÓ, J. (1971) *Science*, **173,** 444.
IKAWA, M. (1964) in: Leone (1964).
IKUMA, H. (1972) *Ann. Rev. Plant Physiol.* **23,** 419.
IMSHENETSKII, A. A. (1961) 5th Int. Congr. Biochem., Moscow.
INGERSOLL, A. P. and LEOVY, C. B. (1971) *Ann. Rev. Astron. Astrophys.* **9,** 147.
INGOLD, C. T. (1969) *The Biology of Fungi*, London.
INGRAHAM, L. L. and PARDEE, A. B. (1967) in: Greenberg (1967), Vol. 1.
IRVING, L. (1939) *Physiol. Rev.* **19,** 112.
ISHAQUE, M. and ALEEM, M. I. H. (1971) in: Buvet and Ponnamperuma (1971).
ISHIMOTO, M. and EGAMI, F. (1957) in: Oparin *et al.* (1957).
ISMAIL-BEIGI, F. and EDELMAN, I. S. (1970) *Proc. Nat. Acad. Sci., Wash.* **67,** 1071.
IZAWA, S. and GOOD, N. E. (1968) *Biochim. Biophys. Acta*, **162,** 380.
JACKSON, J. B. and CROFTS, A. R. (1968) *Biochem. Biophys. Res. Commun.* **32,** 908.
JACKSON, W. A. and VOLK, J. R. (1970) *Ann. Rev. Plant Physiol.* **21,** 385.
JACOBSON, K. B. and KAPLAN, N. O. (1957) *J. Biol. Chem.* **226,** 603.
JACQ, C. and LEDERER, F. (1974) *Eur. J. Biochem.* **41,** 311.
JAHN, B. M. and SHIH, C. Y. (1974) *Geochim. Cosmochim. Acta*, **38,** 873.
JAHN, E. (1924) *Beiträge zur botanischen Protistologie* I. *Die Polyangiden*, Leipzig.
JAMES, W. O. and DAS, V. S. R. (1957) *New Phytol.* **56,** 325.
JANE, F. W. (1955) *New Biology*, Vol. 19, London.
JANSSEN, M. A., HILLS, R. E., THORNTON, D. D. and WELCH, W. J. (1973) *Science*, **179,** 994.
JEFFCOAT, R. and DAGLEY, S. (1973) *Nature New Biol.* **241,** 186.
JENSEN, O. (1909) *Centralbl. Bakt.* Abt. II, **22,** 305.
JENSEN, R. G. and BASSHAM, J. A. (1966) *Proc. Nat. Acad. Sci., Wash.* **56,** 1095.
JINKS, J. L. (1964) *Extrachromosomal Inheritance*, Englewood Cliffs.
JOEL, C. D., TREBLE, D. E. and BALL, E. G. (1964) *Feder. Proc.* **23,** 271.
JOHN, P. and WHATLEY, F. R. (1975) *Nature* **254,** 495.
JOKLIK, W. K. (1974) in: Carlile and Skehel (1974).
JOLIOT, P. (1966) *Brookhaven Symp. Biol.* **19,** 418.
JOLIOT, P. and JOLIOT, A. (1973) *Biochim. Biophys. Acta* **305,** 302.
JOLLÈS, P. and JOLLÈS, J. (1971) *Prog. Biophys. Molec. Biol.* **22,** 97.
JONES, C. W. and VERNON, L. P. (1969) *Biochim. Biophys. Acta*, **180,** 149.
JONES, D. and SNEATH, P. H. A. (1970) *Bact. Rev.* **34,** 40.
JONES, J. D. (1972) *Comparative Physiology of Respiration*, London.
JONES, M. E. and LIPMANN, F. (1960) *Proc. Nat. Acad. Sci., Wash.* **46,** 1194.
JONES, W. J., ACKRELL, B. A. C. and ERICKSON, S. K. (1971) *Biochim. Biophys. Acta*, **245,** 54.
JUKES, T. H. (1974) *Origin of Life*, **5,** 331.
JUNGCK, J. R. and FOX, S. W. (1973) *Naturwiss.* **60,** 425.
JUNGE, C. (1972) *Quart. J. Roy. Meteorol. Soc.* **98,** 771.
KALCKAR, H. (1937) *Enzymologia*, **2,** 47.
KALCKAR, H. (1941) *Chem. Rev.* **28,** 71.
KALCKAR, H. (1966a) in: Kaplan and Kennedy (1966).
KALCKAR, H. (1966b) in: Cairns *et al.* (1966).
KALCKAR, H. (1969) *Biological Phosphorylation*, Englewood Cliffs.
KALCKAR, H. (1973) *Conference on the Historical Development of Bioenergetics, American Academy of Arts and Sciences.*
KALKOWSKI, E. (1908) *Z. Dtsch. Geol. Ges.* **60,** 68.
KAMEN, M. D. (1963) *Primary Processes in Photosynthesis*, New York.
KAMEN, M. D. and GEST, H. (1949) *Science*, **109,** 560.
KAMEN, M. D. and HORIO, T. (1970) *Ann. Rev. Biochem.* **39,** 673.
KAMEN, M. D. and VERNON, L. P. (1955) *Biochim. Biophys. Acta*, **17,** 10.
KANAZAWA, T., KANAZAWA, K., KIRK, M. R. and BASSHAM, J. A. (1970) *Plant and Cell Physiol.* **11,** 149.
KAPLAN, N. O. and KENNEDY, E. P., Eds. (1966) *Current Aspects of Biochemical Energetics*, New York.
KARAKASHIAN, S. J. (1963) *Physiology Zoology*, **36,** 52.

KARAKASHIAN, S. J. (1970) *Ann. N.Y. Acad. Sci.* **175,** 474.
KARAKASHIAN, S. J. and SIEGEL, R. W. (1965) *Exp. Parasitol.* **17,** 103.
KASHA, M. and PULLMAN, B., Eds. (1962) *Horizons in Biochemistry*, New York.
KATCHALSKY, A. (1973) *Naturwiss.* **60,** 215.
KATZ, A., DREWS, G., MAYER, H. and FROMME, I. (1974) *J. Bact.* **120,** 672.
KATZ, E. (1949) in: Franck and Loomis (1949).
KATZ, J. J. and NORRIS, J. R. (1973) *Current Topics Bioenergetics*, **5,** 41.
KAWAGUTI, S. and YAMASU, T. (1965) *Biol. J. Okayama Univ.* **3,** 57.
KAYUSHIN, L. P. and SKULACHEV, V. P. (1974) *FEBS Letters*, **39,** 39.
KE, B. (1973) *Biochim. Biophys. Acta*, **301,** 1.
KE, B., HANSEN, R. E. and BEINERT, H. (1973) *Proc. Nat. Acad. Sci., Wash.* **70,** 2941.
KEETON, W. T. (1972) *Biological Science*, New York.
KEIL, F. (1912) *Beitr. Biol. Pflanz.* **11,** 335.
KEILIN, D. (1953) *Nature*, **172,** 390.
KEILIN, D. (1966) *The History of Cell Respiration and Cytochrome*, Cambridge.
KEILIN, D. and RYLEY, J. F. (1953) *Nature*, **172,** 451.
KEILIN, D. and TISSIÈRES, A. (1953) *Nature*, **172,** 393.
KEISTER, D. L. and YIKE, N. J. (1966) *Biochem. Biophys. Res. Commun.* **24,** 519.
KEISTER, D. L. and YIKE, N. J. (1967) *Arch. Biochem. Biophys.* **121,** 415.
KEISTER, D. L. and MINTON, N. J. (1971) *Arch. Biochem. Biophys.* **147,** 330.
KELLOGG, W. W., CADLE, R. D., ALLEN, E. R., LAZRUS, A. L. and MARTELL, E. A. (1972) *Science*, **175,** 587.
KELLY, D. P. (1967) *Science Prog.* **55,** 35.
KELLY, D. P. (1968) *Arch. Mikrobiol.* **61,** 59.
KELLY, D. P. (1971) *Ann. Rev. Microbiol.* **25,** 177.
KELVIN, LORD (1882) *Phil. Math. Papers*, Cambridge, Vol. 1.
KENNEDY, E. P. and LEHNINGER, A. L. (1948) *J. Biol. Chem.* **172,** 874.
KENYON, D. H. and STEINMAN, G. (1969) *Biochemical Predestination*, New York.
KESSLER, E. (1960) in: Ruhland (1960), Vol. V/1.
KIDSTON, R. and LANG, W. H. (1921) *Trans. Roy. Soc. Edinburgh*, **52,** 855.
KIESOW, L. A. (1967) *Current Topics Bioenerget.* **2,** 96.
KIMBALL, A. P. and ORÓ, J., Eds. (1971) *Prebiotic and Biochemical Evolution*, Amsterdam.
KIMURA, M. (1968a) *Nature*, **217,** 624.
KIMURA, M. (1968b) *Genet. Res.* **11,** 247.
KIMURA, M. and OHTA, T. (1974) *Proc. Nat. Acad. Sci., Wash.* **71,** 2848.
KING, J. L. and JUKES, T. H. (1969) *Science* **164,** 788.
KING, T. E., MASON, H. S. and MORRISON, M., Eds. (1964) *Symposium on Oxidases and Related Redox Systems*, New York.
KIRCHHOFF, J. and TRÜPER, H. G. (1974) *Arch. Microbiol.* **100,** 115.
KIRK, J. T. (1966) in: Goodwin (1966).
KIRK, J. T. and TILNEY-BASSETT, R. A. (1967) *The Plastids*, London.
KIRSTEN, T., KRANKOWSKY, D. and ZÄHRINGER, J. (1963) *Geochim. Cosmochim. Acta*, **27,** 1.
KISAKI, T. and TOLBERT, N. E. (1969) *Plant Physiol.* **44,** 242.
KISTNER, A. (1954) *Kon. Nederl. Akad. Wet.*, Proc. Ser. C, **57** (2).
KLEIBER, M. (1961) *The Fire of Life*, New York.
KLEIN, R. M. (1970) *Ann. N.Y. Acad. Sci.* **175,** 623.
KLEIN, R. M. and CRONQUIST, A. (1967) *Quart. Rev. Biol.* **42,** 105.
KLEMME, J. H. (1968) *Arch. Mikrobiol.* **64,** 29.
KLEMME, J. H. (1969) *Z. Naturforsch.* **24b,** 67.
KLEMME, J. H. and SCHLEGEL, H. G. (1968) *Arch. Mikrobiol.* **63,** 154.
KLINGENBERG, M. (1958) *Arch. Biochim. Biophys.* **75,** 376.
KLINGENBERG, M. (1963) *Angew. Chem.* **75,** 900.
KLINGENBERG, M. and SCHOLLMEYER, P. (1960) *Biochem. Z.* **333,** 335.
KLOB, W., KANDLER, O. and TANNER, W. (1973) *Plant Physiol.* **51,** 825.
KLOTZ, I. M. (1957) *Energetics in Biochemical Reactions*, New York.
KLOTZ, I. M. (1964) *Chemical Thermodynamics*, New York.
KLUGE, M. (1971) in: Hatch *et al.* (1971).
KLUYVER, A. J. (1931) *The Chemical Activities of Microorganismus*, London.
KLUYVER, A. J. (1952) *Chem. and Ind.* 136.
KLUYVER, A. J. (1953) 6th Int. Congr. Microbiol., Rome.
KLUYVER, A. J. and DONKER, H. J. L. (1926) *Chem. d. Zelle und Gew.* **13,** 124.
KLUYVER, A. J. and VAN NIEL, C. B. (1936) *Zentralbl. Bakt.*, Abt. II, **94,** 369.
KLUYVER, A. J. and VAN NIEL, C. B. (1956) *The Microbe's Contribution to Biology*, Cambridge.

KNAF, D. and ARNON, D. I. (1970) *Biochim. Biophys. Acta*, **226,** 400.
KNOBLOCH, K. (1966) *Planta*, **70,** 73, 172.
KNOBLOCH, K., ELEY, J. H. and ALEEM, M. I. H. (1971) *Arch. Mikrobiol.* **80,** 97.
KNOX, R. S. (1969) *Biophys. J.* **9,** 1351.
KOCH, A. L. (1972) *Genetics*, **72,** 297.
KOCKARTS, G. (1973) *Space Sciences Rev.* **14,** 723.
KOHLMILLER, E. F. and GEST, H. (1951) *J. Bact.* **61,** 269.
KOHNE, D. (1970) *Quart. Rev. Biophys.* **3,** 327.
KOHNO, T. and YOURNO, J. (1971) *J. Biol. Chem.* **246,** 2203.
KOK, B. (1967) in: San Pietro *et al.* (1967).
KOK, B. and CHENAIE, G. M. (1966) *Current Topics Bioenerget.* **1,** 2.
KOK, B. and JAGENDORF, A. (1963) *Photosynthetic Mechanisms in Green Plants*, Publ. 1145, U.S. Nat. Acad. Sci. Nat. Res. Council.
KOK, B., MALKIN, S., OWENS, O. and FORBUSH, B. (1966) *Brookhaven Symp. Biol.* **19,** 446.
KOK, B., RURAINSKI, H. J. and OWENS, O. V. (1965) *Biochim. Biophys. Acta*, **109,** 347.
KONDRATIEVA, E. N. (1963) *Photosynthetic Bacteria* (Russ.), Moscow.
KONINGS, A. W. T. (1970) *Biochim. Biophys. Acta*, **223,** 398.
KORMAN, E. F., ADDINK, A. D. F., WAKABAYASHI, T. and GREEN, D. E. (1970) *J. Bioenerget.* **1,** 9.
KORN, E. D. (1969) in: Cole (1968), Vol. 2.
KORNBERG, H. L. (1959) *Ann. Rev. Microbiol.* **13,** 49.
KORNBERG, H. L. (1966) *Essays in Biochem.* **2,** 1.
KORNBERG, H. L. and KREBS, H. A. (1957) *Nature*, **179,** 988.
KORNBERG, H. L. and QUAYLE, J. R. (1970) in: Bartley *et al.* (1970).
KORTSCHAK, H. P., HARTT, C. E. and BURR, G. O. (1965) *Plant Physiol.* **40,** 209.
KRAMPITZ, L. O. (1961) in: Gunsalus and Stanier (1960), Vol. 2.
KRASNOVSKY, A. A. (1960) *Ann. Rev. Plant Physiol.* **11,** 363.
KRASNOVSKY, A. A. (1972) *Biophys. J.* **12,** 749.
KRASNOVSKY, A. A. and VOINOVSKAYA, K. K. (1952) *Dokl. Akad. Nauk SSSR*, **67,** 325.
KREBS, H. A. (1972) *Essays in Biochemistry*, **8,** 1.
KREBS, H. A., GURIN, S. and EGGLESTON, L. V. (1952) *Biochem. J.* **51,** 614.
KREBS, H. A. and JOHNSON, W. A. (1937) *Enzymologia*, **4,** 148.
KREBS, H. A. and KORNBERG, H. L. (1957) *Energy Transformations in Living Matter*, Berlin.
KREBS, H. A. and LOWENSTEIN, J. M. (1960) in: Greenberg (1960), Vol. 1.
KREBS, H. A. and VEECH, R. L. (1970) in: Sund, H., Ed. (1970) *Pyridine Nucleotide-dependent Dehydrogenases*, Berlin.
KREISEL, H. (1969) *Grundzüge eines natürlichen Systems des Pilze*, Jena.
KRINSKY, N. I. (1967) in: Goodwin (1967).
KRINSKY, N. I. (1968) in: Giese (1964), Vol. 3.
KROGH, A. (1941) *The Comparative Physiology of Respiratory Mechanisms*, Philadelphia.
KROGMANN, D. W. (1973) in: Carr and Whitton (1973).
KROON, A. M. and SACCONE, C., Eds. (1974) *The Biogenesis of Mitochondria*, New York.
KUBITSCHEK, H. E. (1974) in: Carlile and Skehel (1974).
KÜNZLER, A. and PFENNIG, N. (1973) *Arch. Mikrobiol.* **91,** 83.
KÜNTZEL, H. (1969) *Nature*, **222,** 142.
KÜNTZEL, H. and NOLL, H. (1967) *Nature*, **215,** 1340.
KUHN, H. (1972) *Angew. Chem.* **80,** 838.
KUIPER, G. P., Ed. (1952) *The Atmosphere of the Earth and Planets*, Chicago.
KUIPER, G. P. (1952) in: Kuiper (1952).
KUIPER, G. P., Ed. (1954) *The Earth as a Planet*, Chicago.
KULAEV, I. S. (1971) in: Buvet and Ponnamperuma (1971).
KULAEV, I. S. and BOBYK, M. A. (1971) *Biokhimiya*, **36,** 426.
KUSHNER, D. J. (1969) *Bact. Rev.* **33,** 302.
KVENVOLDEN, K. A. (1974) *Origin of Life*, **5,** 71.
KVENVOLDEN, K. A. and HODGSON, G. W. (1969) *Geochim. Cosmochim. Acta*, **33,** 1195.
KVENVOLDEN, K. A., LAWLESS, J. G., PERING, K., PETERSON, E., FLORES, J., PONNAMPERUMA, C., KAPLAN, I. R. and MOORE, C. (1970) *Nature*, **228,** 923.
KVENVOLDEN, K. A., LAWLESS, J. G. and PONNAMPERUMA, C. (1971) *Proc. Nat. Acad. Sci., Wash.* **68,** 486.
KVENVOLDEN, K. A., PETERSON, E. and BROWN, F. S. (1970b) *Science*, **169,** 1079.
KVENVOLDEN, K. A., PETERSON, E. and POLLOCK, G. E. (1969) *Nature*, **221,** 141.
LA BERGE, G. L. (1967) *Bull. Geol. Soc. Am.* **78,** 331.
LAETSCH, W. M. (1969) *Science Prog.* **57,** 323.
LAETSCH, W. M. (1971) in: Hatch (1971b).

LAETSCH, W. M. (1974) *Ann. Rev. Plant Physiol.* **25,** 27.
LAMPE, H. H. and DREWS, G. (1972) *Arch. Mikrobiol.* **84,** 1.
LANCE, C. and BONNER, W. D. (1968) *Plant Physiol.* **43,** 756.
LANG, N. J. (1968) *Ann. Rev. Microbiol.* **22,** 15.
LANG, N. J. and WHITTON, B. A. (1973) in: Carr and Whitton (1973).
LAPAN, E. A. and MOROWITZ, H. J. (1972) *Scient. Amer.* **227** (6), 94.
LARDY, H. A. and FERGUSON, S. M. (1969) *Ann. Rev. Biochem.* **38,** 991.
LARIMER, J. W. (1973) *Space Science Rev.* **15,** 103.
LARIMER, J. W. and ANDERS, E. (1970) *Geochim. Cosmochim. Acta,* **31,** 1239.
LA RIVIÈRE, J. W. M. (1959) *Jan Albert Kluyver*, Amsterdam.
LARSEN, H. (1954) *Symp. Soc. Gen. Microbiol.* **4,** 186.
LARSEN, H. (1960) in: Ruhland (1960), Vol. 5/2.
LARSEN, H., YOCUM, C. S. and VAN NIEL, C. B. (1952) *J. Gen. Physiol.* **36,** 161.
LASAGA, A. C., HOLLAND, H. D. and DWYER, M. J. (1971) *Science,* **174,** 53.
LASCELLES, J. (1960) *J. Gen. Microbiol.* **23,** 499.
LASCELLES, J. (1962) in: Gunsalus and Stanier (1960), Vol. 3.
LASCELLES, J. (1964a) *Tetrapyrrole Biosynthesis and its Regulation*, New York.
LASCELLES, J. (1964b) in: Dickens and Neil (1964).
LASCELLES, J. (1968) *Adv. Microb. Physiol.* **2,** 1.
LASCELLES, J. and WERTLIEB, D. (1971) *Biochim. Biophys. Acta,* **226,** 328.
LAUTERBORN, R. (1895) *Z. wiss. Zool.* **59,** 537.
LAWLESS, J. G. (1973) *Geochim. Cosmochim. Acta,* **37,** 2207.
LAWLESS, J. G., FOLSOM, C. E. and KVENVOLDEN, K. A. (1972b) *Scient. Amer.* **226** (6), 38.
LAWLESS, J. G., KVENVOLDEN, K. A., PETERSON, E. and PONNAMPERUMA, C. (1972a) *Nature,* **236,** 66.
LAWLESS, J. G., KVENVOLDEN, K. A., PETERSON, E., PONNAMPERUMA, C. and MOORE, C. (1971) *Science,* **173,** 626.
LAWLESS, J. G. and PETERSON, E. (1975) *Origin of Life,* **6,** 3.
LAWRENCE, A. J. and QUAYLE, J. R. (1970) *J. Gen. Microbiol.* **63,** 371.
LEBEDEFF, A. F. (1908) *Biochem. Z.* **7,** 1, quoted by Kluyver and Van Niel (1956).
LEBEDEFF, A. F. (1921), quoted by Arnon (1961).
LEDERBERG, J. (1952) *Physiol. Rev.* **32,** 403.
LEDERBERG, J. and COWIE, D. B. (1958) *Science,* **127,** 1473.
LEE, R. E. (1972) *Nature,* **237,** 44.
LEECH, R. M. (1968) in: Pridham (1968).
LEEDALE, G. F. (1970) *Ann. N.Y. Acad. Sci.* **175,** 429.
LEEDALE, G. F. (1971) *The Euglenoids*, Oxford.
LEEDALE, G. F. (1974) *Taxon,* **23,** 225.
LEES, H. (1954) *Symp. Soc. Gen. Microbiol.* **4,** 84.
LEES, H. (1960) *Ann. Rev. Microbiol.* **14,** 83.
LEES, H. (1962) *Bact. Rev.* **26,** 165.
LEGALL, J. and POSTGATE, J. R. (1973) *Adv. Microb. Physiol.* **10,** 82.
LEHNINGER, A. L. (1964) *The Mitochondrion*, New York.
LEHNINGER, A. L. (1970) *Biochemistry*, New York.
LEHNINGER, A. L. (1971) *Bioenergetics*, New York.
LÉJOHN, H. B. (1971) *Nature,* **231,** 164.
LEMBERG, R. (1969) *Physiol. Rev.* **49,** 48.
LEMBERG, R. and BARRETT, J. (1973) *Cytochromes*, London.
LEMMON, R. M. (1970) *Chem. Rev.* **70,** 95.
LEMMON, R. M. (1973a) *Contemp. Phys.* **14,** 463.
LEMMON, R. M. (1973b) *Environment Biol. Med.* **2,** 1.
LEMOIGNE, M. (1925) *Ann. Inst. Pasteur,* **39,** 144.
LENFANT, C. (1969) in: Andersen (1969).
LEONE, C. A., Ed. (1964) *Taxonomic Biochemistry, Physiology and Serology*, New York.
LEPP, H. and GOLDICH, S. S. (1964) *Econ. Geol.* **59,** 1025.
LEVINE, N. D. (1972) in: Chen (1967), Vol. 4.
LEVINE, R. P. (1968) *Science,* **162,** 768.
LEVINE, R. P. (1969) *Ann. Rev. Plant Physiol.* **20,** 523.
LEVITT, L. S. (1953) *Science,* **118,** 696.
LEVY, R. L., GRAYSON, M. A. and WOLF, C. J. (1973) *Geochim. Cosmochim. Acta,* **37,** 467.
LEWIN, R. A., Ed. (1962) *Physiology and Biochemistry of Algae*, New York.
LEWIN, R. A. (1969) *J. Gen. Microbiol.* **58,** 189.
LEWIN, R. A. and LOUNSBERY, D. M. (1969) *J. Gen. Microbiol.* **58,** 145.

LEWIS, D. (1970) in: Miller (1970).
LEWIS, E. B. (1951) *Cold Spring Harb. Symp. Quant. Biol.* **16,** 159.
LEWIS, G. N. and LIPKIN, D. (1942) *J. Amer. Chem. Soc.* **64,** 2801.
LEWIS, G. N. and RANDALL, M. (1923) *Chemical Thermodynamics*, New York, quoted after the German edition, Vienna 1927.
LEWIS, J. S. (1971) *Amer. Scient.* **59,** 557.
LEWIS, J. S. (1973) *Space Science Rev.* **14,** 401.
LIAAEN-JENSEN, S. and ANDREWES, A. G. (1972) *Ann. Rev. Microbiol.* **26,** 225.
LICARI, G. R. and CLOUD, P. E. (1968) *Proc. Nat. Acad. Sci., Wash.* **59,** 1053.
LICARI, G. R. and CLOUD, P. E. (1972) *Proc. Nat. Acad. Sci., Wash.* **69,** 2500.
LIEBERMAN, M. and BAKER, J. E. (1965) *Ann. Rev. Plant Physiol.* **16,** 343.
LIEN, S. and GEST. H. (1973a) *J. Bioenerget.* **4,** 423.
LIEN, S. and GEST, H. (1973b) *Arch. Biochem. Biophys.* **159,** 730.
LINDBERG, O., Ed. (1970) *Brown Adipose Tissue*, New York.
LINDENMAYER, A. (1965) in: Ainsworth and Sussman (1965).
LINDENMAYER, A. and SMITH, L. (1964) *Biochim. Biophys. Acta,* **93,** 445.
LINDMARK, D. G. and MÜLLER, M. (1973) *J. Biol. Chem.* **248,** 7724.
LINKE, H. A. B. (1969) *Arch. Microbiol.* **64,** 203.
LINNANE, A. W., HASLAM, J. M., LUKINS, H. B. and NAGLEY, P. (1972) *Ann. Rev. Microbiol.* **26,** 163.
LIPMANN, F. (1941) *Adv. Enzymol.* **1,** 99.
LIPMANN, F. (1946) in: Green (1946).
LIPMANN, F. (1951) in: McElroy and Glass (1951).
LIPMANN, F. (1965) in: Fox (1965a).
LIPMANN, F. (1971a) *Wanderings of a Biochemist*, New York.
LIPMANN, F. (1972b) *Science,* **173,** 875.
LIU, A. Y. and BLACK, C. C. (1972) *Arch. Biochem. Biophys.* **149,** 269.
LJUNGDAHL, L. G. and WOOD, H. G. (1969) *Ann. Rev. Biochem.* **23,** 515.
LLOYD, D. (1974) *The Mitochondria of Microorganisms*, New York.
LOCKE, M., Ed. (1964) *Cellular Membranes in Development*, New York.
LÖB, W. (1913) *Ber. Deutsch. Chem. Ges.* **46,** 684.
LÖW, H. and ALM, B. (1964) First Meet. Europ. Biochem. Soc.
LOHRMANN, R. and ORGEL, L. E. (1968) *Science,* **161,** 64.
LOHRMANN, R. and ORGEL, L. E. (1973) *Nature,* **244,** 418.
LOOMIS, R. S., WILLIAMS, W. A. and HALL, A. E. (1971) *Ann. Rev. Plant Physiol.* **22,** 431.
LOPUCHIN, A. S. (1975) *Origin of Life,* **6,** 45.
LOQUIN, M. (1972) *Synopsis Generalis Fungorum*, Paris.
LORIMER, G. H. and ANDREWS, T. J. (1973) *Nature,* **243,** 369.
LOSADA, M., TREBST, L. V., OGATA, S. and ARNON, D. I. (1960) *Nature,* **186,** 753.
LOSADA, M., WHATLEY, F. R. and ARNON, D. I. (1961) *Nature,* **190,** 606.
LOVELOCK, J. E. and MARGULIS, L. (1974) *Origins of Life* **5,** 93.
LOVENBERG, W., Ed. (1973) *Iron–Sulfur Proteins*, New York.
LOWENSTEIN, J. M. (1967) in: Greenberg (1967, etc.), Vol. 1.
LOWENSTEIN, J. M., Ed. (1969) *Citric Acid Cycle: Control and Compartmentation*, New York.
LOWENSTEIN, O., Ed. (1962, etc.) *Adv. Compar. Physiol. Biochem.*, New York.
LUCK, D. J. (1963) *J. Cell Biol.* **16,** 783.
LUCK, D. J. and RICH, E. (1969). *Proc. Nat. Acad. Sci., Wash.* **52,** 931.
LUNDEGARDH, H. (1961) *Nature,* **192,** 243.
LUNDSGAARD, E. (1930) *Biochem. Z.* **217,** 162.
LURIA, S. E. and DARNELL, J. E. (1967) *General Virology*, New York.
LWOFF, A. (1944) *L'évolution physiologique*, Paris.
LWOFF, A. (1951) in: Lwoff (1951).
LWOFF, A., Ed. (1951) *Biochemistry and Physiology of Protozoa*, New York, Vol. 1.
LWOFF, A. (1957) *J. Gen. Microbiol.* **17,** 239.
LWOFF, A. and TOURNIER, P. (1966) *Ann. Rev. Microbiol.* **20,** 45.
LWOFF, M. (1951) in: Lwoff (1951).
MACGREGOR, A. M. (1940) *Trans. Geol. Soc. S. Africa,* **43,** 9.
MACHTA, L. and HUGHES, E. (1970) *Science,* **168,** 1582.
MÄGDEFRAU, K. (1967) in: Heberer (1967).
MAHLER, H. R. and BASTOS, R. N. (1974) *Proc. Nat. Acad. Sci., Wash.* **71,** 2241.
MALKIN, R. (1971) *Biochim. Biophys. Acta,* **253,** 421.
MALKIN, R., APARICIO, P. J. and ARNON, D. I. (1962) *Proc. Nat. Acad. Sci., Wash.* **71,** 2362.
MALKIN, R. and BEARDEN, A. J. (1971) *Proc. Nat. Acad. Sci., Wash.* **68,** 16.

MALKIN, R. and BEARDEN, A. J. (1973) *Biochim. Biophys. Acta*, **292,** 169.
MALKIN, R. and RABINOWITZ, J. C. (1967) *Ann. Rev. Biochem.* **36,** 113.
MALMSTRÖM, B. G. (1973) *Quart. Rev. Biophys.* **6,** 389.
MAMIKUNIAN, G. and BRIGGS, M. H., Eds. (1965) *Current Aspects of Exobiology*, London.
MANDEL, M. (1969) *Ann. Rev. Microbiol.* **23,** 239.
MANDELES, S. (1972) *Nucleic Acid Sequence Analysis*, New York.
MARGOLIASH, E. (1963) *Proc. Nat. Acad. Sci., Wash*, **50,** 672.
MARGOLIASH, E. (1964) in: Dickens and Neil (1964).
MARGOLIASH, E., BARLOW, G. H. and BYERS, V. (1970) *Nature*, **228,** 773.
MARGOLIASH, E., FITCH, W. M. and DICKERSON, R. E. (1971) in: Schoffeniels (1971).
MARGULIS, L. (1968) *Science*, **161,** 1020 (= L. Sagan).
MARGULIS, L. (1969) *J. Geol.* **77,** 606.
MARGULIS, L. (1970) *Origin of Eukaryotic Cells*, New Haven.
MARGULIS, L. (1971) *Evolution*, **25,** 242.
MARMUR, J. (1962) *Bull. New York Acad. Med.* **38,** 364.
MARMUR, J., FALKOW, S. and MANDEL, M. (1963) *Ann. Rev. Microbiol.* **17,** 329.
MARMUR, J. and LANE, D. (1970) *Proc. Nat. Acad. Sci., Wash.* **46,** 453.
MAROC, J., AZERAD, R., KAMEN, M. D. and LEGALL, J. (1970) *Biochim. Biophys. Acta*, **197,** 87.
MARRS, B., STAHL, C. L., LIEN, S. and GEST, H. (1972) *Proc. Nat. Acad. Sci., Wash.* **69,** 916.
MARSH, H. V., GALMICHE, J. M. and GIBBS, M. (1965) *Plant. Physiol.* **40,** 1013.
MARTIN, G. W. (1968) in: Ainsworth and Sussman (1965) Vol. 3.
MASON, H. S. (1965) *Ann. Rev. Biochem.* **34,** 595.
MATHEWS, F. S., LEVINE, M. and ARGOS, P. (1971) *Nature New Biol.* **233,** 15.
MATTHEWS, C. N. (1971) in: Buvet and Ponnamperuma (1971).
MATTHEWS, C. N. (1975) *Origin of Life*, **6,** 155.
MATTHEWS, C. N. and MOSER, R. E. (1966) *Proc. Nat. Acad. Sci., Wash.* **56,** 1087.
MATTHEWS, C. N. and MOSER, R. E. (1967) *Nature*, **215,** 1230.
MAXWELL, J. R., PILLINGER, C. T. and EGLINTON, G. (1971) *Quart. Rev. Chem. Soc.* **25,** 571.
MAYER, J. F. (1845) *Mechanik der Wärme*, Stuttgart, quoted by Stanier (1961).
MCELHINNY, M. W. (1971) *Science*, **172,** 157.
MCELROY, M. B. and DONAHUE, T. D. (1972) *Science*, **177,** 987.
MCELROY, W. D. and GLASS, B. H., Eds. (1951) *Phosphorus Metabolism*, Baltimore, Vol. 1.
MCELROY, W. D. and GLASS, B. H., Eds. (1956) *Inorganic Nitrogen Metabolism*, Baltimore.
MCELROY, W. D. and SELIGER, H. H. (1962) in: Kasha and Pullman (1962).
MCELROY, W. D. and SELIGER, H. H. (1963) *Adv. Enzymol.* **25,** 119.
MCEWEN, B., ALLFREY, V. and MIRSKY, A. (1963) *J. Biol. Chem.* **238,** 758, 2571, 2579.
MCEWEN, B., ALLFREY, V. and MIRSKY, A. (1964) *Biochim. Biophys. Acta*, **91,** 23.
MCFADDEN, B. A. (1973) *Bact. Rev.* **37,** 289.
MCLAUGHLIN, P. J. and DAYHOFF, M. O. (1970) *Science*, **168,** 1469.
MCLAUGHLIN, P. J. and DAYHOFF, M. O. (1973) *J. Molec. Evol.* **2,** 99.
MCREYNOLDS, J. H., FURLONG, N. B., BIRRELL, J. B., KIMBALL, A. P. and ORÓ, J. (1971) in: Kimball and Oró (1971).
MCSWAIN, B. D. and ARNON, D. I. (1968) *Proc. Nat. Acad. Sci., Wash.* **61,** 989.
MECHALAS, B. J. and RITTENBERG, S. C. (1960) *J. Bact.* **80,** 501.
MEHLER, A. H. (1951) *Arch. Biochem. Biophys.* **33,** 65.
MEINSCHEIN, W. G. (1965) *Science*, **150,** 601.
MEINSCHEIN, W. G. (1969) in: Eglinton and Murphy (1969).
MELANDRI, B., BACCARINI-MELANDRI, A., SAN PIETRO, A. and GEST, H. (1971) *Science.* **174,** 514.
MENDELSOHN, E. (1964) *Heat and Life*, Cambridge, Mass.
MENKE, W. (1961) *Z. Naturforsch.* **16b,** 543.
MENKE, W. (1962) *Ann. Rev. Plant Physiol.* **13,** 27.
MENKE, W. (1966) *Brookhaven Symp. Biol.* **19.**
MENKE, W. (1966a) in: Goodwin (1966).
MERESCHKOWSKY, C. (1905) *Biol. Centralb.* **25,** 593.
METZNER, H., Ed. (1966) *Die Zelle-Struktur und Funktion*, Stuttgart.
METZNER, H., Ed. (1969) *Progress in Photosynthesis Research*, Tübingen.
METZNER, H. (1973) *Naturwiss.* **60,** 507.
MEYER, D. J. and JONES, C. W. (1973) *Eur. J. Biochem.* **36,** 144.
MEYER, R. E. (1973) *J. Theoret. Biol.* **38,** 647.
MEYERHOF, O. (1930) *Die chemischen Vorgänge im Muskel*, Berlin.
MEYERHOF, O. (1937) *Ergebn. Physiol.* **39,** 10.
MICHEL, P. H. (1973) *The Cosmology of Giordano Bruno*, London.

MIDDLEHURST, B. M. and KUIPER, G. P. (1963) *The Moon, Meteorites and Planets*, Chicago.
MILHAUD, G., AUBERT, J. P. and MILLET, J. (1958) *C.R. Acad. Sci. (Paris)*, **246,** 1766.
MILLBANK, J. W. (1969) *Arch. Mikrobiol.* **68,** 32.
MILLER, P. L., Ed. (1970) *Control of Organelle Development*, Symp. 24, Soc. Exp. Biol., London.
MILLER, S. L. (1953) *Science*, **117,** 528.
MILLER, S. L. (1955) *J. Amer. Chem. Soc.* **77,** 2351.
MILLER, S. L. (1957a) *Biochim. Biophys. Acta*, **23,** 480.
MILLER, S. L. (1957b) *Ann. N.Y. Acad. Sci.* **69,** 260.
MILLER, S. L. (1974) *Origin of Life*, **5,** 139.
MILLER, S. L. and HOROWITZ, N. H. (1966) in: Pittendrigh (1966).
MILLER, S. L. and ORGEL, L. E. (1974) *The Origins of life on the Earth*, Englewood Cliffs.
MILLER, S. L. and PARRIS, M. (1964) *Nature*, **204,** 1248.
MILLER, S. L. and PARRIS, M. (1971) in: Kimball and Oró (1971).
MILLER, S. L. and UREY, H. C. (1959) *Science*, **130,** 245.
MILLS, D. R., KRAMER, F. R. and SPIELGELMAN, S. (1973) *Science*, **180,** 916.
MILTON, D. J. (1974) *Science*, **183,** 880.
MIRCHA, C. J. and DEVAY, J. E. (1971) *Can. J. Bot.* **17,** 1353.
MITCHELL, P. (1961) *Nature*, **191,** 144.
MITCHELL, P. (1966) *Biol. Rev.* **41,** 445.
MITCHELL, P. (1967) in: Florkin and Stotz (1962), Vol. 22.
MITCHELL, P. (1968) *Chemiosmotic Coupling and Energy Transduction*, Bodmin, England.
MITCHELL, P. (1969) in: Cole (1968), Vol. 2.
MITCHELL, P. (1970) *Symp. Soc. Gen. Microbiol.* **20,** 121.
MITCHELL. P. (1972) *J. Bioenerget.* **3,** 5.
MITCHELL, P. (1973) *J. Bioenerget.* **4,** 63.
MOLISCH, H. (1907) *Die Purpurbakterien nach neuen Untersuchungen*, Jena.
MONOD, J. (1942) *Recherches sur la croissance des cultures bactériennes*, Paris.
MOORBATH, S., O'NIONS, R. K. and PANKHURST, R. J. (1973) *Nature*, **245,** 138.
MOORE, L. R. (1969) in: Eglinton and Murphy (1969).
MORIARTY, D. J. W. and NICHOLAS, D. J. D. (1970) *Biochim. Biophys. Acta*, **216,** 130.
MORITA, S. (1968) *Biochim. Biophys. Acta*, **153,** 241.
MORPURGO, G., SERLUPI-CRESCENZI, G., TECCE, G., VALENTE, F. and VENETACCI, D. (1964) *Nature*, **201,** 897.
MORRIS, I. (1974) *Science Prog.* **61,** 99.
MORTENSON, L. E. (1963) *Ann. Rev. Microbiol.* **17,** 115.
MORTENSON, L. E., VALENTINE, R. C. and CARNAHAN, J. E. (1962) *Biochem. Biophys. Res. Comm.* **7,** 448.
MÜHLETHALER, K. (1967) in: Goodwin (1967).
MÜHLETHALER, K. (1971) in: Gibbs (1971).
MUELLER, G. (1967) in: Bernal (1967).
MÜLLER, M. (1969) *Ann. N.Y. Acad. Sci.* **168,** 292.
MÜLLER, M., HOGG, J. F. and DEDUVE, C. (1968) *J. Biol. Chem.* **243,** 5385.
MÜNNICH, K. O. (1963) *Naturwiss.* **50,** 211.
MUIR, M. D. and HALL, D. O. (1974) *Nature*, **252,** 376.
MULLER, H. J. (1929) *Proc. Int. Congr. Plant Sci.* **1,** 897.
MULLER, H. J. (1955) *Science*, **121,** 1.
MURRAY, B. C. (1973) *Scient. Amer.* **228** (1), 48.
MURRAY, R. G. E. (1962) *Symp. Soc. Gen. Microbiol.* **12,** 119.
MYERS, J. (1971) *Ann. Rev. Plant Physiol.* **22,** 289.
NAGY, B. (1970) *Geochim. Cosmochim. Acta*, **34,** 525.
NAGY, B. and NAGY, L. A. (1969) *Nature*, **223,** 1226.
NAGY, L. A. (1975) *Science*, **183,** 514.
NAKAMURA, H. (1939) *Acta Biochem. (Japan)*, **11,** 109.
NASON, A. (1962) *Bact. Rev.* **26,** 16.
NASON, A. (1963) in: Boyer *et al.* (1959), Vol. 7.
NASON, A. and TAKAHASHI, H. (1958) *Ann. Rev. Microbiol.* **12,** 203.
NASS, M. M. K. (1967) in: Vogel *et al.* (1967).
NASS, M. M. K. (1969a) *Science*, **165,** 25.
NASS, M. M. K. (1969b) *Science*, **165,** 1128.
NASS, M. M. K. and NASS, S. (1963) *J. Roy. Microsc. Soc.* **81,** 209.
NASS, M. M. K., NASS, S. and AFZELIUS, B. A. (1965) *Exptl Cell Res.* **37,** 516.
NASS, S. (1969c) *Int. Rev. Cytol.* **25,** 55.
NASS, S. and NASS, M. M. K. (1963) *J. Cell Biol.* **19,** 593, 613.

NEALSON, K. H., EBERHARD, A. and HASTINGS, J. W. (1972) *Proc. Nat. Acad. Sci., Wash.* **69,** 1073.
NELSON, E. B., TOLBERT, N. E. and HESS, J. L. (1969) *Plant Physiol.* **44,** 55.
NEWBURN, R. L. and GULKIS, S. (1973) *Space Science Rev.* **14,** 179.
NEWCOMB, E. H. and FREDERICK, S. E. (1971) in: Hatch *et al.* (1971).
NEWTON, J. W. and KAMEN, M. D. (1961) in: Gunsalus and Stanier (1960), Vol. 2.
NICHOLAS, D. J. D. (1963) in: Steward (1960), Vol. 3.
NICOL, J. A. C. (1962b) *Proc. Roy. Soc.* A, **265,** 355.
NIEDERPRUEM, D. J. (1965) in: Ainsworth and Sussman (1965), Vol. 1.
NIKLOWITZ, W. and DREWS, G. (1956) *Arch. Mikrobiol.* **24,** 134.
NIKLOWITZ, W. and DREWS, G. (1957) *Arch. Mikrobiol.* **27,** 150.
NISHIMURA, M. and CHANCE, B. (1963) *Biochim. Biophys. Acta,* **66.**
NOBEL, P. S., (1970) *Plant Cell Physiology*, San Francisco.
NODDACK, W. (1937) *Angew. Chem.* **30,** 505.
NOLAN, C. and MARGOLIASH, E. (1968) *Ann. Rev. Biochem.* **37,** 727.
NOLL, H. (1970) in: Miller (1970).
NOVIKOFF, A. B. (1961) in: Brachet and Mirsky (1961), Vol. 2.
NOZAKI, M., TAGAWA, K. and ARNON, D. I. (1961) *Proc. Nat. Acad. Sci., Wash.* **47,** 1334.
NOZAKI, M., TAGAWA, K. and ARNON, D. I. (1963) in: Gest *et al.* (1963).
NÜHRENBERG, B., LESEMANN, D. and PIRSON, A. (1968) *Planta* **79,** 162.
NURSALL, J. R. (1959) *Nature,* **183,** 1170.
OBERLIES, F. and PRASHNOWSKY, A. A. (1968) *Naturwiss.* **55,** 25.
O'BRIEN, R. W. and MORRIS, J. G. (1971) *J. Gen. Microbiol.* **68,** 307.
OEHLER, J. H., and SCHOPF, J. W. (1971) *Science,* **174,** 1229.
OELZE, J. and DREWS, G. (1970) *Biochim. Biophys. Acta,* **203,** 189.
OELZE, J. and DREWS, G. (1972) *Biochim. Biophys. Acta,* **265,** 209.
ÖPIK, H. (1968) in: Pridham (1968).
OESPER, R. (1951) in: McElroy and Glass (1951).
ÖSTERBERG, R. and ORGEL, L. E. (1972) *J. Molec. Evol.* **1,** 241.
OESTERHELT, D. and STOECKENIUS, W. (1971) *Nature New Biol.* **233,** 152.
OESTERHELT, D. and STOECKENIUS, W, (1973) *Proc. Nat. Acad. Sci., Wash.* **60,** 2833.
OGINSKY, E. L. and UMBREIT, W. W. (1959) *An Introduction to Bacterial Physiology*, San Francisco.
OGSTON, A. G. and SMITHIES, O. (1948) *Physiol. Rev.* **28,** 283.
OHNO, S. (1970) *Evolution by Gene Duplication*, London.
OLSON, J. M. (1970) *Science,* **168,** 438.
O'NIONS, R. K. and PANKHURST, R. J. (1972) *Nature,* **237,** 446.
OPARIN, A. I. (1924) *The Origin of Life* (Russian), Moscow, translated in Bernal (1967).
OPARIN, A. I. (1938) *The Origin of Life*, London.
OPARIN, A. I., Ed. (1959) *Origin of Life on Earth*, Oxford.
OPARIN, A. I. (1961) *Life: Its Nature, Origin and Development*, New York.
OPARIN, A. I. (1964) *The Chemical Origin of Life*, Springfield.
OPARIN, A. I. (1965a) in: Fox (1965a).
OPARIN, A. I. (1965b) *Adv. Enzymol.* **27,** 347.
OPARIN, A. I. (1968) *Genesis and Evolutionary Development of Life*, New York.
OPARIN, A. I. (1971) in: Kimball and Oró (1971).
ORGEL, L. E. (1968) *J. Molec. Biol.* **38,** 381.
ORGEL, L. E. (1973) *The Origins of Life*, London.
ORGEL, L. E. and SULSTON, J. E. (1971) in: Kimball and Oró (1971).
ORME, T. W., REVSIN, B. and BRODIE, A. F. (1969), *Arch. Biochem. Biophys.* **134,** 172.
ORMEROD, J. G. (1956) *Biochem. J.* **64,** 373.
ORMEROD, J. G. and GEST, H. (1962) *Bact. Rev.* **26,** 51.
ORNSTON, L. N. (1974) *Bact. Rev.* **35,** 87.
ORÓ, J. (1961) *Nature,* **190,** 389.
ORÓ, J. (1965a) in: Fox (1965a).
ORÓ, J. (1965b) in: Mamikunian and Briggs (1965).
ORÓ, J. (1972) *Space Life Sciences,* **3,** 507.
ORÓ, J. and STEPHEN-SHERWOOD, E. (1974) *Origins of Life,* **5,** 159.
ORÓ, J., GIBERT, J., LICHTENSTEIN, H., WIKSTRÖM, S. and FLORY, D. A. (1971a) *Nature,* **230,** 105.
ORÓ, J. and KAMAT, S. S. (1961) *Nature,* **190,** 442.
ORÓ, J. and KIMBALL, A. P. (1962) *Arch. Biochem. Biophys.* **96,** 293.
ORÓ, J., NAKAPARKSIN, S., LICHTENSTEIN, H. and GIL-AV, E. (1971b) *Nature,* **230,** 107.
ORÓ, J. and NOONER, D. W. (1967) *Nature,* **213,** 1082.
ORÓ, J., NOONER, D. W., ZLATKIN, A., WIKSTRÖM, S. A. and BARGHOORN, E. S. (1965) *Science,* **148,** 77.

OSAWA, S., ALLFREY, V. G. and MIRSKY, A. E. (1957) *J. Gen. Physiol.* **40,** 491.
OSMOND, C. B. (1971) in: Hatch *et al.* (1971).
OSMOND, C. B. and HARRIS, B. (1971) *Biochim. Biophys. Acta,* **234,** 270.
OTTOW, J. C. G. and GLATHE, H. (1971) *Soil Biol. Biochem.* **3,** 43.
OVERSBY, V. M. and RINGWOOD, A. E. (1971) *Nature,* **234,** 463.
PACE, D. M. and IRELAND, R. L. (1945) *J. Gen. Physiol.* **28,** 547.
PAECHT-HOROWITZ, M. (1971) in: Buvet and Ponnamperuma (1971).
PAECHT-HOROWITZ, M. (1973) *Angew. Chem.* **85,** 422.
PAECHT-HOROWITZ, M. (1974) *Origin of Life,* **5,** 173.
PAECHT-HOROWITZ, M., BERGER, J. and KATCHALSKY, A. (1970) *Nature,* **228,** 636.
PAECHT-HOROWITZ, M. and KATCHALSKY, A. (1973) *J. Molec. Evol.* **2,** 91.
PAINTER, R. B. (1970) in: Giese (1964), Vol. 5.
PALLADE, G. E. (1964) *Proc. Nat. Acad. Sci., Wash.* **52,** 613.
PALMER, J. M. and HALL, D. O. (1972) *Prog. Biophys. Molec. Biol.* **24,** 127.
PANGBORN, J., MARR, A. G. and ROBRISH, S. A. (1962) *J. Bact.* **84,** 669.
PARK, R. B. (1966) in: Vernon and Seely (1966).
PARK, R. B. and SANE, P. V. (1971) *Ann. Rev. Plant Physiol.* **22,** 395.
PARSON, W. W. (1974) *Ann. Rev. Microbiol.* **28,** 41.
PARSON, W. W. and CODGELL, R. J. (1975) *Biochim. Biophys. Acta,* **416,** 105.
PASCHER, A. (1929) *Z. wiss. Bot.* **71,** 386.
PASCHINGER, H., PASCHINGER, J. and GAFFRON, H. (1974) *Arch. Microbiol.* **96,** 341.
PASTEUR, L. (1862) See, for example, Oparin (1961).
PASTEUR, L. (1875) *C.R. Acad. Sci., Paris,* **80,** 452; quoted after Krebs (1972).
PASTEUR, L. (1876) *Etudes sur la bière,* Paris.
PATT, T. E., COLE, G. C., BLAND, J. and HANSON, R. S. (1974) *J. Bact.* **120,** 955.
PATTEE, H. H. (1965) *Adv. Enzymol.* **27,** 381.
PATTERSON, C. (1956) *Geochim. Cosmochim. Acta,* **10,** 230.
PATTERSON, C., TILTON, G. and INGHRAM, M. (1953) *Bull. Geol. Soc. Amer.* **64,** 1461.
PATTERSON, C., TILTON, G. and INGHRAM, M. (1955) *Science,* **121,** 69.
PAULING, L. (1970) *Chem. in Brit.* **6,** 468.
PAVLOVSKAYA, T. E. (1971) Int. Symp. *Origin of Life and Evolutionary Biochemistry,* Varna.
PAVLOVSKAYA, T. E. and PASSYNSKY, A. G. (1957) in: Oparin *et al.* (1957).
PAYNE, W. J. (1970) *Ann. Rev. Microbiol.* **24,** 17.
PAYNE, W. J. (1973) *Bact. Rev.* **37,** 409.
PECK, H. D. (1962) *Bact. Rev.* **26,** 67.
PECK, H. D. (1966) *Some Evolutionary Aspects of Inorganic Sulfur Metabolism,* Lecture, University of Maryland.
PECK, H. D. (1968) *Ann. Rev. Microbiol.* **22,** 489.
PECK, H. D. (1974) in: Carlile and Skehel (1974).
PEETERS, T. and ALEEM, M. I. H. (1970) *Arch. Mikrobiol.* **71,** 319.
PEETERS, T., LIU, M. S. and ALEEM, M. I. H. (1970) *J. Gen. Microbiol.* **64,** 29.
PELROY, R. A., RIPPKA, R. and STANIER, R. Y. (1972) *Arch. Mikrobiol.* **87,** 303.
PENNIALL, R., SAUNDERS, J. P. and LIU, S. M. (1964) *Biochemistry,* **3,** 1454, 1459.
PESCHEK, G. A. (1975) Thesis, Vienna.
PFENNIG, N. (1967) *Ann. Rev. Microbiol.* **21,** 285.
PFENNIG, N. (1969) Personal communication.
PFENNIG, N. (1970) *J. Gen. Microbiol.* **61,** II.
PFLUG, H. D. (1965) *Palaeontographica,* Abt. A, **125,** 46.
PFLUG, H. D. (1967a) *Natuwrwiss.* **54,** 236.
PFLUG, H. D. (1967b) *Rev. Palaeobot. Palynol.* **5,** 9.
PFLUG, H. D. (1971) *Naturwiss.* **58,** 348.
PFLUGER, U. N. and BACHOFEN, R. (1971) *Arch. Mikrobiol.* **77,** 36.
PHILLIPS, P. G., REVSIN, B., DRELL, E. G. and BRODIE, A. F. (1970) *Arch. Biochem. Biophys.* **139,** 59.
PHILLIPS, R. C., GEORGE, P. and RUTMAN, R. J. (1969) *J. Biol. Chem.* **244,** 3330.
PICHINOTY, F., AZOULAY E., COUCHOUD-BEAUMONT, P. and DEMINOR, L. (1969) *Ann. Inst. Pasteur,* **116,** 27.
PICHINOTY, F. and CHIPPAUX, M. (1969) *Ann. Inst. Pasteur,* **117,** 145.
PIERSON, B. K. and CASTENHOLZ, R. W. (1971) *Nature New Biol.* **233,** 25.
PIERSON, B. K. and CASTENHOLZ, R. W. (1974) *Arch. Microbiol.* **100,** 5.
PIRIE, N. W. (1953) *Discovery,* **14,** 238.
PIROZYNSKI, K. A. and MALLOCH, D. W. (1975), to be published.
PIRT, S. J. (1965) *Proc. Roy. Soc.* B, **163,** 224.

PITTENDRIGH, C. S. (1958) in: Roe, A. and Simpson, G. G., Eds. (1958) *Behavior and Evolution*, New Haven.
PITTENDRIGH, C. S., VISHNIAC, W. and PEARMAN, J. P. T., Eds. (1966) *Biology and the Exploration of Mars*, Publ. 1296, Nat. Acad. Sci. (NRC), Washington, D.C.
PLATTNER, H., SALPETER, M. M., SALTZGABER, J. and SCHATZ, G. (1970) *Proc. Nat. Acad. Sci., Wash.* **66,** 1252.
PLATTNER, H. and SCHATZ, G. (1969) *Biochemistry*, **8,** 339.
PODOSEK, F. A. (1970) *Geochim. Cosmochim. Acta*, **34,** 341.
PODOSEK, F. A. and HUNEKE, J. C. (1971) *Earth Planet. Sci. Lett.* **12,** 74.
POLLOCK, M. R. (1969) *Prog. Biophys. Molec. Biol.* **19,** 273.
PON, N. G. (1964) in: Florkin and Mason (1960), Vol. 7.
PONNAMPERUMA, C. (1968) in: Giese (1964), Vol. 3.
PONNAMPERUMA, C. (1971) *Quart. Rev. Biophys.* **4,** 77.
PONNAMPERUMA, C. (1972a) *The Origins of Life*, London.
PONNAMPERUMA, C., Ed. (1972b) *Exobiology*, Amsterdam.
PONNAMPERUMA, C. and KLEIN, H. P. (1970) *Quart. Rev. Biol.* **45,** 235.
PONNAMPERUMA, C. and MACK, R. (1965) *Science*, **148,** 1221.
PONNAMPERUMA, C. and PETERSON, E. (1965) *Science*, **147,** 1572.
PONNAMPERUMA, C., SAGAN, C. and MARINER, R. (1963) *Nature*, **199,** 222.
POOLE, J. H. J. (1951) *Sci. Proc. Roy. Dub. Acad.* **25,** 201.
PORFIREV, V. B. (1971) *Problem of the Inorganic Origin of Oil* (Russ.), Kiev.
POSTGATE, J. R. (1952) *Research*, **6,** 189.
POSTGATE, J. R. (1959) *Ann. Rev. Microbiol.* **13,** 505.
POSTGATE, J. R. (1963) *J. Bact.* **85,** 1450.
POSTGATE, J. R. (1965), *Bact. Rev.* **29,** 425.
POSTGATE, J. R. (1968a) *Science J.* **4** (3), 69.
POSTGATE, J. R. (1968b) *Proc. Roy. Soc. Lond.* B, **171,** 67.
POSTGATE, J. R. (1969) in: Nickless, G. (Ed.) *Inorganic Sulphur Chemistry*, Amsterdam.
POSTGATE, J. R. (1970) *Nature*, **226,** 25.
POSTGATE, J. R., Ed. (1971) *The Chemistry and Biochemistry of Nitrogen Fixation*, London.
POSTGATE, J. R. (1974) in: Carlile and Skehel (1974).
PRIDHAM, J. B., Ed. (1968) *Plant Cell Organelles*, London.
PRIGOGINE, I. and GLANSDORFF, P. (1971) *Thermodynamic Theory of Structure, Stability and Fluctuations*, London.
PRIGOGINE, I. and NICOLIS, G. (1971) *Quart. Rev. Biophys.* **4,** 107.
PRIGOGINE, I., NICOLIS, G. and BABLOYANTZ, A. (1972) *Physics Today*, **25** (11), 23; (12), 38.
PRINGSHEIM, E. G. (1949) *Bact. Rev.* **13,** 47.
PRINGSHEIM, E. G. (1958) in: S. Prat, Ed., *Studies in Plant Physiology*, Prague.
PRINGSHEIM, E. G. (1963a) *Naturwiss.* **50,** 146.
PRINGSHEIM, E. G. (1963b) *Farblose Algen*, Stuttgart.
PRINGSHEIM, E. G. (1964) *Naturwiss.* **51,** 154.
PRINGSHEIM, E. G. and WIESSNER, W. (1960) *Nature*, **188,** 919.
PRINGSHEIM, E. G. and WIESSNER, W. (1961) *Arch. Mikrobiol.* **40,** 231.
PROVASOLI, L., HUTNER, S. H. and SCHATZ, A. (1948) *Proc. Soc. Exp. Biol.* **69,** 279.
PRUSINER, S. and POE, M. (1968) *Nature*, **220,** 235.
PULLMAN, A. and PULLMAN, B. (1967) in: Florkin and Stotz (1962), Vol. 22.
PULLMAN, B. and PULLMAN, A. (1960) *Radiation Res.* Suppl. **2,** 160.
PULLMAN, B. and PULLMAN, A. (1962) *Nature*, **196,** 1137.
PULLMAN, B. and PULLMAN, A. (1963) *Quantum Biochemistry*, New York.
PULLMAN, M. E. and SCHATZ, G. (1967) *Ann. Rev. Biochem.* **36,** 539.
PUNNETT, T. (1966) *Brookhaven Symp. Biol.* **19,** 375.
PUNNETT, T. (1971) *Science*, **171,** 284.
PUNNETT, T. and DERRENBACKER, E. C. (1966) *J. Gen. Microbiol.* **44,** 105.
QUAYLE, J. R. (1961) *Ann. Rev. Microbiol.* **15,** 119.
QUAYLE, J. R. (1972) *Adv. Microb. Physiol.* **7,** 119.
QUAYLE, J. R. and KEECH, D. B. (1959) *Biochem. J.* **72,** 623, 631.
RABINOWITCH, E. (1945, etc.) *Photosynthesis and Related Processes*, New York, Vol. 1.
RABINOWITCH, E. (1959) *Disc. Faraday Soc.* **27,** 161.
RABINOWITCH, E. (1961) *Proc. Nat. Acad. Sci., Wash.* **47,** 1296.
RABINOWITCH, E. and GOVINDJEE (1969) *Photosynthesis*, New York.
RABINOWITCH, E. and WEISS, J. (1937) *Proc. Roy. Soc.* A, **162,** 251.
RABINOWITZ, J., CHANG, C. and PONNAMPERUMA, C. (1968) *Nature*, **218,** 442.
RABINOWITZ, J., CHANG, C. and PONNAMPERUMA, C. (1971) in: Kimball and Oró (1971).

RACKER, E. (1965) *Mechanisms in Bioenergetics*, New York.
RACKER, E. (1973) *Conference on the Historical Development of Bioenergetics, American Academy of Arts and Sciences.*
RACKER, E. (1970) *Essays in Bioenerget.* **6,** 1.
RACKER, E. and STOECKENIUS, W. (1974) *J. Biol. Chem.* **249,** 662.
RADCLIFFE, B. C. and NICHOLAS, D. J. D. (1970) *Biochim. Biophys. Acta*, **205,** 273.
RAFF, R. A. and MAHLER, H. R. (1972) *Science*, **177,** 575.
RAFF, R. A. and MAHLER, H. R. (1973) *Science*, **180,** 517.
RAFF, R. A. and RAFF, E. C. (1970) *Nature*, **228,** 1003.
RAHE, J. (1974) *Naturwiss.* **61,** 45.
RAMDOHR, P. (1958) *Abh. Akad. Wiss. Berlin. Kl. Chem. Geol. Biol.* p. 35.
RAMIREZ, J. M., DEL CAMPO, F. F. and ARNON, D. I. (1968) *Proc. Nat. Acad. Sci., Wash.* **59,** 606.
RAMIREZ, J. and SMITH, L. (1968) *Biochim. Biophys. Acta*, **153,** 466.
RAMSAY, J. G. (1963) *Trans. Geol. Soc. S. Afr.* **66,** 353.
RASOOL, S. I. (1968) *Astronautics and Aeronautics*, October issue.
RASOOL, S. I. (1972) in: Ponnamperuma (1972b).
RASOOL, S. I. and MCGOVERN, W. E. (1966) *Nature*, **212,** 1225.
RASOOL, S. I. and DEBERGH, C. (1970) *Nature*, **226,** 1037.
RAVEN, J. A. (1970a) *J. Exptl Bot.* **21,** 1.
RAVEN, P. H. (1970b) *Science*, **169,** 641.
RAVIN, A. W. (1960) *Bact. Rev.* **24,** 201.
RAZIN, S. (1973) *Adv. Microb. Physiol.* **10,** 2.
REANNEY, D. C. (1974) *J. Theor. Biol.* **48,** 243.
RECHLER, M. M. and BRUNI, C. B. (1971) *J. Biol. Chem.* **246,** 1802.
REDDY, C. A., BRYANT, M. P. and WOLIN, M. J. (1972) *J. Bact.* **109,** 539.
REEVES, R. B. (1963) *Amer. J. Physiol.* **205,** 33.
REEVES, S. G. and HALL, D. O. (1973) *Biochim. Biophys. Acta*, **314,** 66.
REID, C., ORGEL, L. and PONNAMPERUMA, C. (1967) *Nature*, **216,** 939.
REINERT, J. and URSPRUNG, H., Eds. (1971) *Origin and Continuity of Cell Organelles*, Berlin.
REMANE, K. (1967) in: Heberer (1967).
RENNER, O. (1929) *Handb. d. Vererbungswissenschaften*, Vol. 2A, Berlin.
REVELLE, R. and FAIRBRIDGE, R. W. (1957), quoted in Fairbridge (1972).
REVELLE, R. and SUESS, H. E. (1957) *Tellus*, **9,** 18.
RIBBONS, D. W., HARRISON, J. E. and WADZINSKI, A. M. (1970) *Ann. Rev. Microbiol.* **24,** 135.
RIBBONS, D. W. and NORRIS, J. R., Eds. (1969, etc.) *Methods in Microbiology*, London.
RICH, A. (1962) in: Kasha and Pullman (1962).
RICHARDSON, K. E. and TOLBERT, N. E. (1961) *J. Biol. Chem.* **236,** 1285.
RICHARDSON, M. (1974) *Science Prog.* **61,** 41.
RICHMOND, M. H. and WIEDERMAN, B. (1974) in: Carlile and Skehel (1974).
RICHMOND, R. C. (1970) *Nature*, **225,** 1025.
RIDLEY, S. M. and LEECH, R. M. (1970) *Nature*, **227,** 463.
RIENITS, K. G., HARDT, H. and AVRON, M. (1974) *Eur. J. Biochem.* **43,** 291.
RIESKE, J. S. (1967) in: Gottlieb and Shaw (1967).
RIGBY, P. W., BURLEIGH, B. D. and HARTLEY, B. S. (1974) *Nature* **251,** 200.
RINGWOOD, A. E. (1959) *Geochim. Cosmochim. Acta*, **15,** 257.
RINGWOOD, A. E. (1960) *Geochim. Cosmochim. Acta*, **20,** 241.
RINGWOOD, A. E. (1966a) *Geochim. Cosmochim. Acta*, **30,** 41.
RINGWOOD, A. E. (1966b) *Rev. Geophys.* **4,** 113.
RIPPKA, R., WATERBURY, J. and COHN-BAZIRE, G. (1974) *Arch. Microbiol.* **100,** 419.
RIS, H. (1961) *Can. J. Genet. Cytol.* **3,** 95.
RIS, H. and PLAUT, W. (1962) *J. Cell. Biol.* **13,** 383.
RITTENBERG, S. C. (1969) *Adv. Microb. Physiol.* **3,** 159.
RITTENBERG, S. C. (1971) *J. Gen. Microbiol.* **69,** III.
ROBBINS, P. W. and LIPMANN, F. (1958) *J. Biol. Chem.* **233,** 686.
ROBERTSON, J. D. (1960) *J. Physiol.* **153,** 58.
ROBERTSON, J. D. (1964) in: Locke (1964).
ROBERTSON, J. D. (1967) in: Allen (1967).
ROBINSON, G. W. (1964) *Ann. Rev. Physic. Chem.* **15,** 311.
ROBINSON, N., Ed. (1966) *Solar Radiation*, Amsterdam.
ROBINSON, R. (1964) in: Colombo, J. and Hobson, G. D., Eds. (1964) *Advances in Organic Geochemistry.*
ROBINSON, R. (1956) *Nature*, **212,** 1291.
ROBINSON, W. E. (1969) in: Eglinton and Murphy (1969).

ROCHLEDER, F. (1854) *Phytochemie*, Leipzig.
ROELOFSEN, P. A. (1935) Thesis, Utrecht.
ROGERS, H. J. and PERKINS, H. R. (1968) *Cell Walls and Membranes*, London.
ROHLFING, D. L. and FOX, S. W. (1969) *Adv. Catalysis*, **20,** 373.
RONOV, A. B. (1968) *Sedimentology*, **10,** 40.
ROODYN, D. B. and WILKIE, D. (1968) *The Biogenesis of Mitochondria*, London.
ROSE, A. H. (1968) *Chemical Microbiology*, London.
ROSSIGNOL-STRICK, M. and BARGHOORN, E. S. (1971) *Space Life Sciences*, **2,** 144.
ROY, A. B. and TRUDINGER, P. A. (1970) *The Biochemistry of Inorganic Compounds of Sulphur*, Cambridge.
RUBEN, S. (1943) *J. Amer. Chem. Soc.* **65,** 279.
RUBEY, W. W. (1951) *Geol. Soc. Amer. Bull.* **62,** 1124.
RUBEY, W. W. (1955) in: A. Poldervaart, Ed., *The Crust of the Earth*, Geological Society of America Special Paper No. 62.
RUBEY, W. W. (1964) in: Brancazio and Cameron (1964).
RUBEY, W. W. (1974), quoted by Cloud (1974b).
RUDERMAN, M. A. (1974) *Science*, **184,** 1079.
RUEDA, A. (1973) *Space Life Sciences*, **4,** 469.
RUHLAND, W., Ed. (1960, etc.) *Handbuch der Pflanzenphysiologie*, Berlin.
RUPERT, C. S. (1964) in: Giese (1964), Vol. 2.
RUSSELL, D. and TUCKER, W. (1971) *Nature*, **229,** 553.
RUSSELL, G. K. and GIBBS, M. (1968) *Plant Physiol.* **43,** 649.
RUSSELL, N. and MENZEL, H. (1933) *Proc. Nat. Acad. Sci., Wash.* **19,** 997.
RUTTEN, M. G. (1962) *The Geological Aspects of the Origin of Life on Earth*, Amsterdam.
RUTTEN, M. G. (1969) *Proc. 15th Int. Inter-Univ. Geol. Congr.*, Leicester.
RUTTEN, M. G. (1970) *Space Life Sciences*, **1,** 1.
RUTTEN, M. G. (1971) *The Origin of Life by Natural Causes*, Amsterdam.
RUUD, J. T. (1954) *Nature*, **173,** 848.
RYLEY, J. F. (1967) in: Florkin and Scheer (1967), Vol. 1.
RYTHER, J. H. (1969) *Science*, **116,** 72.
RYTHER, J. H. (1970) *Nature*, **227,** 374.
SAGAN, C. (1957) *Evolution*, **11,** 40.
SAGAN, C. (1961) *Radiation Research*, **15,** 174.
SAGAN, C. (1965) in: Fox (1965a).
SAGAN, C. (1972) *Nature*, **238,** 77.
SAGAN, C. (1973a) *J. Theor. Biol.* **39,** 195.
SAGAN, C. (1973b) *The Cosmic Connection*, New York.
SAGAN, C. (1974) *Origin of Life*, **5,** 497.
SAGAN, C. and KHARE, B. N. (1971) *Science*, **173,** 417.
SAGAN, L. (1967) *J. Theor. Biol.* **14,** 225 (= L. Margulis).
SAGER, R. (1972) *Cytoplasmic Genes and Organelles*, New York.
SALSER, W. A. (1974) *Ann. Rev. Biochem.* **43,** 923.
SALTON, M. R. J. (1960) in: Gunsalus and Stanier (1960), Vol. 1.
SALTON, M. R. J. (1964) *The Bacterial Cell Wall*, Amsterdam.
SALTON, M. R. J. (1967) *Ann. Rev. Microbiol.* **21,** 417.
SANCHEZ, R. A., FERRIS, J. P. and ORGEL, L. E. (1966) *Science*, **154,** 784.
SANCHEZ, R. A., FERRIS, J. P. and ORGEL, L. E. (1967) *J. Molec. Biol.* **30,** 223.
SAN PIETRO, A. (1967) in: San Pietro *et al.* (1967).
SAN PIETRO, A., GREER, F. A. and ARMY, T. J., Eds. (1967) *Harvesting the Sun*, New York.
SANYAL, S. K., KVENVOLDEN, K. A. and MARSDEN, S. S. (1971) *Nature*, **232,** 325.
SAPSHEAD, L. M. and WIMPENNY, J. W. T. (1972) *Biochim. Biophys. Acta*, **267,** 388.
SATO, R. (1956) in: McElroy and Glass (1956).
SATO, T. and TAMIYA, H. (1937) *Cytologia*, Fujii Jubilee Volume, p. 1133, Tokyo.
SAVILE, D. B. O. (1968) in: Ainsworth and Sussman (1965, etc.), Vol. 3.
SAWYER, J. S. (1972) *Nature*, **239,** 23.
SAXENA, J. and ALEEM, M. I. H. (1972) *Arch. Mikrobiol.* **84,** 317.
SCHACHMAN, H. K., PARDEE, A. B. and STANIER, R. Y. (1952) *Arch. Biochem. Biophys.* **38,** 245.
SCHATZ, G. (1965) *Biochim. Biophys. Acta*, **96,** 342.
SCHATZ, G. (1967) *Angew. Chem.* **79,** 1088.
SCHATZ, G. and MASON, T. L. (1974) *Ann. Rev. Biochem.* **43,** 51.
SCHEIBEL, L. W., SAZ, H. J. and BUEDING, E. J. (1968) *J. Biol. Chem.* **243,** 2229.
SCHIDLOWSKI, M. (1965) *Nature*, **205,** 895.
SCHIDLOWSKI, M. (1966) *N. Jahrb. Mineral. Abh.* **105,** 183.

SCHIDLOWSKI, M. (1970) *Paläontol. Z.* **44,** 128.
SCHIFF, J. A. and EPSTEIN, H. T. (1966) in: Goodwin (1966).
SCHIFF, J. A. and EPSTEIN, H. T. (1967) in: Locke (1967).
SCHIFF, J. A. and HODSON, R. C. (1973) *Ann. Rev. Plant Physiol.* **23,** 381.
SCHILDKRAUT, C., MARMUR, J. and DOTY, P. (1961) *J. Molec. Biol.* **5,** 595.
SCHIMPER, A. F. W. (1885) *Jb. wiss. Botan.* **16,** 1.
SCHIMPL, A., LEMMON, R. M. and CALVIN, M. (1965) *Science,* **147,** 149.
SCHLEGEL, H. G. (1966) in: Lowenstein (1962), Vol. 2.
SCHLEGEL, H. G. (1969) in: Perlman, D., Ed., *Fermentation Advances,* New York.
SCHLEGEL, H. G. (1972) *Allgemeine Mikrobiologie,* Stuttgart.
SCHLEGEL, H. G. (1974) *Tellus,* **26,** 1.
SCHLEGEL, H. G. and EBERHARDT, U. (1972) *Adv. Microb. Physiol.* **7,** 205.
SCHLEGEL, H. G. and GOTTSCHALK, G. (1962) *Angew. Chem.* **74,** 342.
SCHLEGEL, H. G. and LAFFERTY, R. M. (1971) *Adv. Biochem. Eng.* **1,** 143.
SCHLEGEL, H. G., LAFFERTY, R. and KRAUSS, I. (1970) *Arch. Mikrobiol.* **71,** 283.
SCHMID, G. H. and GAFFRON, H. (1966) *Brookhaven Symp. Biol.* **19,** 380.
SCHMID, G. H. and GAFFRON, H. (1969) *Prog. Photosynthesis Res.* **2,** 857.
SCHMID, G. H. and GAFFRON, H. (1971) *Photochem. and Photobiol.* **14,** 451.
SCHMIDT, L. S., YEN, H. C. and GEST, H. (1974) *Arch. Biochem. Biophys.* **165,** 229.
SCHNEPF, E. and BROWN, R. M. (1971) in: Reinert and Ursprung (1971).
SCHOBERTH, S. and GOTTSCHALK, G. (1969) *Arch. Mikrobiol.* **65,** 318.
SCHÖN, G. and DREWS, G. (1966) *Arch. Mikrobiol.* **54,** 199.
SCHOENHEIMER, R. (1942) *The Dynamic State of Body Constituents,* Boston.
SCHOFFENIELS, E. (1964) in: Florkin and Mason (1969), Vol. 7.
SCHOFFENIELS, E. (1967) *Cellular Aspects of Membrane Permeability,* Oxford.
SCHOFFENIELS, E., Ed. (1971) *Biochemical Evolution and the Origin of Life* (*Molecular Evolution* II), Amsterdam.
SCHOLANDER, P. F. (1940) *Hvalråd Skrift.* **22,** 1, quoted by Steen (1971).
SCHOLES, P. B. and SMITH, L. (1968) *Biochim. Biophys. Acta,* **153,** 363.
SCHOPF, J. W. (1967) *McGraw-Hill Yearbook of Science and Technology,* New York.
SCHOPF, J. W. (1968) *J. Paleontol.* **42,** 651.
SCHOPF, J. W. (1969) *Grana Palynologica,* **9,** 147.
SCHOPF, J. W. (1970a) *J. Paleontol.* **44,** 1.
SCHOPF, J. W. (1970b) *Biol. Rev.* **45,** 319.
SCHOPF, J. W. (1974) *Origin of Life,* **5,** 119.
SCHOPF, J. W. and BARGHOORN, E. S. (1967) *Science,* **156,** 508.
SCHOPF, J. W. and BARGHOORN, E. S. (1969) *J. Paleontol.* **43,** 111.
SCHOPF, J. W., BARGHOORN, E. S., MASER, M. D. and GORDON, R. O. (1965) *Science,* **149,** 1365.
SCHOPF, J. W. and BLACIC, J. M. (1971) *J. Paleontol.* **45,** 925.
SCHOPF, J. W., HAUGH, B. N., MOLNAR, R. E. and SATTERTHWAIT, D. F. (1973) *J. Paleontol.* **47,** 1.
SCHOPF, J. W., KVENVOLDEN, K. A. and BARGHOORN, E. S. (1968) *Proc. Nat. Acad. Sci., Wash.* **59,** 639.
SCHOPF, J. W., OEHLER, D. Z., HORODYSKI, R. J. and KVENVOLDEN, K. A. (1971) *J. Paleontol.* **45,** 477.
SCHRÖDINGER, E. (1938) Public lecture, Vienna.
SCHRÖDINGER, E. (1945) *What is Life?,* Cambridge.
SCHÜRMANN, P., BUCHANAN, B. B. and ARNON, D. I. (1971) *Biochim. Biophys. Acta,* **267,** 111.
SCHULMAN, M., GHAMBEER, R. K., LJUNGDAHL, L. G. and WOOD, H. G. (1973) *J. Biol. Chem.* **248,** 6255.
SCHULMAN, M., PARKER, D., LJUNGDAHL, L. G. and WOOD, H. G. (1972) *J. Bact.* **109,** 633.
SCHUSTER, P. (1972) *Chemie in uns. Zeit* **6,** 1.
SCHWARTZ, A. (1971) in: Buvet and Ponnamperuma (1971).
SCHWARTZ, A. (1972) *Biochim. Biophys. Acta,* **281,** 477.
SCHWARTZ, A. and PONNAMPERUMA, C. (1968) *Nature,* **218,** 443.
SCHWARTZ, A. and PONNAMPERUMA, C. (1971) in: Kimball and Oró (1971).
SCHWARTZ, A. C. and SPORKENBACH, J. (1975) *Arch. Microbiol.* **102,** 261.
SCHWARTZ, A. W., VAN DER VEEN, M., BISSELING, T. and CHITTENDEN, G. J. F. (1975) *Origin of Life,* **6,** 163.
SCHWARTZMAN, D. W. (1973) *Geochim. Cosmochim. Acta,* **37,** 2479.
SCHWENDINGER, R. B. (1969) in: Eglinton and Murphy (1969).
SCOTT, W. M., MODZELESKI, V. E. and NAGY, B. (1970) *Nature,* **225,** 1129.
SCOTTEN, H. L. and STOKES, J. L. (1962) *Ann. Rev. Microbiol.* **42,** 353.
SEELY, G. R. (1973) *J. Theoret. Biol.* **40,** 189.
SENEZ, J. (1962) *Bact. Rev.* **26,** 95.
SELIGER, H. H. and MCELROY, W. D. (1965) *Light: Physical and Biological Action,* New York.

SEWELL, D. L. and ALEEM, M. I. H. (1969) *Biochim. Biophys. Acta*, **172,** 467.
SHAFIA, F., BRINSON, K. B., HEINZMAN, M. W. and BRADY, J. M. (1972) *J. Bact.* **111,** 56.
SHANMUGAM, K. T. and ARNON, D. I. (1972) *Biochim. Biophys. Acta*, **256,** 487.
SHANMUGAM, K. T. and VALENTINE, R. C. (1975) *Science* **187,** 919.
SHAPIRO, J. (1973) *Science*, **179,** 383.
SHAPIRO, L., AGABIAN-KESHISHIAN, N. and BENDIS, I. (1971) *Science*, **173,** 884.
SHAPLEY, H. (1958) *Of Stars and Men*, Boston.
SHAW, G. (1970) in: Harborne (1970).
SHIN, M., TAGAWA, K. and ARNON, D. I. (1963) *Biochem. Z.* **338,** 84.
SHKLOVSKY, I. S. and SAGAN, C. (1966) *Intelligent Life in the Universe*, San Francisco.
SIEBERT, G. (1968) in: Florkin and Stotz (1962), Vol. 23.
SIEDOW, J. N., YOCUM, C. F. and SAN PIETRO, A. (1973a) *Curr. Topics Bioenerget.* **5,** 107.
SIEDOW, J. N., CURTIS, V. A. and SAN PIETRO, A. (1973b) *Arch. Biochem. Biophys.* **158,** 889, 898.
SIEGEL, B. Z. and SIEGEL, S. M. (1970) *Proc. Nat. Acad. Sci., Wash.* **67,** 1005.
SIEGEL, S. M., RENWICK, G., DALY, O., GIUMARRO, C., DAVIS, G. and HALPERN, L. (1965) in: Mamikunian and Briggs (1965).
SIEKEVITZ, P. (1957) *Scient. Amer.* **197** (1), 57.
SIES, H. (1974) *Angew. Chem.* **86,** 789.
SILLÉN, L. G. (1965) *Ark. Kemi.*, **24,** 431.
SILLÉN, L. G. (1966) *Ark. Kemi.*, **25,** 159.
SILLÉN, L. G. (1967) *Science*, **156,** 1189.
SILVER, L. T. (1963), quoted by Cloud (1968b).
SILVER, W. S. and POSTGATE, J. R. (1973) *J. Theoret. Biol.* **40,** 1.
SIMMONDS, P. G., SHULMAN, G. P. and STEMBRIDGE, C. H. (1969) *J. Chrom. Sci.* **7,** 36.
SIMONIS, W. and URBACH, W. (1973) *Ann. Rev. Plant Physiol.* **24,** 89.
SIMPSON, G. G. (1964) *Science*, **146,** 1535.
SIMPSON, G. G. (1968) *Science*, **162,** 140.
SINGER, C. E. and AMES, B. N. (1970) *Science*, **170,** 822.
SINGER, R. (1973) *Mycologia*, **65,** 1378.
SINGER, R. (1975) *The Agaricales in Modern Taxonomy*, Lehre (Germany).
SINGER, S. F. (1968a) *Geophys. J. Roy. Astr. Soc.* **15,** 205.
SINGER, T. P. (1968b) *Biological Oxidations*, New York.
SINGER, T. P. (1971) in: Schoffeniels (1971).
SINHA, K. S. and TILTON, G. R. (1973) *Geochim. Cosmochim. Acta*, **37,** 1823.
SIREVÅG, R. (1974) *Arch. Microbiol.* **98,** 3.
SIREVÅG, R. and ORMEROD, J. G. (1970a) *Science*, **169,** 186.
SIREVÅG, R. and ORMEROD, J. G. (1970b) *Biochem. J.* **120,** 399.
SISTROM, W. E., GRIFFITHS, M. and STANIER, R. Y. (1956) *J. Cell. Comp. Physiol.* **48,** 473.
SITTE, P. (1965) *Bau und Feinbau der Pflanzenzelle*, Stuttgart.
SKULACHEV, V. P. (1971) *Current Topics in Bioenergetics*, **4,** 127.
SKULACHEV, V. P. (1972) *J. Bioenerget.* **3,** 25.
SKULACHEV, V. P. *et al.* (1963) *Biokhimiya*, **28,** 70.
SLACK, C. R. (1968) *Biochem. Biophys. Res. Commun.* **30,** 438.
SLATER, E. C. (1953) *Nature*, **178,** 975.
SLATER, E. C. (1968) in: Florkin and Stotz (1962), Vol. 14.
SLATER, E. C. (1971) *Quart. Rev. Biophys.* **4,** 35.
SLATER, J. H. and MORRIS, I. (1973) *Arch. Mikrobiol.* **92,** 235.
SLATYER, R. O. and TOLBERT, N. E. (1971) *Science*, **173,** 1162.
SLEIGH, M. (1973) *The Biology of Protozoa*, New York.
SLONIMSKI, P. P. (1953) *La formation des enzymes respiratoires chez la levure*, Paris.
SLONIMSKI, P. P. (1974) *Symp. Brit. Soc. Cell Biol.*, see *Nature*, **249,** 13.
SMILLIE, R. M. and EVANS, W. R. (1963) in: Gest *et al.* (1963).
SMILLIE, R. M., RIGOPOULOS, N. and KELLY, H. (1962) *Biochim. Biophys. Acta*, **56,** 612.
SMILLIE, R. M. and SCOTT, N. S. (1969) in: Hahn (1969), Vol. 1.
SMITH, A. J. (1973a) in: Carr and Whitton (1973).
SMITH, A. J. and HOARE, D. S. (1968) *J. Bact.* **95,** 844.
SMITH, A. J., LONDON, L. and STANIER, R. Y. (1967) *J. Bact.* **94,** 972.
SMITH, D. C. (1973a) *Symbiosis of Algae with Invertebrates*, Oxford.
SMITH, D. C. (1973b) *The Lichen Symbiosis*, Oxford.
SMITH, G. (1962) *Synp. Soc. Gen. Microbiol.* **12,** 111.
SMITH, J. E. and GALBRAITH, J. C. (1971) *Adv. Microb. Physiol.* **5,** 41.
SMITH, J. H. C. and FRENCH, C. S. (1963) *Ann. Rev. Plant Physiol.* **14,** 181.

SMITH, J. W., SCHOPF, J. W. and KAPLAN, I. R. (1970) *Geochim. Cosmochim. Acta*, **34,** 659.
SMITH, L. (1954) *Bact. Rev.* **18,** 106.
SMITH, L. (1961) in: Gunsalus and Stanier (1960), Vol. 2.
SMITH, L. (1968) in: Singer (1968b).
SMITH, L. and BALTSCHEFFSKY, M. (1959) *J. Biol. Chem.* **234,** 1575.
SMITH, L. and RAMIREZ, J. (1959) *Arch. Biochem. Biophys.* **79,** 233.
SMITH, M. H. (1969) *Nature*, **223,** 1129.
SMITH, P. F., Ed. (1971) *The Biology of Mycoplasmas*, New York.
SMOLY, J., KUYLENSTIERNA, B. and ERNSTER, L. (1970) *Proc. Nat. Acad. Sci., Wash.* **66,** 125.
SNEATH, P. H. A. (1962) *Symp. Soc. Gen. Microbiol.* **12,** 289.
SNEATH, P. H. A. (1964) in: Leone (1964).
SNEATH, P. H. A. (1974) in: Carlile and Skehel (1974).
SNEATH, P. H. A. and SOKAL, R. R. (1973) *Numerical Taxonomy*, San Francisco.
SÖHNGEN, N. L. (1910) *Rec. Trav. Chim. Pays-Bas*, **29,** 238.
SOJKA, G. A., DIN, G. A. and GEST, H. (1967) *Nature*, **216,** 1021.
SORIANO, S. (1973) *Ann. Rev. Microbiol.* **27,** 155.
SORIANO, S. and LEWIN, R. A. (1965) *Antonie van Leeuwenhoek J. Microbiol. Serol.* **31,** 66.
SOROKIN, Y. I. (1957) in: Oparin *et al.* (1957).
SOROKIN, YU, I. (1961) *Mikrobiologiya*, **49,** 307.
SOROKIN, YU, I. (1966a) *Nature*, **210,** 551.
SOROKIN, YU. I. (1966b) *Z. Allg. Mikrobiol.* **6,** 69.
SPAMPANATO, V. (1921) *Vita di Giordano Bruno*, Messina.
SPANNER, D. C. (1964) *Introduction to Thermodynamics*, London.
SPENCER, G. (1844) *Phil. Mag.*, 3rd ser., **24,** 90; quoted by Pirie (1953) and Blum (1955).
SPIEGELMAN, S. (1971) *Quart. Rev. Biophys.* **4,** 213.
SPRIGG, R. C. (1949) *Trans. Roy. Soc. South Australia*, **73,** 72.
SPRUIT, C. J. P. (1962) in: Lewin (1962).
STADTMAN, E. R. (1966) in: Kaplan and Kennedy (1966).
STADTMAN, R. C. (1967) *Ann. Rev. Microbiol.* **21,** 121.
STADTMAN, T. C. and BARKER, H. A. (1951) *J. Bact.* **61,** 67.
STANIER, R. Y. (1961) *Bact. Rev.* **25,** 1.
STANIER, R. Y. (1964) in: Gunsalus and Stanier (1960), Vol. 5.
STANIER, R. Y. (1970) in: Charles and Knight (1970).
STANIER, R. Y. (1973) in: Carr and Whitton (1973).
STANIER, R. Y. (1974) in: Carlile and Skehel (1974).
STANIER, R. Y., WACHTER, D., GASSER, C. and WILSON, A. C. (1970) *J. Bact.* **102,** 351.
STANIER, R. Y. and COHEN-BAZIRE, G. (1957) *Symp. Soc. Gen. Microbiol.* **7,** 56.
STANIER, R. Y., DOUDOROFF, M. and ADELBERG, E. (1966) *General Microbiology*, London (2nd ed. 1970).
STANIER, R. Y., DOUDOROFF, M., KUNISAWA, R. and CONTOPOULOU, R. (1959) *Proc. Nat. Acad. Sci., Wash.* **45,** 1246.
STANIER, R. Y., KUNISAWA, R., MANDEL, M. and COHEN-BAZIRE, G. (1971) *Bact. Rev.* **35,** 176.
STANIER, R. Y. and VAN NIEL, C. B. (1941) *J. Bact.* **42,** 437.
STANIER, R. Y. and VAN NIEL, C. B. (1962) *Arch. Mikrobiol.* **42,** 17.
STANLEY, S. M. (1973) *Proc. Nat. Acad. Sci., Wash.* **70,** 1486.
STEEMANN-NIELSEN, A. (1960) *Ann. Rev. Plant Physiol.* **11,** 341.
STEEN, J. B. (1971) *Comparative Physiology of Respiratory Mechanisms*, London.
STEIN, W. D. (1967) *The Movement of Molecules Across Cell Membranes*, New York.
STEINBÖCK, O. (1937) *Arch. exp. Zellforsch.* **19,** 343.
STEINMAN, G. (1971) in: Kimball and Oró (1971).
STEINMAN, G., KENYON, D. H. and CALVIN, M. (1966) *Biochim. Biophys. Acta*, **124,** 339.
STEINMAN, G., LEMMON, R. M. and CALVIN, M. (1964) *Proc. Nat. Acad. Sci., Wash.* **52,** 27.
STEINMAN, G., LEMMON, R. M. and CALVIN, M. (1965) *Science*, **147,** 1574.
STEPHENSON, M. (1949) *Bacterial Metabolism*, London.
STERN, K. (1933) *Pflanzenthermodynamik*, Berlin.
STEVENS, E. D. (1973) *J. Theoret. Biol.* **38,** 597.
STEWARD, F. C., Ed. (1960, etc.) *Plant Physiology*, New York.
STEWART, W. D. P. (1973) *Ann. Rev. Microbiol.* **27,** 283.
STEWART, W. D. P. and PEARSON, H. W. (1970) *Proc. Roy. Soc.* B, **175,** 293.
STOKES, J. E. and HOARE, D. S. (1969) *J. Bact.* **100,** 890.
STOPPANI, A. O. M. and DEBOISO, J. F. (1973) *Experientia*, **29,** 1495.
STOPPANI, A. O. M., FULLER, R. C. and CALVIN, M. (1955) *J. Bact.* **69,** 491.
STOREY, B. T. (1970) *J. Theoret. Biol.* **28,** 233.

STOREY, B. T. (1972) *Plant Physiol.* **49,** 314.
STOREY, B. T. and BAHR, J. T. (1969) *Plant Physiol.* **44,** 115.
STOREY, K. B. and HOCHACHKA, P. W. (1974) *J. Biol. Chem.* **249,** 1417, 1423.
STOUTHAMER, A. H. (1969) in: Ribbons and Norris (1969), Vol. 1.
STOUTHAMER, A. H. and BETTENHAUSSEN, C. (1973) *Biochim. Biophys. Acta,* **301,** 53.
STREICHER, S. L. and VALENTINE, R. C. (1973) *Ann. Rev. Biochem.* **42,** 279.
STRØM, T., FERENCI, T. and QUAYLE, J. R. (1974) *Biochem. J.* **144,** 465.
STUART, T. S. and GAFFRON, H. (1971) *Planta,* **100,** 228.
STUART, T. S. and GAFFRON, H. (1972) *Plant Physiol.* **50,** 136.
STUBBE, W. (1971) in: Reinert and Ursprung (1971).
STUDIER, M. H., HAYATSU, R. and ANDERS, E. (1965) *Science,* **149,** 1455 (1965).
STUDIER, M. H.., HAYATSU, R. and ANDERS, E. (1966) *Science,* **152,** 102 (1966).
STUDIER, M. H., HAYATSU, R. and ANDERS, E. (1972) *Geochim. Cosmochim. Acta.* **36,** 189.
STUTZ, E. and NOLL, H. (1967) *Proc. Nat. Acad. Sci., Wash.* **57,** 774.
SUESS, H. E. (1949) *J. Geol.* **57,** 600.
SUESS, H. E. (1962) *J. Geophys. Res.* **67,** 2029.
SULSTON, J., LOHRMANN, R., ORGEL, L. E. and MILES, H. T. (1968) *Proc. Nat. Acad. Sci., Wash.* **59,** 726; **60,** 409.
SUSSMAN, A. S. (1974) *Taxon,* **23,** 301.
SUZUKI, I. (1974) *Ann. Rev. Microbiol.* **28,** 85.
SWAIN, F. M. (1969a) *Ann. Rev. Microbiol.* **23,** 455.
SWAIN, F. M. (1969b) in: Eglinton and Murphy (1969).
SWAIN, T., Ed. (1963) *Chemical Plant Taxonomy,* London.
SWAIN, T., Ed. (1966) *Comparative Phytochemistry,* London.
SWAIN, T. (1971) in: Schoffeniels (1971).
SWAIN, T., Ed. (1973) *Chemistry in Evolution and Systematics,* London.
SWAIN, T. (1974) in: Florkin and Stotz (1962), Vol. 29A.
SWEETMAN, A. J. and GRIFFITHS, D. E. (1971) *Biochem. J.* **121,** 117.
SYBESMA, C. (1969) *Biochim. Biophys. Acta,* **172,** 177.
SZABO, A. S. and AVERS, C. J. (1969) *Ann. N.Y. Acad. Sci.* **168,** 302.
TAGAWA, K. and ARNON, D. I. (1962) *Nature,* **195,** 537.
TAGAWA, K., TSUJIMOTO, H. Y. and ARNON, D. I. (1963) *Proc. Nat. Acad. Sci., Wash.,* **49,** 567.
TAKAHASHI, H., TANIGUCHI, S. and EGAMI, F. (1963) in: Florkin and Mason (1960), Vol. 5.
TALLING, J. F. (1961) *Ann. Rev. Plant Physiol.* **12,** 133.
TANIGUCHI, S. and ITAGAKI, E. (1960) *Biochim. Biophys. Acta,* **44,** 263.
TANIGUCHI, S. and KAMEN, M. D. (1964) *Arch. Biochem. Biophys.* **105,** 367.
TANIGUCHI, S., SATO, R. and EGAMI, F. (1956) in: McElroy and Glass (1956).
TANNER, W., LÖFFLER, M. and KANDLER, O. (1969) *Plant Physiol.* **44,** 422.
TARVIN, D. and BUSWELL, A. M. (1934) *J. Amer. Chem. Soc.* **56,** 1751.
TASKER, P. W. G. (1961) *Trans. Roy. Soc. Trop. Med. Hyg.* **55,** 36.
TATSUMOTO, M., KNIGHT, J. R. and DOE, B. R. (1972) *Geochim. Cosmochim. Acta. Suppl.* **2,** 1521.
TAYLOR, D. L. (1970) *Int. Rev. Cytol.* **27,** 29.
TAYLOR, D. L. (1973) *Ann. Rev. Microbiol.* **27,** 171.
TAYLOR, F. J. R. (1974) *Taxon,* **23,** 229.
TEORELL, T. (1967) *Ber. Bunsen-Ges. physik. Chem.* **71,** 814.
TERENIN, A. N. (1959) in: Oparin (1959).
TEWARI, K. K. (1971) *Ann. Rev. Plant. Physiol.* **22,** 141.
THAUER, R. K., JUNGERMANN, K., HENNINGER, H., WENNING, J. and DECKER, K. (1968) *Eur. J. Biochem.* **4,** 173.
THIELE, H. H. (1968) *Arch. Mikrobiol.* **60,** 124.
THIEMANN, W., Ed. (1974) *Symposium on Generation and Amplification of Asymmetry in Chemical Systems. J. Mol. Evol.* **4,** 1.
THOMSON, W. (1852) see Kelvin, Lord (1882).
THORE, A., KEISTER, D. L. and SAN PIETRO, A. (1969) *Arch. Mikrobiol.* **67,** 378.
TING, I. P. (1971) in: Hatch *et al.* (1971).
TODD, A. R. (1968) *Chemie in unserer Zeit,* **2,** 1.
TOLBERT, N. E. (1963) in: *Photosynthetic Mechanisms of Green Plants,* Nat. Acad. Sci.—Nat. Res. Council Publ. **1145,** 648.
TOLBERT, N. E. (1971a) *Ann. Rev. Plant Physiol.* **22,** 45.
TOLBERT, N. E. (1971b) in: Hatch *et al.* (1971).
TOLBERT, N. E., NELSON, E. B. and BRUIN, W. J. (1971) in: Hatch *et al.* (1971).
TOLBERT, N. E., OESER, A., YAMAZAKI, R. K., HAGEMAN, R. H. and KISAKI, T. (1969) *Plant Physiol.* **44,** 135.

TOLBERT, N. E. and ZILL, L. P. (1956) *J. Biol. Chem.* **222,** 895.
TOMLINSON, N. and BARKER, H. A. (1954) *J. Biol. Chem.* **209,** 585.
TOSTESON, D. C. (1971) in: Schoffeniels (1971).
TOWE, K. M. (1970) *Proc. Nat. Acad. Sci., Wash.* **65,** 781.
TRAGER, W. (1934) *Biol. Bull.* **66,** 182.
TRAGER, W. (1974) *Science,* **183,** 269.
TREBST, A. (1974) *Ann. Rev. Plant Physiol.* **25,** 423.
TREBST, A. and HAUSKA, G. (1974) *Naturwiss.* **61,** 308.
TRENCH, R. K. (1971) *Proc. Roy. Soc.* B, **177,** 225.
TRUDINGER, P. A. (1956) *Biochem. J.* **64,** 274.
TRUDINGER, P. A. (1961) *Biochem. J.* **78,** 673, 680.
TRUDINGER, P. A. (1967) *Rev. Pure Appl. Chem.* **17,** 1.
TRUDINGER, P. A. (1969) *Adv. Microb. Physiol.* **3,** 111.
TRÜPER, H. G. (1964) *Arch. Mikrobiol.* **49,** 23.
TRÜPER, H. G. and PECK, H. D. (1970) *Arch. Mikrobiol.* **73,** 125.
TRÜPER, H. G. and ROGERS, L. A. (1971) *J. Bact.* **108,** 1112.
TUPPY, H. and BERKMAYER, G. D. (1969) *Eur. J. Biochem.* **8,** 237.
TURCOTTE, D. L., NORDMANN, J. C. and CISNE, J. L. (1974) *Nature,* **251,** 124.
TUTTLE, J. H. and JANNASCH, H. W. (1973) *J. Bact.* **115,** 732.
TYLER, S. A. and BARGHOORN, E. S. (1954) *Science,* **119,** 606.
TYNDALL, J. (1874), quoted by Pirie (1953) and Bernal (1967).
UFFEN, R. L. (1973) *J. Bact.* **116,** 874, 1086.
UFFEN, R. L., SYBESMA, C. and WOLFE, R. S. (1971) *J. Bact.* **108,** 1348.
UMBREIT, W. W. (1962) *Bact. Rev.* **26,** 145.
UREY, H. C. (1952a) *The Planets: Their Origin and Development,* New Haven.
UREY, H. C. (1952b) *Proc. Nat. Acad. Sci., Wash.* **38,** 351.
UREY, H. C. (1953a) *Proc. Roy. Soc.* A, **219,** 281.
UREY, H. C. (1953b), quoted by Anders *et al.* (1973).
UREY, H. C. (1957) in: Ahrens *et al.* (1957), Vol. 2.
UREY, H. C. (1959) *Handbuch der Physik, Berlin,* **52,** 363.
UREY, H. C. (1962) *Geochim. Cosmochim. Acta,* **26,** 1.
UREY, H. C. and LEWIS, J. S. (1966) *Science,* **152,** 106.
USSING, H. H. (1954) *Symp. Soc. Exptl Biol.* **8,** 407.
USSING, H. H. (1960) *Hdb. exptl Pharmakol. Erg.-Werk, Berlin,* Vol. 13.
USSING, H. H. (1967) *Ber. Bunsen-Ges. physik. Chem.* **71,** 807.
UTTER, M. F. and WOOD, H. G. (1951) *Adv. Enzymol.* **12,** 41.
UZZELL, T. and SPOLSKY, C. (1974) *Amer. Scient.* **62,** 334.
VALENTINE, R. C. (1964) *Bact. Rev.* **28,** 497.
VAN DAM, K. and MEYER, A. J. (1971) *Ann. Rev. Biochem.* **40,** 115.
VAN DEMARK, P. J. and SMITH, P. F. (1964) *J. Bact.* **88,** 121.
VAN DER BEEK, E. G. and STOUTHAMER, A. H. (1973) *Arch. Mikrobiol.* **89,** 327.
VAN GEMERDEN, H. (1967) Thesis, Leiden.
VAN GEMERDEN, H. (1968) *Arch. Mikrobiol.* **64,** 118.
VAN GEMERDEN, H. (1971) *J. Gen. Microbiol.* **69,** VII.
VAN HOUTEN, F. B. (1973) *Ann. Rev. Earth Planet. Sci.* **1,** 39.
VAN NIEL, C. B. (1936) *Arch. Mikrobiol.* **7,** 373.
VAN NIEL, C. B. (1941) *Adv. Enzymol.* **1,** 263.
VAN NIEL, C. B. (1944) *Bact. Rev.* **8,** 1.
VAN NIEL, C. B. (1946) *Cold Spring Harbor Symp. Quant. Biol.* Vol. 11.
VAN NIEL, C. B. (1949a) *Bact. Rev.* **13,** 161.
VAN NIEL, C. B. (1949b) in: Franck and Loomis (1949).
VAN NIEL, C. B. (1953) *J. Cell Comp. Physiol.* **41,** Suppl. p. 9.
VAN NIEL, C. B. (1955) in: *A Century of Progress in the Natural Sciences,* California Academy of Sciences, San Francisco.
VAN NIEL, C. B. (1962) *Ann. Rev. Plant Physiol.* **13,** 1.
VAN NIEL, C. B. (1963) in: Gest *et al.* (1963).
VAN NIEL, C. B. (1966) in: Kaplan and Kennedy (1966).
VAN NIEL, C. B. (1969) Personal communication.
VANVALEN, L. (1971) *Science,* **171,** 439.
VERHOEVEN, W. (1956) in: McElroy and Glass (1956).
VERNADSKY, W. (1929) *La Biosphère,* Paris.
VERNON, L. P. (1964) *Ann. Rev. Plant. Physiol.* **15,** 73.

VERNON, L. P. (1967) in: San Pietro *et al.* (1967).
VERNON, L. P. (1968) *Bact. Rev.* **32,** 243.
VERNON, L. P. and AVRON, M. (1965) *Ann. Rev. Plant Biochem.* **34,** 269.
VERNON, L. P. and SEELY, G. R., Eds. (1966) *The Chlorophylls*, New York.
VERON, M., FALCOZ-KELLY, F. and COHEN, G. N. (1972) *Eur. J. Biochem.* **28,** 520.
VINOGRADOV, A. P. (1961) *Geochemistry*, London.
VISHNIAC, W. (1971) *Soc. Gen. Microbiol. Symp.* **21,** 355.
VISHNIAC, W. and SANTER, M. (1957) *Bact. Rev.* **21,** 195.
VISHNIAC, W. and TRUDINGER, P. (1962) *Bact. Rev.* **26,** 168.
VOGEL, H. J., LAMPEN, J. O. and BRYSON, V., Eds. (1967) *Organizational Biosynthesis*, New York.
VOSE, J. R. and SPENCER, M. (1969) *Can. J. Biochem.* **47,** 443.
WAEHNELDT, T. V. and FOX, S. W. (1967) *Biochim. Biophys. Acta*, **134,** 1.
WÄNKE, H. and KÖNIG, H. (1959) *Z. Naturforsch.* A. **14,** 860.
WAGNER, R. P. (1969) *Science*, **163,** 1026.
WALD, G. (1957) *Ann. N.Y. Acad. Sci.* **69,** 352.
WALD, G. (1964) *Proc. Nat. Acad. Sci., Wash.* **52,** 595.
WALD, G. (1965) in: Bowen (1965).
WALD, G. (1966) in: Kaplan and Kennedy (1966).
WALKER, D. A. (1962) *Biol. Rev.* **37,** 215.
WALKER, D. A. (1966) *Endeavour*, **25,** 21.
WALLACE, W. and NICHOLAS, D. J. D. (1969) *Biol. Rev.* **44,** 359.
WALLES, B. (1971) in: Gibbs (1971).
WALLIN, J. E. (1927) *Symbionticism and the Origin of Species*, Baltimore.
WALTER, M. R., BAULD, J. and BROCK, T. D. (1972) *Science*, **178,** 402.
WANLESS, R. K., LOVERIDGE, W. D. and STEVENS, R. D. (1970) *Geochim. Cosmochim. Acta*, Suppl. 1, p. 1729.
WARBURG, O. (1920) *Biochem. Z.* **103,** 188.
WARBURG, O. (1966) in: Kaplan and Kennedy (1966).
WARBURG, O., KRIPPAHL, G., GEWITZ, H. S. and VÖLKER, W. (1959) *Z. Naturforsch.* **14b,** 712.
WARREN, K. D., Ed. (1967) *Formation and Fate of Cell Organelles*, New York.
WASSERBURG, G. J. (1964) in: Brancazio and Cameron (1964).
WASSINK, E. C. (1963) in: Florkin and Mason (1960), Vol. 5.
WATTS, R. L. (1970) in: Harborne (1970).
WATTS, R. L. and WATTS, D. C. (1968) *J. Theor. Biol.* **20,** 227.
WEIER, T. E. (1963) *Amer. J. Bot.* **50,** 604.
WEIER, T. E., STOCKING, C. R. and SHUMWAY, L. K. (1966) *Brookhaven Symp. Biol.* **19,** 353.
WEINBERG, A. (1972) *Science*, **177,** 27.
WEISS, E. (1973) *Bact. Rev.* **37,** 359.
WEIZSÄCKER VON, C. F. (1937) *Phys. Z.* **38,** 623.
WELTE, D. H. (1967) *Naturwiss.* **54,** 325.
WEST, K. R. and WISKICH, J. T. (1968) *Biochem. J.* **109,** 527.
WEST, K. R. and WISKICH, J. T. (1973) *Biochim. Biophys. Acta*, **292,** 197.
WEST, M. W., GILL, E. D. and KVENVOLDEN, K. A. (1975) *Origin of Life*, **6,** 285.
WEST, M. W., GILL, E. D. and PONNAMPERUMA, C. (1972) *Space Life Sciences*, **3,** 293.
WEST, M. W., GILL, E. D. and SHERWOOD, B. R. (1973) *Space Life Sciences*, **4,** 309.
WEST, M. W., GILL, E. D., SHERWOOD, B. and KVENVOLDEN, K. A. (1974) *Origin of Life*, **5,** 507.
WEST, M. W. and PONNAMPERUMA, C. (1970) *Space Life Sciences*, **2,** 225.
WETTSTEIN, R. VON (1935) *Handbuch der Systematischen Botanik*, Vienna.
WHATLEY, F. R., TAGAWA, K. and ARNON, D. I. (1963) *Proc. Nat. Acad. Sci., Wash.* **49,** 266.
WHIPPLE, F. L. (1974) *Scient. Amer.* **230** (2), 48.
WHITE, C. D. and SINCLAIR, P. R. (1971) *Adv. Microb. Physiol.* **5,** 173.
WHITTAKER, R. H. (1959) *Quart. Rev. Biol.* **34,** 210.
WHITTAKER, R. H. (1969) *Science*, **163,** 150.
WHITTAKER, R. H. (1970) *Communities and Ecosystems*, New York.
WHITTAM, R. (1964) in: Hoffman, J. F., Ed. (1964) *The Cellular Functions of Membrane Transport*, New Jersey, quoted by Ismail-Beigi and Edelman (1970).
WHITTENBURY, R. (1963) *J. Gen. Microbiol.* **32,** 375.
WHITTENBURY, R. (1964) *J. Gen. Microbiol.* **35,** 13.
WHITTENBURY, R. (1971) *J. Gen. Microbiol.* **69,** V.
WHITTENBURY, R., DAVIES, S. L. and DAVEY, J. F. (1970a) *J. Gen. Microbiol.* **61,** 219.
WHITTENBURY, R., PHILLIPS, K. C. and WILKINSON, J. F. (1970b) *J. Gen. Microbiol.* **61,** 205.
WHITTINGHAM, C. P. (1970) *Prog. Biophys. Molec. Biol.* **21,** 125.
WHITTINGHAM, C. P., COOMBS, J. and MARKER, A. F. (1966) in: Goodwin (1966), Vol. 2.

WHITTON, B. A. (1973) in: Carr and Whitton (1973).
WHITTOW, G. C., Ed. (1970, 1971) *Comparative Physiology of Thermoregulation*, New York.
WICKRAMASINGHE, R. H. (1973) *Space Life Sciences*, **4**, 341.
WIELAND, H. (1922) *Ergebn. Physiol.* **20**, 477.
WIERINGA, K. T. (1940) *Antonie van Leeuwenhoek J. Microbiol. Serol.* **6**, 251, quoted by Kluyver and Van Niel (1956).
WIESSNER, W. (1963) *Arch. Mikrobiol.* **45**, 33.
WIESSNER, W. (1965) *Nature*, **205**, 56.
WIESSNER, W. and GAFFRON, H. (1964) *Nature*, **201**, 725.
WIGGLESWORTH, V. (1972) *Insect Respiration*, Oxford.
WILDT, R. (1937) (1942) *Rev. Mod. Phys.* **14**, 157.
WILDY, P. (1962) *Symp. Soc. Gen. Microbiol.* **12**, 145.
WILKIE, D. (1968) in: Pridham (1968).
WILKIE, D. (1970a) *Chem. in Brit.* **6**, 11.
WILKIE, D. (1970b) in: Charles and Knight (1970).
WILKINSON, J. F. (1971) *Symp. Soc. Gen. Microbiol.* **21**, 15.
WILSON, A. C. and KAPLAN, N. O. (1964) in: Leone (1964).
WILSON, D. F., DUTTON, P. L. and WAGNER, M. (1973) *Current Topics Bioenergetics*, **5**, 234.
WILSON, E. B. (1925) *The Cell in Development and Heredity*, New York.
WILSON, L. G. (1962) *Ann. Rev. Microbiol.* **13**, 201.
WILSON, L. G. and BANDURSKI, R. W. (1958) *J. Amer. Chem. Soc.* **80**, 5576.
WILSON, M. A. and CASCARANO, J. (1970) *Biochim. Biophys. Acta*, **216**, 54.
WILSON, S. B. (1970) *Biochim. Biophys. Acta*, **223**, 383.
WIMPENNY, J. W. T. and NECKLEN, D. K. (1971) *Biochim. Biophys. Acta*, **253**, 352.
WINOGRADSKY, S. N. (1888) *Beiträge zur Morphologie und Physiologie der Bakterien*, Leipzig.
WINOGRADSKY, S. (1949) *Microbiologie du sol*, Paris.
WINTER, H. W. and ARNON, D. I. (1970) *Biochim. Biophys. Acta*, **263**, 352.
WITT, H. T. (1971) *Quart. Rev. Biophys.* **4**, 365.
WITT, H. T. (1972) *J. Bioenerget.* **3**, 47.
WITT, H. T., MÜLLER, A. and RUMBERG, B. (1961) *Nature*, **191**, 170.
WITT, H. T., RUMBERG, B., SCHMIDT-MENDE, P., SIGGEL, U., SKERRA, B., VATER, J. and WEIKARD, J. (1965) *Angew. Chem. Int. Ed.* **4**, 799.
WOESE, C. (1967) *The Genetic Code*, New York.
WOESE, C. R. (1969) in: Hahn (1969), Vol. 1.
WOESE, C. R. (1973a) *Naturwiss.* **60**, 447.
WOESE, C. R. (1973b) *J. Molec. Evol.* **2**, 205.
WOLFE, R. S. (1971) *Adv. Microb. Physiol.* **6**, 107.
WOLK, P. (1973) *Bact. Rev.* **37**, 32.
WOOD, H. G., KATZ, J. and LANDAU, B. R. (1963) *Biochem. Z.* **338**, 809.
WOOD, H. G. and STJERNHOLM, R. L. (1962) in: Gunsalus and Stanier (1960), Vol. 3.
WOOD, H. G. and UTTER, M. F. (1965) in: Campbell and Greville (1965), Vol. 1.
WOOD, H. G. and WERKMAN, C. H. (1942) *Adv. Enzymol.* **3**, 135.
WOOD, J. A. (1968) *Meteorites and the Origin of Planets*, New York.
WOOD, W. A. (1961) in: Gunsalus and Stanier (1960), Vol. 2.
WOODCOCK, C. L. F. and BOGORAD, L. (1971) in: Gibbs (1971).
WOODS, D. D. (1938) *Biochem. J.* **32**, 2000.
WOODS, D. D. and LASCELLES, J. (1954) *Symp. Soc. Gen. Microbiol.* **4**, 1.
WRIGHT, B., Ed. (1963) *Control Mechanisms in Respiration and Fermentation*, New York.
WU, T. T., FITCH, W. M. and E. MARGOLIASH (1974) *Ann. Rev. Biochem.* **43**, 539.
WU, T. T., LIU, E. C. and TANAKA, S. (1968) *J. Bact.* **96**, 447.
YAMANAKA, T. (1964) *Nature*, **204**, 253.
YAMANAKA, T. (1967) *Nature*, **213**, 1183.
YAMANAKA, T. (1972) *Adv. Biophys.* **3**, 227.
YAMANAKA, T. (1973) *Space Life Sciences*, **4**, 490.
YAMANAKA, T. and OKUNUKI, K. (1964) *J. Biol. Chem.* **239**, 1813.
YAMAZAKI, R. K. and TOLBERT, N. E. (1970) *J. Biol. Chem.* **245**, 5137.
YANG, C. C. and ORÓ, J. (1971) in: Buvet and Ponnamperuma (1971).
YČAS, M. (1969) *The Biological Code*, Amsterdam.
YČAS, M. (1972) *Nature*, **238**, 163.
YOCH, D. C. and ARNON, D. I. (1970) *Biochim. Biophys. Acta*, **196**, 180.
YOCH, D. C. and VALENTINE, R. C. (1972) *Ann. Rev. Microbiol.* **26**, 139.
YOCUM, C. F. and SAN PIETRO, A. (1969) *Biochim. Biophys. Res. Commun.* **36**, 614.

York, D. and Farquhar, R. M. (1972) *The Earth's Age and Geochronology*, Oxford.
Yotsuyanagi, Y. (1962) *J. Ultrastruct. Res.* **7,** 121, 141.
Young, A. T. (1974) *Science*, **183,** 407.
Young, R. S. (1973) *Space Life Sciences*, **4,** 505.
Yourno, J., Kohno, T. and Roth, J. R. (1970) *Nature*, **228,** 820.
Yu, C. A., Gunsalus, I. C., Katagiri, M., Suhara, K. and Takemori, S. (1974) *J. Biol. Chem.* **249,** 94.
Zamenhof, S. and Eichhorn, H. H. (1967) *Nature*, **206,** 456.
Zeikus, J. G. and Wolfe, R. S. (1973) *J. Bact.* **113,** 461.
Zeikus, J. G. and Wolin, R. S. (1972) *J. Bact.* **109,** 707.
Zelitch, I. (1964) *Ann. Rev. Plant Physiol.* **15,** 121.
Zelitch, I. (1967) in: San Pietro *et al.* (1967).
Zelitch, I. (1968) *Plant Physiol.* **43,** 1829.
Zelitch, I. (1971) *Photosynthesis, Photorespiration and Plant Productivity*, New York.
Zimen, K. E. and Altenhein, F. K. (1973) *Z. Naturforsch.* **28a,** 1747.
Zuckerkandl, E. (1963) in: Washburn, S. L., Ed. (1963) *Perspectives in Molecular Anthropology.*
Zuckerkandl, E., Jones, R. T. and Pauling, L. (1960) *Proc. Nat. Acad. Sci., Wash.* **46,** 1349
Zuckerkandl, E. and Pauling, L. (1962) in: Kasha and Pullman (1962).
Zuckerkandl, E. and Pauling, L. (1965) *J. Theoret. Biol.* **8,** 357.

SUBJECT INDEX

LATIN NAMES OF ORGANISMS